MW00606555

MISTLETOE

Medicinal and Aromatic Plants – Industrial Profiles

Individual volumes in this series provide both industry and academia with in-depth coverage of one major medicinal or aromatic plant of industrial importance.

Edited by Dr Roland Hardman

Volume 1
Valerian, edited by Peter J. Houghton

Volume 2
Perilla, edited by He-Ci Yu, Kenichi Kosuna and Megumi Haga

Volume 3
Poppy, edited by Jenö Bernáth

Volume 4
Cannabis, edited by David T. Brown

Volume 5
Neem, by H.S. Puri

Volume 6
Ergot, edited by Vladimír Křen and Ladislav Cvak

Volume 7
Caraway, edited by Éva Németh

Volume 8
Saffron, edited by Moshe Negbi

Volume 9
Tea Tree, edited by Ian Southwell and Robert Lowe

Volume 10
Basil, edited by Raimo Hiltunen and Yvonne Holm

Volume 11
Fenugreek, edited by Georgios Petropoulos

Volume 12
Ginkgo biloba, edited by Teris A. Van Beek

Volume 13
Black Pepper, edited by P.N. Ravindran

Volume 14
Sage, edited by Spiridon E. Kintzios

Volume 15
Ginseng, edited by W.E. Court

Volume 16
Mistletoe, edited by Arndt Büssing

This book is part of a series. The publisher will accept continuation orders which may be cancelled at any time and which provide for automatic billing and shipping of each title in the series upon publication. Please write for details.

MISTLETOE

The Genus *Viscum*

Edited by

Arndt Büssing
Department of Applied Immunology
University Witten Herdecke
Herdecke, Germany

harwood academic publishers
Australia • Canada • France • Germany • India • Japan
Luxembourg • Malaysia • The Netherlands • Russia • Singapore
Switzerland

Printed in Singapore.

Amsteldijk 166
1st Floor
1079 LH Amsterdam
The Netherlands

British Library Cataloguing in Publication Data

A catalogue record for this book is available from the British Library.
ISBN: 90-5823-092-9
ISSN: 1027-4502

Dedicated to my parents,
my wife Claudia,
and our children, Oliver and Annika,
for their attendance.

CONTENTS

PREFACE TO THE SERIES

There is increasing interest in industry, academia and the health sciences in medicinal and aromatic plants. In passing from plant production to the eventual product used by the public, many sciences are involved. This series brings together information which is currently scattered through an ever increasing number of journals. Each volume gives an in-depth look at one plant genus, about which an area specialist has assembled information ranging from the production of the plant to market trends and quality control.

Many industries are involved such as forestry, agriculture, chemical food, flavour, beverage, pharmaceutical, cosmetic and fragrance. The plant raw materials are roots, rhizomes, bulbs, leaves, stems, barks, wood, flowers, fruits and seeds. These yield gums, resins, essential (volatile) oils, fixed oils, waxes, juices, extracts and spices for medicinal and aromatic purposes. All these commodities are traded worldwide. A dealer's market report for an item may say "Drought in the country of origin has forced up prices."

Natural products do not mean safe products and account of this has to be taken by the above industries, which are subject to regulation. For example, a number of plants which are approved for use in medicine must not be used in cosmetic products.

The assessment of safe to use starts with the harvested plant material which has to comply with an official monograph. This may require absence of, or prescribed limits of, radioactive material, heavy metals, aflatoxin, pesticide residue, as well as the required level of active principle. This analytical control is costly and tends to exclude small batches of plant material. Large scale contracted mechanised cultivation with designated seed or plantlets is now preferable.

Today, plant selection is not only for the yield of active principle, but for the plant's ability to overcome disease, climatic stress and the hazards caused by mankind. Such methods as *in vitro* fertilisation, meristem cultures and somatic embryogenesis are used. The transfer of sections of DNA is giving rise to controversy in the case of some end-uses of the plant material.

Some suppliers of plant raw material are now able to certify that they are supplying organically-farmed medicinal plants, herbs and spices. The Economic Union directive (CVO/EU No 2092/91) details the specifications for the **obligatory** quality controls to be carried out at all stages of production and processing of organic products.

Fascinating plant folklore and ethnopharmacology leads to medicinal potential. Examples are the muscle relaxants based on the arrow poison, curare, from species of *Chondrodendron*, and the antimalarials derived from species of *Cinchona* and *Artemisia*. The methods of detection of pharmacological activity have become increasingly reliable and specific, frequently involving enzymes in bioassays and avoiding the use of laboratory animals. By using bioassay linked fractionation of crude plant juices or extracts, compounds can be specifically targeted which, for example, inhibit blood platelet aggregation, or have antitumour, or antiviral, or any other required activity. With the assistance of robotic devices, all the members of a genus may be readily screened. However, the plant material must be **fully** authenticated by a specialist.

The medicinal traditions of ancient civilisations such as those of China and India have a large armamentarium of plants in their pharmacopoeias which are used throughout South East Asia. A similar situation exists in Africa and South America. Thus, a very high percentage of the world's population relies on medicinal and aromatic plants for their medicine. Western medicine is also responding. Already in Germany all medical practitioners have to pass an examination in phytotherapy before being allowed to practise. It is noticeable that throughout Europe and the USA, medical, pharmacy and health related schools are increasingly offering training in phytotherapy.

Multinational pharmaceutical companies have become less enamoured of the single compound magic bullet cure. The high costs of such ventures and the endless competition from me too compounds from rival companies often discourage the attempt. Independent phytomedicine companies have been very strong in Germany. However, by the end of 1995, eleven (almost all) had been acquired by the multinational pharmaceutical firms, acknowledging the lay public's growing demand for phytomedicines in the Western World.

The business of dietary supplements in the Western World has expanded from the Health Store to the pharmacy. Alternative medicine includes plant based products. Appropriate measures to ensure the quality, safety and efficacy of these either already exist or are being answered by greater legislative control by such bodies as the Food and Drug Administration of the USA and the recently created European Agency for the Evaluation of Medicinal Products, based in London.

In the USA, the Dietary Supplement and Health Education Act of 1994 recognised the class of phytotherapeutic agents derived from medicinal and aromatic plants. Furthermore, under public pressure, the US Congress set up an Office of Alternative Medicine and this office in 1994 assisted the filing of several Investigational New Drug (IND) applications, required for clinical trials of some Chinese herbal preparations. The significance of these applications was that each Chinese preparation involved several plants and yet was handled as a **single** IND. A demonstration of the contribution to efficacy, of **each** ingredient of **each** plant, was not required. This was a major step forward towards more sensible regulations in regard to phytomedicines.

My thanks are due to the staff of Harwood Academic Publishers who have made this series possible and especially to the volume editors and their chapter contributors for the authoritative information.

Roland Hardman

PREFACE

Mistletoe is still a controversial plant. Growing between heaven and earth, never touching the ground, and not accepting the seasons. Even discussing its clinical impact results in polarisation: Rejected by clinical oncologists but used by practitioners and cancer patients. It is applied as a remedy to treat a broad spectrum of different diseases, such as epilepsy, diabetes, hypertension, arthrosis, hepatitis, HIV infection, labour pains, and cancer.

What is fact and what is fiction in the mistletoe story? The following chapters may give a glimpse that the book is still open, and that the last and final chapter of mistletoe research remains to be written.

Thirty spokes join together in the hub.
It is because of what is not there that the cart is useful.
Clay is formed into a vessel.
It is because of its emptiness that the vessel is useful.
Cut doors and windows to make a room.
It is because of its emptiness that the room is useful.
Therefore, what is present is used for profit.
But it is in absence that there is usefulness.

Lao Tzu, The Tao Te Ching

CONTRIBUTORS

Elida M.C. Alvarez
Member of the Research Career
CONICET
Cátedra de Inmunología-IDEHU
Facultad de Farmacia y Bioquímica
Universidad de Buenos Aires
Junín 956 (1113), Buenos Aires
Argentina

Maria Barberaki
Department of Plant Physiology
Faculty of Agricultural Biotechnology
Agricultural University of Athens
Iera Odos 75
11855 Athens
Greece

Hans Becker
Institute of Pharmacognosy and
Analytical Phytochemistry
University of the Saarland
P.O. Box 151150
66041 Saarbrücken
Germany

Peter A. Berg
Medical Clinic
Department of Internal Medicine II
University of Tübingen
Otfried-Müller-Strasse 10
72076 Tübingen
Germany

Josef Beuth
Institute for Scientific Evaluation for
Naturopathy
University of Cologne
Robert-Koch Strasse 10
50931 Cologne
Germany

Arndt Büssing
Krebsforschung Herdecke
Department of Applied Immunology
University Witten/Herdecke
Communal Hospital
58313 Herdecke
Germany

Teresa B. Fernández
Cátedra de Inmunología-IDEHU
Facultad de Farmacia y Bioquímica
Universidad de Buenos Aires
Junín 956 (1113), Buenos Aires
Argentina

Gianfranco Grazi
Institut Hiscia
Verein für Krebsforschung
Kirschweg 9
4144 Arlesheim
Switzerland

Alberto A. Gurni
Cátedra de Farmacobotánica
Facultad de Farmacia y Bioquímica
Universidad de Buenos Aires
Junín 956 (1113), Buenos Aires
Argentina

Silvia E. Hajos
Member of the Research Career
CONICET
Cátedra de Inmunología-IDEHU
Facultad de Farmacia y Bioquímica
Universidad de Buenos Aires
Junín 956 (1113), Buenos Aires
Argentina

Spiridon Kintzios
Department of Plant Physiology
Faculty of Agricultural Biotechnology
Agricultural University of Athens
Iera Odos 75
11855 Athens
Greece

Donald W. Kirkup
Herbarium, Royal Botanic Gardens
Kew, Richmond
Surrey TW10 3AE
England

Elmar Lorch
Helixor Heilmittel
Hofgut Fischermühle
P.O. Box 8
72344 Rosenfeld
Germany

Won-Bong Park
College of Natural Science
Seoul Women's University
Seoul, 139-774
Korea

Uwe Pfüller
Institute of Phytochemistry
University Witten/Herdecke
Stockumer Strasse 10
58453 Witten
Germany

Roger M. Polhill
Herbarium, Royal Botanic Gardens
Kew, Richmond
Surrey TW10 3AE
England

Hartmut Ramm
Institut Hiscia
Verein für Krebsforschung
Kirschweg 9
4144 Arlesheim
Switzerland

Rafael A. Ricco
Cátedra de Farmacobotánica
Facultad de Farmacia y Bioquímica
Universidad de Buenos Aires
Junín 956 (1113), Buenos Aires
Argentina

Markus Scheibler
Institut Hiscia
Verein für Krebsforschung
Kirschweg 9
4144 Arlesheim
Switzerland

Michael Schietzel
Krebsforschung Herdecke
Communal Hospital
University Witten/Herdecke
58313 Herdecke
Germany

Gerburg M. Stein
Krebsforschung Herdecke
Communal Hospital
University Witten/Herdecke
58313 Herdecke
Germany

Carlos A. Taira
Member of the Research Career
CONICET
Cátedra de Farmacología
Facultad de Farmacia y Bioquímica
Universidad de Buenos Aires
Junín 956 (1113), Buenos Aires
Argentina

Wilfried Tröger
Helixor Heilmittel
Hofgut Fischermühle
P.O. Box 8
7234 Rosenfeld
Germany

Konrad Urech
Institut Hiscia
Verein für Krebsforschung
Kirschweg 9
4144 Arlesheim
Switzerland

Beatriz G. Varela
Cátedra de Farmacobotánica
Facultad de Farmacia y Bioquímica
Universidad de Buenos Aires
Junín 956 (1113), Buenos Aires
Argentina

Marcelo L. Wagner
Cátedra de Farmacobotánica and
Museo de Farmacobotánica "Juan A. Domínguez"
Facultad de Farmacia y Bioquímica
Universidad de Buenos Aires
Junín 956 (1113), Buenos Aires
Argentina

Delbert Wiens
Department of Biology
University of Utah
Salt Lake City, Utah 84112
USA

1. INTRODUCTION: HISTORY OF MISTLETOE USES

ARNDT BÜSSING

Krebsforschung Herdecke, Department of Applied Immunology, University Witten/Herdecke, Communal Hospital, 58313 Herdecke, Germany

MISTLETOE: THE MYTHICAL PLANT

Mistletoes belong to the families *Loranthaceae* and *Viscaceae*, which both are taxonomically related to each other, and share the order *Santalales*. The family of *Viscaceae* has seven genera (*Arceuthobium, Dendrophthora, Ginalloa, Korthalsella, Notothixos, Phoradendron, Viscum*) and several hundred species world-wide. The European white-berry mistletoe (*Viscum album* L.) is an evergreen, dioecious plant growing half-parasitically on its host. *V. album* is a small shrub with linear lanceolate leathery leaves which persist for several seasons. The yellowish-green flowers grow in the sprout axil and develop the translucent, whitish berries in the late fall and early winter. Theophrastos (371–286 BC) described mistletoe as an evergreen plant growing on pine and fir trees, fed to animals. He recognised that mistletoe does not grow on the earth, but is spread to trees by birds whose excretions contain "seeds" from the berries.

Unlike other plants, mistletoe does not follow a 12 month vegetation period, never touches the earth, and blooms during winter (see Ramm *et al.*, this book). The plant was considered sacred by the Celtic Druids, because in the dead of winter, when branches of the oak tree were bare, the mistletoe is still green and flourished without having roots on the earth. To them, the plant represented ever-lasting life. Plinius G.P. Secundus (23–79 AC) reported the Druids to ceremoniously remove mistletoe from oak trees with a golden sickle on the 6th day after new moon. They believed the plant was an antidote for poisons and ensured fertility, and to possess miraculous properties to cure each illness as an *omnia sanans* (Historia naturalis, liber XVI, 95).

The mythical plant mistletoe was used in ancient times together with aromatic substances as an incense (Rätsch, 1997; Fischer-Rizzi, 1999). Scenting of houses, animals and man with the blend of these herbs was suggested to protect against lightning, spells and bad dreams (Marzell, 1923), or to get in contact with "elementary power of nature" and to find the "inner stability" (Fischer-Rizzi, 1999). Burning of the "light-grown" mistletoe may release the "captured elementary power of light" (F. Wollner, personal communication).

Even today, the evergreen mistletoe is a symbol of fertility and good luck, and kissing under branches of mistletoe during the Christmas tide is popular in many

European countries and North America. This exchange of kisses is interpreted as a promise to marry and a prediction of happiness and long life. In the Middle ages and later, branches of mistletoe were hung from ceilings to ward off evil spirits, and were located over house and stable doors to prevent the entrance of witches and ghosts (Tabernaemontanus, 1731; Marzell, 1923). In contrast, a Tanganyika species of *Loranthus* was used to put spells on somebody (Watts and Breyer-Brandwisk, 1962).

According to Nordic mythology as described in the "Edda" (Snorri Shurlason, 1200 AC), which is a collection of ancient Viking poems, the god of shamans (Odin) and the goddess of love and beauty (Free) bound all being of earth from ever harming their son Balder. However, the tiny mistletoe did not take root in the earth, and therefore, was not bound to the oath. Balder was killed at Loki's instigation by a twig of mistletoe (*mistilteinn*) shot by his blind brother Hödur. Interestingly, von Tubeuf (1923) clearly stated that mistletoe is unknown in the northern parts of Scandinavia and Island, and thus, the legend must be rooted in other areas of Europe, or probably the Near East.

Mistletoe as a Remedy

The intentions of mistletoe uses were manifold and conflicting in several cases. According to the Greek physician and author Dioskorides (15–85 AC), Hippocrates (460–377 BC) used the mistletoe to treat diseases of the spleen and complaints associated with menstruation. Plinius (23–79 AC) reported mistletoe from oak trees, when applied as a chewed pulp, to be beneficial for epilepsy, infertility, and ulcers (Historia naturalis, liber XXIV, 12). Around 150 AC, the Platonist Celsus reported the use of mistletoe in the treatment of swellings or tumours (De Medicina, liber V, 18 and 23). Although the conditions have never been accurately defined, the use of mistletoe in medicine was referred also by the Alexandrian physician and surgeon Paulus Aegineta (625–690 AC), and the Persian philosopher and physician Avicenna (Ibn-Sina, 980–1037 AC) (Foy, 1904). However, as suggested by Marzell (1923), "oak mistletoes" described and used by most ancient scientists might not be identical with the white-berry mistletoe (*V. album*), which is green even in winter, but might be the yellow-berry *Loranthus europaeus* which in turns looses the leaves during winter. It seems unlikely that also the Celtic druids may have used *Loranthus*, which is common on oak trees, as the exceptional botanical properties of *V. album* predisposes its use as a miraculous plant.

During the middle ages, mistletoe was recommended as a treatment for epilepsy because it never fell to the ground. In fact, Paracelsus (Theophrastus Bombastus von Hohenheim, 1493–1541) recommended that epileptics should wear oak mistletoe on their right hand, and rosaries were made of mistletoe (*mistlin paternoster*) during the 15th century to prevent the disease (Marzell, 1923). In the 12th century, the abbess and composer Hildegard von Bingen (1098–1179) wrote treaties about natural history and medicinal uses of plants, animals and stones, and described mistletoe as a treatment for diseases of spleen and liver. In the European herbals of the 16th century, mistletoe was reported to warm, to soften, to astringent, and to be more sharp than bitter (von Tubeuf, 1923). Epilepsy and diseases of the kidneys and spleens were treated with mistletoe from oaks; poultices and plasters with mistletoe and other plants were used to treat ulcers, fractures of the bone, and labour-pains. In 1729, the English physician Colbatch reported that not alone mistletoe from oak trees but also from other deciduous trees was an effective treatment for epilepsy.

In 1731, the physician and botanist Jacobus Theodorus Tabernaemontanus reported in a collection of herbal remedies that mistletoe never touching the ground is used for childhood epilepsy, when applied as a pulverised drug, or even by wearing it as a silver-amulet. Mistletoe was also applied for deworming children, to treat labour-pains, gout, and affections of lung and liver (Tabernaemontanus, 1731). However, when applied in wine, mistletoe was used to treat leprosy. When applied as a plaster, mistletoe was suggested to be beneficial in the treatment of mumps and fractures, while the binding of their leaves to the palms and sole will heal hepatitis (Tabernaemontanus, 1731).

During the 18th century, mistletoe was applied for "weakness of the heart" and oedema. These indications have been recorded in the homeopathic *materia medica* until today (Boericke, 1992). By the end of the 19th century, mistletoe was rejected by the scientists as a folklore remedy. The only remaining acceptable application was the mistletoe-containing ointment, Viscin, which was a yellowish bird-lime. Viscin was reported to be effective for eczema, ulcers of the feet, burns, and granulating wounds (Riehl, 1900; Klug, 1906). In a German encyclopaedia from 1934, it was stated that mistletoe did not contain clinically relevant compounds (Oestergaards Lexicon, 1934). However, an encyclopaedia from 1962 reports the historical use of mistletoe as a cure for epilepsy, convulsions, and worms (Duden Lexicon, 1962). The scientific interest on mistletoe awakened in the 20th century, as Gaultier (1907, 1910) investigated the effect of oral or subcutaneous applications of fresh *Viscum album* L. extracts on blood pressure in man and in animals.

Use of mistletoe as a remedy was not restricted to Europe, but is also developed in other parts of the world, or were transformed to similar plants. Argentine mistletoe (*Ligaria cuneifolia*) is used in local folk medicine to treat hypertension (Domínguez, 1928; Ratera and Ratera, 1980; Martinez-Crovetto, 1981). In fact, depending on the host tree, mistletoe may reduce or even increase blood pressure (Domínguez, 1928; Ratera and Ratera, 1980). Argentine mistletoe was further used as an external remedy to stabilise fractures of the bones, and as a lime for birds and insects (Ratera and Ratera, 1980). In veterinary medicine, Argentine mistletoe was used as a sedative (Arenas, 1982).

The Northern American mistletoe (*Phoradendron* subspecies) was used by the Native Americans as an abortifacient, and by farmers and veterinarians for "clearing cattle" (Hanzlik and French, 1924). Indeed, uterine contractions were reported in pregnant and non-pregnant women and animals after administration of mistletoe (Howard, 1892; Hanzlik and French, 1924). Howard (1892) also reported the use of mistletoe in menorrhagia, post-partum haemorrhage, and in haemoptysis.

In Japanese folk medicine, mistletoe (*Taxillus kaempferi*) was a remedy to treat hypotension (Nanba, 1980), while mistletoe (Sangjisheng; *Ramulis Loranthi et Visci*: *Viscum coloratum* (Kom.) Nakai, *Loranthus parasitikus* (L.) Merr., *Loranthus yadoriki* Sieb.) was used in Traditional Chinese Medicine to treat hypertension, spasms of the heart, rheumatic pain, threatened abortion and locally to treat frostbite (Paulus and Ding Yu-he, 1987).

In West-India, a tea prepared from mistletoe leaves is traditionally used to treat diabetes (Peters, 1957), while a preparation of *Viscum articulatum* is given in fever attended with aching limbs (Chopra *et al.*, 1956).

In Africa, *Viscum aethiopicum* was a remedy to treat diarrhoea, and *Loranthus* and *Viscum* subspecies was used by the Zulu as an enema for stomach troubles in children (Watts and Breyer-Brandwisk, 1962). To treat diabetes mellitus, *Loranthus bengwensis* L. has been widely used in Nigerian folk medicine (Obatomi *et al.*, 1994). The Xhosa used a decoction of a *Viscum* subspecies in lumbago and sore throat, while a *Loranthus* subspecies was used as a poultice for orchitis in Southern Rhodesia (Watts and Breyer-Brandwisk, 1962). A Tanganyika species of *Loranthus* is used in witchcraft, and another species of *Loranthus* as a poison by the Zezuru (Watts and Breyer-Brandwisk, 1962).

In 1920, *Viscum album* L. was introduced as a cancer treatment by Rudolf Steiner (1861–1925; Steiner, 1985), founder of anthroposophy. He recommended a drug extract produced in a complicated manufacturing process combining sap from mistletoe harvested in the winter and summer (Steiner, 1989). Clinical evaluations of mistletoe as an adjuvant cancer treatment have expanded. During the 1960s, Vester and Nienhaus (1965) isolated carcinostatic protein fractions which were recognised later as viscotoxins and mistletoe lectins.

Table 1 Short history of mistletoe uses in Europe.

5th century BC	Diseases of the spleen and complaints associated with menstruation
1st century AC	Cure every illness as an *omnia sanans*, antidote for poisons, infertility
12th century AC	Epilepsy, diseases of the liver and spleen, infertility for women, ulcers
16th century AC	Epilepsy, diseases of the kidneys and spleen
18th century AC	Epilepsy, diseases of lung and liver, labour pains, "weakness of the heart", oedema
19th century AC	Rejection of mistletoe uses by the scientists as a folklore remedy
20th century AC	Hypertension, arthrosis and cancer

Until today, several groups of researchers are still writing exciting new chapters of the mistletoe story. This book may give a glimpse that the last chapter of the mistletoe story remains to be written.

ACKNOWLEDGEMENTS

I am grateful to Sabine Rieger and Ursula Haid, Rosenfeld, for the finding of all the ancient writings, and to Fred Wollner, Lemgo, and Sigrun Scherneck, Ulrichstein, for their advises in the mythical use of mistletoe as an incense.

REFERENCES

Arenas, P. (1982) Recolección y agricultura entre los indíginas Maká del Chaco Boreal. *Parodiana*, **1**, 171–243.

Boericke, W. (1992) *Manual der Homöopathischen Materia Medica*. Heidelberg, Karl F. Haug Verlag.

Chopra, R.N., Nayar, J.L., Chopra, I.C. (1956) *Glossary of Indian Medicinal Plants*. Council of Scientific & Industrial Research. New Delhi, India.

Colbatch (1776) *Abhandlung von dem Mistel und dessen Kraft wider die Epilepsie*. Altenburg.

Domínguez, J.A. (1928) *Contribuciones a la Materia Médica Argentina*. Peuser ed., Buenos Aires

Duden Lexikon in 3 Bänden. Vol. 2. (1962) Dudenverlag des Bibliographischen Instituts, Mannheim.

Fischer-Rizzi, S. (1999) Botschaft an den Himmel. Anwendung, Wirkung und Geschichten von duftendem Räucherwerk. Heinrich Gugeldubel Verlag, München (1996) and Wilhelm Heyne Verlag, München, pp, 15 and 63–64.

Foy, G. (1887) Mistletoe. *Med. Press and Circular*, pp. 588.

Gaultier, M.R. (1907) Action hypotensive de l' extrait aqueux de gui. *La Semaine Médicale*, **43**, 513.

Gaultier, R. (1910) Etudes physiologiques sur le Gui (*Viscum album*). *Arch Internat. De Pharmacodynam et de Thérapie*, **20**, 96–116.

Hanzlik, P.J., French, W.O. (1924) The pharmacology of *Phoradendron flavescens* (American mistletoe). *The Journal of Pharmacology and Experimental Therapy*, **23**, 269–306.

Howard, H.P. (1892) Mistletoe as an oxytoxic. *Medical News*, **60**, 547–548.

Klug (1906) Viscolan, eine neue Salbengrundlage. *Deutsche Medizinische Wochenschrift*, **51**, 2071–2072.

Martínez Crovetto, R.(1981) *Las plantas utilizadas en Medicina Popular en el Noroeste de Corrientes (República Argentina)*. Miscelanea N° 69. S.M. Tucumán: Fundación Miguel Lillo ed.

Marzell, H, (1923) Die Mistel in der Volkskunde. In K. von Tubeuf, (ed.), *Monographie der Mistel*. R. Oldenbourg Verlag, München, Berlin, pp. 28–37.

Nanba, T. (1980) *Genshokuwakanyakuzukan*. Hoikusya, Tokyo, p. 172.

Obatomi D.K., Bikomo, E.O., and Temple, V.J. (1994) Anti-diabetic properties of the African mistletoe in streptozotocin-induced diabetic rats. *Journal of Ethnopharmacology*, **43**, 13–17.

Oestergaards Lexikon in zwanzig Bänden. Vol. XIII. (1934) Peter J. Oestergaard Verlag, Berlin-Schöneberg.

Paulus, E, Ding Ye-he (1987) *Handbuch der traditionellen chinesischen Medizin*. Haug Verlag, Heidelberg, pp. 241–242.

Peters, G. (1957) Übersichten Insulin-Ersatzmittel pflanzlichen Ursprungs. *Deutsche Medizinische Wochenschrift*, **82**, 320–322.

Rätsch, C. (1997) Enzyklopädie der psychoaktiven Pflanzen. ATV-Verlag, Zürich, pp. 82.

Ratera, E.L., Ratera, M.O. (1980) *Plantas de la flora Argentina empleadas en Medicina Popular*. Hemisferio Sur ed., Buenos Aires, pp. 82–85.

Riehl, G. (1900) Ueber Viscin und dessen therapeutische Verwendung. *Deutsche Medizinische Wochenschrift*, **41**, 653–655.

Steiner, R. (1985) *Geisteswissenschaft und Medizin* (GA 312). 13. Vortrag 2. April 1920. Rudolf Steiner Verlag, Dornach, pp. 252–255.

Steiner, R. (1989) *Physiologisch-therapeutisches auf Grundlage der Geisteswissenschaft.* (GA 314). Besprechungen mit praktizierenden Ärzten, 22. April 1924. Rudolf Steiner Verlag, Dornach, pp. 294–295.
Tabernaemontanus, J.T. (1731) *Kräuterbuch.* Johann Ludwig König, Basel, pp. 1376–1377.
von Tubeuf, K. (1923) *Monographie der Mistel.* R. Oldenbourg Verlag, Berlin, München.
Vester, F., Nienhaus, J. (1965) Cancerostatic protein components from *Viscum album. Experientia*, **21** (suppl 4), 197–199.
Watts, J.M., Breyer-Brandwisk, M.G. (1962) *The Medicinal & Poisonous Plants of Southern & Eastern Africa.* 2nd Edition. E. & S. Livingstone Ltd., Edinburgh, London, pp. 731–732.

2. VISCUM IN THE CONTEXT OF ITS FAMILY, VISCACEAE, AND ITS DIVERSITY IN AFRICA

DONALD W. KIRKUP[1], ROGER M. POLHILL[1] and DELBERT WIENS[2]

[1]*Herbarium, Royal Botanic Gardens, Kew, Richmond, Surrey TW10 3AE, England*
[2]*Department of Biology, University of Utah, Salt Lake City, Utah 84112, U.S.A.*

INTRODUCTION

More has been written about *Viscum album* L., the common European species, than any other mistletoe. Indeed for many centuries, *Viscum album*, and to a lesser extent *Loranthus europaeus* L., the only other European mistletoe, epitomised what was known about this intriguing group of plants. As other parts of the world were explored many more shrubby plants were discovered that grew exclusively on other woody plants and had an intimate connection, the haustorium, that enabled them to extract solutes and water from their hosts. Most of them had fruits similar to the common European mistletoes, with a highly viscous middle layer to the fruit wall that stuck the seed to a branch when deposited there by a bird. Distinctive features were found in some genera. Some species from southern South America and Australia are root parasites, growing into lianes or small trees. The Australian Christmas tree, *Nuytsia floribunda* R. Br., is notable as being a small tree, but also for having dry, winged fruits. The dwarf mistletoes attributed to the genus *Arceuthobium*, are serious parasites of conifers in the northern hemisphere. They are dispersed by tiny seeds squirted considerable distances by water pressure built up in the small fruits. Overall, however, the mistletoes were considered to share sufficient features for all to be included in a single family, Loranthaceae, by most botanists until the middle of this century. In the census by Engler & Krause in the second edition of Engler & Prantl's Die natürlichen Pflanzenfamilien in 1935 some 1,000 species were included in the family.

It is now apparent that many of the features shared by mistletoes are also found in other families of the order of sandalwoods, the Santalales. The Loranthaceae and Viscaceae are now considered to be of separate origin within that order. The special adaptations for an aerial existence seem to have arisen separately in the two groups.

FAMILY RELATIONSHIPS

The family Loranthaceae is thought to be most closely related to the Olacaceae and the Viscaceae to the Santalaceae. The separation of the families gained credence in the 1960s, with an appreciation of significant differences in the embryology and basic chromosome numbers (Dixit 1962; Barlow 1964). Most accounts in the century up to then ranked Viscaceae as a subfamily of Loranthaceae. Later that decade, Kuijt (1968, 1969) removed certain genera from Viscaceae as the Eremolepidaceae, noting differences in the inflorescence structure. That made the morphological separation more clear cut. The Santalaceae have a single whorl of tepals like Viscaceae in this restricted sense, but Loranthaceae have both calyx and corolla, which suggests a relationship nearer to the more basal family Olacaceae, which has the calyx differentiated to varying degrees.

The main differences between the families are indicated in Table 1, adapted from Kuijt (1969) and Calder (1983). The flowers of Viscaceae are generally less than 3 mm across and adapted to pollination by small insects. Except in the primitive genus *Ginalloa*, the anthers are much modified, opening by one or more small pores. In Loranthaceae the flowers are mostly pollinated by birds and tend to be large and showy. The pollen grains are generally more elaborate and distinctly 3-lobed in Loranthaceae.

Both families have much modified ovaries and peculiar embryology. In Viscaceae there are no ovules and the ovary shows little internal differentiation. A small central placenta, the mamelon, occurs in some genera, but is much reduced or lacking in *Viscum*. Two to several sporogenous cells differentiate out of the mamelon. The embryo lacks a suspensor, except in *Viscum*, where it is very short. The arrangement is generally similar in Loranthaceae, but in some primitive genera there are vestiges of what might be interpreted as a vascularised central placenta and four ovarian locules, without, however, any trace of ovules. The embryos are more elaborate than in Viscaceae and often extend up the style. The common features of a marked reduction in the female gametophyte are striking, but structures

Table 1 Major features distinguishing Viscaceae from Loranthaceae, adapted from Kuijt (1969) and Calder (1983).

Loranthaceae	*Viscaceae*
1. Flowers 5 mm or, with calyx and corolla differentiated, 4–7-merous, bisexual	1. Flowers mostly minute (< 3 mm), with a single whorl of tepals, 2–4-merous, unisexual
2. Anthers opening by slits	2. Anthers opening by pores, much modified, except in *Ginalloa*
3. Pollen trilobate, rarely triangular or spherical	3. Pollen spherical
4. Embryo-sacs several, of Polygonum type	4. Embryo-sacs 2, of Allium type
5. Embryo-suspensor long and multicellular	5. Embryo-suspensor short or absent
6. Endosperm compound, usually achlorophyllous	6. Endosperm simple, chlorophyllous
7. Medium to large chromosomes with a base number of $x = 12$, reduced to $x = 8$ and $x = 9$ in advanced taxa	7. Large chromosomes, with basic numbers of $x = 10$, 11, 12, 13 or 14(primary base number of 14).

apparently homologous with the mamelon do occur in both Santalaceae and Olacaceae. Kuijt (1968, 1969) supposes that what may be envisaged as the retention of juvenile features in the ovary may be correlated with the independent transformation of the root-system into a haustorium necessary for a parasitic mode of existence.

The two families appear to have separate geographic origins and a different cytological history (Barlow 1983a, 1990). The more primitive genera of Loranthaceae occur in South America, New Zealand and Australia. A primary base number of x = 12 occurs in the relictual genera, with divergence to x = 8 in the remaining New World genera and x = 9 in most of the Old World genera, with one apparently derived group reverting to x = 12. In Viscaceae the basic number of x = 14 is consistent through the family, though some genera, notably *Notothixos* and *Viscum*, have reduced numbers, down to x = 10 in some species. The family seems to have its origins in south-east Asia, from where it has dispersed mainly in the tropics and the northern hemisphere.

Nickrent and Soltis (1995) have recently reported comparative molecular analyses of the chloroplast gene *rbc*L and the ribosomal gene 18S rDNA. They included representatives from the seven genera of Viscaceae, *Antidaphne* and *Eubrachion* from Eremolepidaceae, *Santalum* and *Osyris* from Santalaceae, *Misodendrum* from Misodendraceae, *Gaiodendron* from the Loranthaceae, *Opilia* from the Opiliaceae and *Schoepfia* from the Olacaceae. The *rbc*L analysis showed three clades (1) Loranthaceae, Misodendraceae, Olacaceae and Opiliaceae, (2) Eremolepidaceae and Santalaceae (the genera intermixed), and, linked to the latter, (3) Viscaceae. The 18S rDNA sequences similarly linked Loranthaceae with Opiliaceae and Olacaceae, and gave strong support for Viscaceae linked to Santalaceae plus Eremolepidaceae, but Misodendraceae was linked with Santalaceae rather than with Loranthaceae. Nickrent and co-workers have found higher than average substitution rates in rRNA and rDNA in parasitic plants generally, which makes them useful tools for discrimination at the family level in Santalales and even generic level in Viscaceae (Nickrent and Franchina 1990; Nickrent and Soltis 1995; Nickrent 1996).

GENERA OF VISCAEAE

Viscaceae is a family of seven genera and about 400 species. The world distribution of the genera is mapped by Barlow (1983a). All the evidence suggests that this is a close-knit group. The circumscription of genera has been relatively uncontroversial, but there is no clear indication of the relationship between them (Barlow 1997). The adaptations to a parasitic mode of existence are mostly common to all the genera. The structure of the flowers and fruits is relatively simple, without a great deal of variation in the family. The anthers open in various ways, but this does not seem closely correlated with other features. The inflorescences have become modified to a considerable extent, the components often condensed and reduced to minute structures that are somewhat difficult to interpret. The leaves are sometimes reduced to no more than scales and specialised prophylls and cataphylls are commonplace but uninformative. Unlike Loranthaceae, the chromosome complement provides little information. Recent analyses by Nickrent (1996) using the chloroplast gene *rbc*L and the ribosomal gene 18S rDNA, reaffirm a close-knit group without a clear resolution of relationships within it. Both genes do indicate, however, a link between *Arceuthobium* and *Notothixos* and between *Ginalloa* and *Korthalsella. Viscum* tended to be basal to the whole family in the 18SrDNA cladograms and linked as a polytomy with *Notothixos* and *Arceuthobium* in the *rbc*L trees. The two American genera *Dendrophthora* and *Phoradendron* were closely associated, as is evident from their morphology (Kuijt 1959, 1961).

Viscum has about 100 species. About twenty of these are found in mainland tropical Asia and considerable diversity occurs there. Most species are now found in Africa, which has 45 species, and Madagascar, which has a further 30 species. The few species belonging to the *Viscum album* group have adapted to more temperate regions in Eurasia and several of the Asian groups extend sparsely southwards to eastern Australia. The genus has spread into all sorts of wooded habitats and some species have become very specialised parasites of particular hosts, sometimes occurring only on other mistletoes, principally Loranthaceae.

Korthalsella has minute seeds that seem to have permitted long-distance dispersal on the feet and feathers of birds. The genus is now widely distributed from tropical Africa and Indian Ocean islands to Japan, Australia, New Zealnd and the Pacific. The plants are small, inconspicuous and often overlooked. They are difficult to classify and some of the species formerly recognised may prove to be no more than races of a few widespread species. The size of the genus is estimated at somewhere between 7 and 25 species (Barlow, 1997).

The dwarf mistletoes, belonging to the genus *Arceuthobium*, are also often small and with complexities in their taxonomy, but have been studied to a much greater extent because of their economic importance (Hawksworth and Wiens 1996). Unlike almost all other mistletoes, they have become parasites of pine trees. The 40 species range from the Sino-Himalayan region to the Mediterranean, North and eastern Africa and the Canary Islands, then concentrated in North America in Mexico and northern California, with a few species extending north and then eastwards in the Great Lakes region. As indicated above, the genus has a very specialised dispersal technique, the tiny seeds ejected by water pressure up to 10 m or so. The seeds are prone to lodge among pine needles and after rain tend to slide into the axils where they germinate. The genus may well have spread westwards from the Himalayan region and entered North America when the continents were still joined in the Eocene, about 50 million years ago (Hawksworth and Wiens, 1996; Lavin and Luckow, 1993).

Ginalloa has 9 species from Sri Lanka to the Philippines and south-eastwards to Papua New Guinea and the Solomon Islands (Barlow 1997). The species occur mostly in rain-forest communities and there is little host specificity. *Nothothixos*, with eight species, has a similar distribution, extending a little further, from Sri Lanka to eastern Australia and the Santa Cruz Islands. Some species of *Noththixos*, towards the edge of the range, are adapted to drier habitats and tend to be more host specific (Barlow1983b, 1997).

Phoradendron, with about 200 species, and *Dendrophthora*, with about 50 species, are wholly American and widely distributed in the warmer parts of both continents. They are closely related to each other and show trends that can be related to those found in *Notothixos* and *Ginalloa* (Kuijt 1959). They may have migrated on angiosperm hosts about the same time as *Arceuthobium*.

VISCUM ALBUM GROUP

The greater part of this book is concerned with the biology and uses of *Viscum album*. Populations extend further into temperate regions than any other species of the genus. The species has a number of remarkable characters, some of which seem to

Table 2 Morphological groups of Asian and Australian species of *Viscum*. Adapted and updated from Danser (1941).

Group	*Characters*	*Species*
V. album group	Dioecious. Inflorescences terminal.	*V. album* L.; *V. dryophilum* Rech.f.; *V. alni-formosanae* Hayata; *V. fargesii* Lecomte; *V. nudum* Danser; *V. cruciatum* Sieber ex Boiss.
V. mysorense group	Monoecious. Inflorescences mostly lateral, with terminal female flower and subsidiary cymules male or female	*V. mysorense* Gamble; *V. articulatum* Burm.f.; *V. nepalense* Sprengel; *V. liquidambaricum* Hayata; *V. stenocarpum* Danser; *V. angulatum* Heyne; *V. ramosissimum* Roxb. ex DC.; *V. loranthi* Elmer
V. ovalifolium group	Monoecious. Flowers in triads, middle female, laterals male (or female).	*V. ovalifolium* DC.; *V. wrayi* Gamble; *V. acaciae* Danser; *V. indosinense* Danser; *V. scurruloideum* Barlow; *V. exile* Barlow
V. orientale group	Monoecious. Flowers in triads augmented by adventitious flowers, middle female, laterals male or female	*V. orientale* Willd.; *V. heyneanum* DC.; *V. monoicum* DC.; *V. multinerve* Hayata
V. capitellatum group	As *V. orientale* group but middle flower male (or all female)	*V. trilobatum* Talbot; *V. capitellatum* Smith; *V. yunnanense* H.S. Kiu, *V. katikanum* Barlow; *V. whitei* Blakely; *V. bancroftii* Blakely

be adaptations to more severe climates. The most obvious feature is the strictly terminal inflorescences, the male and female flowers borne on different plants. Growth is continued by axillary branches, bearing inconspicuous scale-like prophylls at the base and one pair of leaves at the tip. This gives the characteristic dichotomous branching pattern to the fertile parts of the plant. The petiolar region of the leaf has a cavity, which forms a pocket to encase the buds before the leaves expand. There are several other species in Eurasia with exactly the same sort of inflorescence and branching pattern (Table 2), but with modifications in the vegetative parts, notably the Chinese species, *V. fargesii*, with narrow leaves, and *V. nudum*, with the leaves all reduced to small scales (Danser 1941). *V. alniformosanae* in Taiwan and the more recently described *V. dryophilum*, from Afghanistan and Pakistan, are segregates of more questionable status. The variation of *V. album* will be described in more detail in subsequent chapters, but includes a reasonably distinctive subspecies, subsp. *meridianum* (Danser) D.G. Long, in the Himalayan region (Grieson & Long, 1983) and well-known forms with coloured fruits further east (Park, this book).

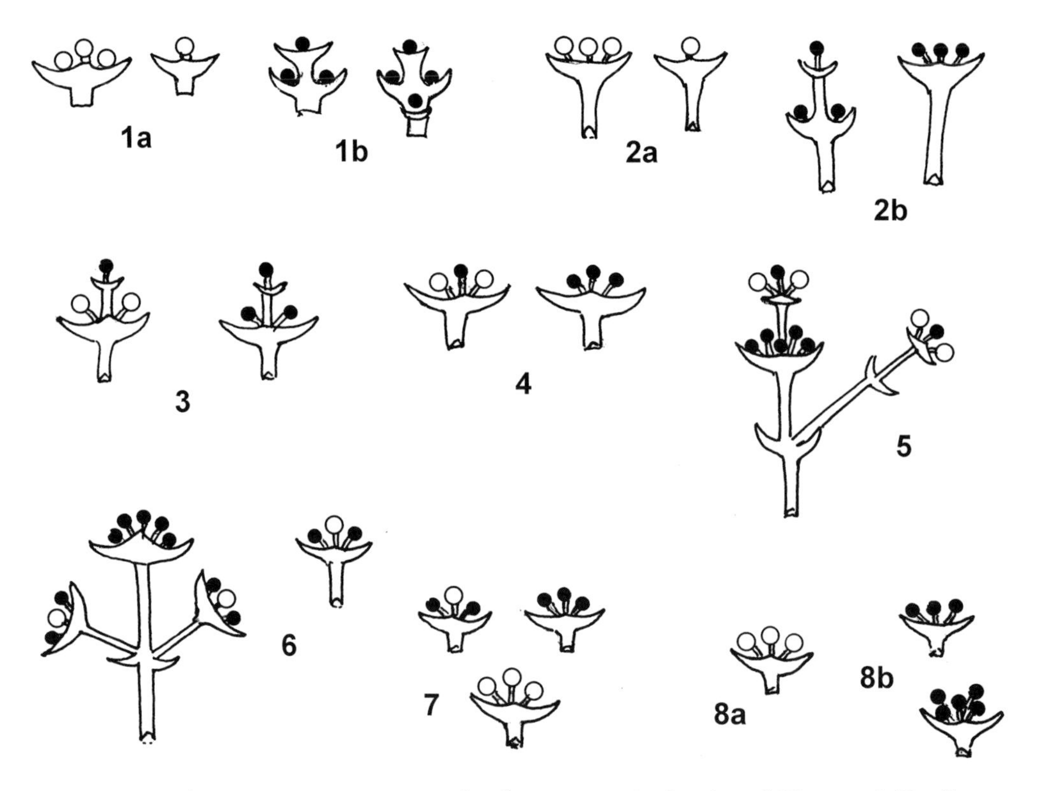

Figure 1 Schematic Representation of Inflorescences in Species of *Viscum*. 1 *V. album* (a male; b female plant); 2 *V. cruciatum* (a male, b female plant); 3 *V. mysorense*; 4 *V. ovalifolium*; 5 *V. orientale*; 6 *V. capitellatum*; 7 *V. triflorum*; 8 *V. congolense* (a male, b female plant). O = male flower; solid O = female flower.

Viscum cruciatum, from North Africa and Spain eastwards to Afghanistan, can also be included in the *V. album* group, but the inflorescence and branching pattern are not quite so modified. In *V. album* and its closest allies the male peduncle is short and swollen, with a pair of bracts subtending three flowers Figure 1.1a). In *V. cruciatum* the peduncle is longer and more flattened, with one to three sessile flowers in the bracteal cup (Figure 1.2a). There are terminal inflorescences in bifurcations, but also lateral inflorescences there and at many nodes below, often up to four together. The branching pattern is correspondingly less obviously dichotomous. The male flowers are, however, up to 6–8 mm long, the largest known in the genus (the tropical species rarely have flowers more than 3 mm, those of *V. album* are about 4 mm long).

The female inflorescences of *V. album* have shorter peduncles than the male, bearing 3–5 flowers, the terminal usually subtended by a pair of bracts, below which further bracts subtend the other flowers (Figure 1.1b). In *V. cruciatum* the peduncle is extended in the female like the male and there are only three flowers, the central one sometimes with a short pedicel and then sometimes with its own small bracts (Figure 1.2b). In all species of the group there is a tendency for supplementary inflorescences to develop from the axils of the bracts and thus appear to be superposed.

The variations of inflorescence structure in the *V. album* group provide some clues to their derivation. All the other species of *Viscum* in Asia are monoecious, with male and female flowers generally borne in the same inflorescence. Sometimes there is a shift in sex ratios with a high proportion of female flowers. On the basis of their inflorescences, Danser (1941) divided the monoecious Asian species into four groups (Table 2). The *V. mysorense* group has axillary inflorescences with a sessile cymule of a single flower, subtended by a bracteal cup, with subsidiary cymules developing lateral to the first one. The first-formed flower is female and the lateral flowers are female or male (Figure 1.3). *V. mysorense* is a species of southern India and there are about half a dozen further species with leaves reduced to scales, including the common *V. articulatum* and *V. nepalense*. The group extends from India to China and thinly southwards to Australia.

The *Viscum ovalifolium* group has the inflorescence reduced to a cymule of three flowers, the central female, the laterals male, subtended by a bracteal cup and borne in the axils or on short lateral shoots (Figure 1.4). Subsidiary cymules often develop around the first one and some cymules may have flowers of only one sex. The small group is restricted to the South-East. Asia and Australia. *V. ovalifolium* extends from Myanmar (Burma) to Hong Kong and southwards to Queensland. It is polymorphic (Barlow 1997) and allied to two other species of the group, *V. wrayi* from Peninsular Malaysia, Borneo and Sumatra, and *V. exile* from the Celebes. The species occur mostly in lowland forests below 1300 m. *V. scurruloideum*, a little known species from western Java, can probably be included in this group.

The *V. orientale* group is restricted to continental Asia and its immediate islands. It is similar to the *V.ovalifolium* group, but the inflorescences are enlarged by the development of adventitious flowers in the bracteal cup (Figure 1.5). The sex ratios are often skewed towards a high proportion of females. *V. orientale* and *V. heynea-*

num grow in the drier parts of Sri Lanka and India, *V. monoicum* extends from India to China and *V. multinerve* occurs in China and Taiwan.

Table 3 Groups of African species of *Viscum*. Main features and list of species.

Group 1. *Viscum triflorum* group. Young internodes slightly flattened, broadened upwards, ribbed. Leafy. Monoecious. Dichasia mostly lateral, mostly 3-flowered, central male, lateral or all female. Bracteal cups pedunculate. Berries sessile, smooth. n = 14. Widespread in diverse forest types. *V. triflorum* DC., *V. petiolatum* Polh. & Wiens.

Group 2. *Viscum congolense* group. Young internodes as *V. triflorum* group. Leafy, the base with a slight cavity. Dioecious, with skewed sex ratios. Dichasia mostly lateral, often more than 3-flowered, especially in female. Bracteal cups pedunculate. Berries subsessile, smooth to minutely tuberculate. 2n = 22 in female, 23 in male. Forests of Congolian region and East Africa. *V. congolense* De Wild., *V. fischeri* Engl., *V. luisengense* Polh. & Wiens.

Group 3. *Viscum rotundifolium* group. Young internodes subterete, ribbed. Leafy. Monoecious. Dichasia 3-flowered, central male, in sessile to pedunculate bracteal cups. Berries stipitate, smooth. n = 14. South African shrublands., *V. rotundifolium* L.f., *V. pauciflorum L.f.*, *V. schaeferi* Engl. & K. Krause.

Group 4. *Viscum tuberculatum* group. Stems terete to angular, ribbed. Leafy. Monoecious. Dichasia 3-flowered, unisexual or the central flower female, in sessile bracteal cups. Berries sessile, tuberculate when young. n = 12 or 23. Eastern Rift zone and SE. African coastal region, in drier forests and associated bushland. *V. tuberculatum* A. Rich., *V. obovatum* Harv.

Group 5. *Viscum decurrens* group. Internodes strongly flattened, ribbed. Leafy. Monoecious, with sex ratio skewed to female. Dichasia usually unisexual, 3(–5)-flowered in pedunculate bracteal cups (longer in female). Berries stipitate, smooth. Congolian forest, only on *Symphonia*. *V. decurrens* (Engl.) Baker & Sprague.

Group 6. *Viscum album* group. Young internodes rounded to only slightly flattened. Leafy, the base with a distinct cavity. Dioecious, with skewed sex ratios. Dichasia terminal and lateral or more commonly all terminal, the branching then strongly dichotomous, 3-flowered or males single, sometimes superposed. Bracteal cups pedunculate. Flowers unusually large. Berries stipitate, smooth. n = 10. Atlas Mts. of North Africa., *V. album* L., *V. cruciatum* Sieber ex Boiss..

Group 7. *Viscum obscurum* group. Internodes angled and flattened at first, becoming rounded. Leafy. Dioecious. Dichasia of male (1–)3(–5)-flowered in sessile bracteal cups; female 1-flowered in sessile to pedunculate cups. Berries sessile to generally stipitate, smooth. n = 15, 14, 12 or 11. South African woodlands and shrublands. *V. obscurum* Thunb., *V. oreophilum* Wiens, *V. crassulae* Eckl. & Zeyh., *V. subserratum* Schltr.

Group 8. *Viscum longiarticulatum* group. Internodes strongly flattened, slightly ribbed. Leafy, but deciduous. Dioecious. Bracteal cups shortly pedunculate. Male flowers in 3s; female single. Berries sessile, tuberculate when young. East African basement-complex mountains (Usambaras). *V. longiarticulatum* Engl..

Table 3 *continued.*

Group 9. *Viscum shirense* group. Stems strongly flattened, often ribbed. Leafless. Dioecious. Dichasia with male flowers in 3s or single; females single. Berries smooth or tuberculate, stipitate. n = 14. Eastern and southern African forests, woodlands and shrublands. *V. engleri* Tieghem, *V. anceps* E. Mey. ex Sprague, *V. combreticola* Engl., *V. shirense* Sprague, *V. goetzei* Engl., *V. cylindricum* Polh. & Wiens, *V. congdoni* Polh. & Wiens.

Group 10. *Viscum menyharthii* group. Stems rounded, except sometimes on youngest internodes. Leafless. Dioecious. Dichasia with male flowers in 3s, rarely 1 or more than 3; female single. Bracteal cups sessile. Berries sessile to shortly stipitate, smooth or tuberculate. n = 14. Zambezian and East African woodlands. *V. menyharthii* Engl. & Schinz, *V. chyuluense* Polh. & Wiens, *V. littorum* Polh. & Wiens, *V. verrucosum* Harv., *V. subverrucosum* Polh. & Wiens, *V. calvinii* Polh. & Wiens, *V. gracile* Polh. & Wiens, *V. griseum* Polh. & Wiens, *V. continuum* E. Mey. ex Sprague, *V. hildebrandtii* Engl., *V. tenue* Engl..

Group 11. *Viscum bagshawei* group. Stems flattened and ribbed to terete and smooth, occasionally (*V. minimum*) very reduced and largely endophytic. Leaves rudimentary or lacking. Monoecious, with male flowers often sparse, or dioecious (*V. capense*, *V. dielsianum*). Dichasia usually 1-flowered (3-flowered in *V. minimum*) in sessile or rarely pedunculate (*V. grandicaule*) bracteal cups. Anthers often fused into synandria in monoecious species. Berries sessile to stipitate, smooth to tuberculate. n = 14 (*V. minimum*), 28 (*V. bagshawei*, *V. schimperi*) or 10 (*V. capense*, *V. hoolei*). Widespread in forest and drier habitats, but often with restricted host ranges., *V. dielsianum* Dinter ex Neusser, *V. capense* L.f., *V. hoolei* (Wiens) Polh. & Wiens, *V. minimum* Harv., *V. grandicaule* Polh. & Wiens, *V. schimperi* Engl., *V. bagshawei* Rendle, *V. loranthicola* Polh. & Wiens, *V. iringense* Polh. & Wiens

Finally in Asia, there is a small group of species ranging from Sri Lanka to Australia, similar to the *V. orientale* group, but with the middle flower of the cymule male, the laterals female, rather than vice versa (Figure 1.6). There are apparently six species in the *V. capitellaum* group and they are mostly epiparasites of other mistletoes. *V. capitellatum* occurs in India and Sri Lanka, *V. trilobatum* in southern India, *V. yunnanense* in southern China, *V. katikanum* in Papua New Guinea and *V. whitei* in northern and north-eastern Australia. *V. bancroftii*, has the leaves reduced to scales, but is sympatric with and possibly even conspecific with *V. whitei* (Barlow 1983c).

Among the Asian groups of species, the *Viscum album* group is not closely matched and all are are monoecious. Among present-day species, the *V. album* group is most similar to a small group of dioecious species in tropical Africa, the *V. congolense* group (Table 3; Figure 1.8 and Figure 2). The three species of this complex have discrete ecogeographic ranges, *V. congolense* in the Guineo-Congolian lowland forests of equatorial Africa, *V. fischeri* and *V. luisengense* in the montane forests of East Africa. Like the *V. album* group, these species are dioecious and have skewed, strongly female biased, sex ratios. In Africa the affinities of the *V. congolense* group have been associated with *V. triflorum*, a protean species widespread on

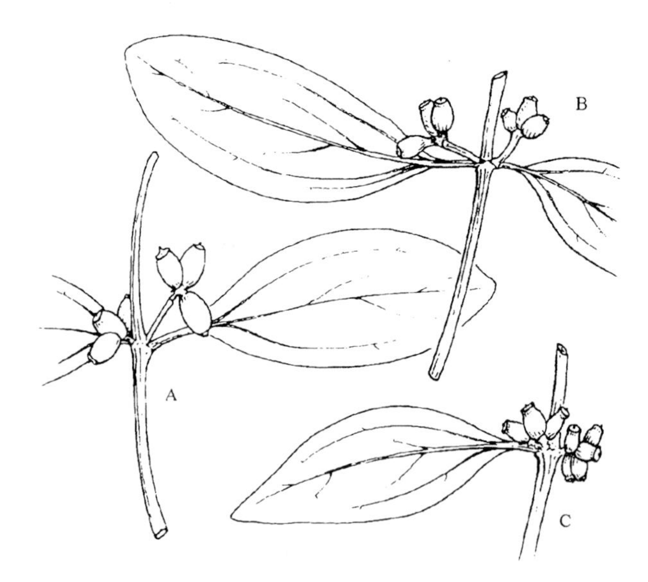

Figure 2 Fruiting branches, × 1, of 1,A, *Viscum congolense;* 2,B, *V. fischeri;* 3,C, *V. luisengense.* Drawn by Christine Grey-Wilson and reproduced from Polhill and Wiens (1998).

the continent and offshore islands. *V. triflorum* has the flowers generally in triads in a bracteal cup, with the central flower male and the laterals female or the triads unisexual (Figure 1.7 and Figure 3), as in the other groups of monoecious species in Africa. The *V. congolense* group has inflorescences comparable to the *V. triflorum*

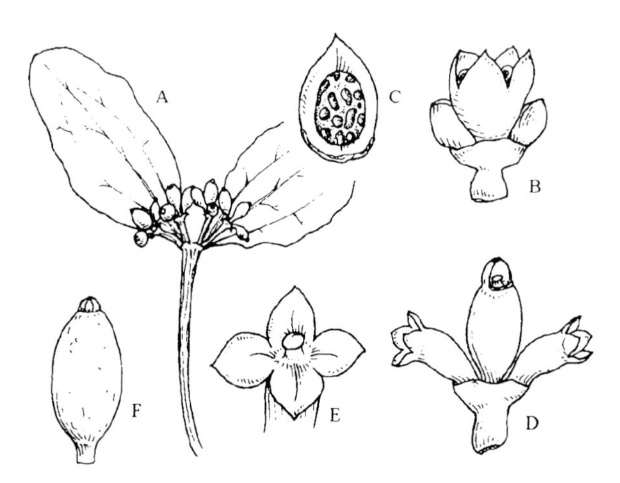

Figure 3 *Viscum triflorum.* 1, A, fertile node × 1; 2, B, staminate triad × 6; 3,C, tepal and porate anther × 8; 4,D, pistillate triad × 6; 5,E, detail of female flower × 12; 6,F, fruit × 4. Drawn by Christine Grey-Wilson and reproduced from Polhill and Wiens (1998).

group except that they are always unisexual and on separate plants and for a tendency to have more than three flowers in the bracteal cup, especially on female plants. The adventitious flowers and the unbalanced sex ratios compare well with the *V. orientale* group of continental Asia and suggest that perhaps the *V. congolense* group may have more affinity with Asian elements than previously supposed.

These putative affinities receive strong support from the cytology. *V. triflorum* has a basic chromosome number of n = 14, characteristic of the family as a whole. Progressive dysploid increase and reduction is commonplace in the genus. Dioecious species in the genus characteristically have sex-associated and floating chromosome translocations that are virtually absent in monoecious species (Wiens and Barlow 1979, Barlow 1981, 1983a). This suggests that the translocations are primarily associated with the origin and establishment of dioecy, by bringing non-allelic male- and female-determining factors into genetic linkage (Barlow 1997). The smallest and simplest translocations occur in other African dioecious species with x = 14 (Wiens and Barlow, 1979) and these belong to several morphological groups mentioned in the discussion below.

The three African species of the *Viscum congolense* group are dioecious with skewed, strongly female biased sex ratios, and a unique sex-determining system in which the male sex-determining factors are located on a translocation chain of chromosomes instead of rings, as in other dioecious species of the genus. These have been described for *V. fischeri* (Barlow and Wiens 1976) but the nature of the chromosome chains and general features of the meiotic genomes in *Viscum congolense* and *V. luisengense* have yet to be described. Male plants of *V. fischeri* have 2n = 23 and constantly produce seven bivalents and a multivalent chain of nine chromosomes at meiosis. Regular assortment results in transmission of 11- and 12-chromosome genomes via the pollen. Female plants have the chromosome number 2n = 22 and are homozygous for the 11-chromosome genome. The multivalent chain in the males is a consequence of reciprocal translocations, one of which was Robertsonian and one of which involved the chromosome carrying the sex-determination factors. There is a constant female-predominant sex ratio of approximately 1:2, possibly maintained by gamete selection; the genes involved may have been linked with the sex-determination mechanism through the translocation system

V. album and *V. criuciatum* have x = 10, a sex-associated ring of 8 is most common, 10 occurs and 12 is rare (Barlow *et al.* 1978, Aparicio 1993). The process of enlargement of the translocation complexes through incorporation of additional translocations has thus occurred along with the dysploid reduction in chromosome number.

On present evidence it would seem that the *Viscum album* group diverged at about the same time as the *Viscum congolense* group at a fairly early stage in the evolution of dioecy, which initiated a significant secondary radiation of the genus in Africa and Madagascar. The Asian groups seem relatively discrete, but there is little obvious sequential arrangement among them or with the African elements. The spread of the *Viscum triflorum* group may have occurred relatively early. *V. triflorum* has a very scattered distribution in the African forests and also occurs at considerable distance from the main distribution on off-shore islands. Much of the radiation in Madagascar probably stems from the *V. triflorum* group (Wiens 1975; Balle 1964a, b) and has resulted in numerous dioecious species, some with x = 13. On the mainland *V. triflorum* occurs in forests of very different types, both wetter and drier, lowland and montane. This sort of distribution is characteristic of species in many groups of plants, including Loranthaceae, that seem to have established in Africa prior to the major rifting and uplift of the continent around and somewhat prior to the Miocene (Polhill and Wiens 1998). By contrast the species of the *V. congolense* group are restricted to different modern ecosystems, *V. congolense* in the equatorial lowland forests of the Guineo-Congolian region, *V. fischeri* and *V. luisengense* in the Afromontane region along the eastern Rift Valley, *V. fischeri* in the drier rain-shadow forests in the Kenya highlands and *V. luisengense* in the wetter types of forest in southern Tanzania. Both have rather small present-day distributions. It seems possible that the progenitors of this group migrated from continental Asia around the Miocene, some 15 million years ago, when the land-surface of Africa had been transformed by major uplifting and rifting and land-connections were re-established with Asia after a separation of some 25 million years by the Tethys Sea (Grove 1983). The *Viscum album* group could well have originated in the general Sino-Himalayan region about this time or a little earlier and spread with the Laurasian floristic components outwards to Japan and North Africa and subsequently into northern regions. It is not possible to speculate further from what we now know, but techniques in molecular systematics are rapidly developing to trace affinities at species level and a more rigorous assessment of the position of the *V. album* group should soon be feasible.

VISCUM IN AFRICA

As indicated above, *Viscum* has speciated to its greatest extent in Africa, which has 45 species, with a further 30 in Madagascar. The African species have been revised recently by Polhill and Wiens (1998). They can be divided provisionally into eleven groups. The more obvious characters on which the species have been recognised for floristic purposes are shown in Table 4 and a data matrix for the species and these characters are given in Table 5. A more rigorous evolutionary investigation would no doubt reveal new characters, but even a preliminary assessment may be helpful in giving some idea of the radiation of the genus over the continent. Much insight is provided by chromosomal features. A considerable coverage of basic information has been achieved, but there are still substantial gaps in the total number of species examined and more significantly in detailed comparison of the karyotypes needed to score characters for evolutionary analysis. The biological characters of sex distribution within populations and host specificity have also been omitted.

The other data has been analysed using the computer program Hennig86 (Farris 1989). Owing to the time that would have been required to run the analysis using empirical methods, faster, hueristic algorithms were employed instead. These algorithms estimate a minimal length tree by a single pass through the data, the branches

Table 4 Groups of species in Africa. List of characters for analysis.

	Final weights	*Character*	*Character state*
0	2	Young internodes cross-section	0 = slightly flattened; 1 = angled or terete; 2 = markedly flattened (tape-like)
1	3	Young internodes surface	0 = ribbed; 1 = not ribbed
2	6	Leaves	0 = well developed; 1 = rudimentary; 2 = scales only
3	10	Cavity at base of leaf	0 = absent; 1 = slight; 2 = marked
4	0[1]	Sex ratios	0 = subequal; 1 = markedly skewed to female
5	10	Inflorescences	0 = all or many terminal; 1 = all or mostly lateral
6	4	Flowers in dichasium	0 = mostly 3 (rarely 1 or more than 3); 1 = mostly 3 or more (especially female); 2 = 1 in male, 3 in female; 3 = several in male, 1 in female; 4 = 1 in male and female
7	10	Anthers	0 = separate; 2 = some or all fused in synandria
8	0	Style	0 = present; 1 = very reduced or absent
9	4	Stigma	0 = small; 1 = expanded; 2 = bifurcate
10	0	Berry colour	0 = white; 1 = yellow or orange; 2 = red
11	0	Stipe of berry	0 = absent; 1 = present
12	0[1]	Chromosome number	0 = n: 14; 1 = 2n: 32 in female, 23 in male; 2 = n: 10; 3 = n: 15; 4 = n: 12; 5 = n: 11; 6 = n: 23; 7 = n: 28
13	0[1]	Translocations	0 = none; 1 = present
14	0[1]	Host range	0 = wide; 1 = limited to 1–few hosts; 2 = only on legumes
15	2	Gender	0 = monoecious; 1 = dioecious
16		Sex of dichasia	0 = dichasia with male & female flowers; 0 = dichasia unisexual
17	0	Immature berry surface	0 = smooth; 1 = tuberculate,
18	0	Mature berry surface	0 = smooth; 1 = tuberculate
19	1	Peduncle of male bracteal cup	0 = present; 1 = lacking
20	1	Peduncle of female bracteal cup	0 = present; 1 = lacking

[1] Not used in analysis.

of that initial tree are then swapped to find all other trees of the same length. Meaningful trees are obtained only with some considerable weighting of selected characters as shown in Table 4. The final weights were obtained by an iterative procedure which involves calculating the fit of each character to the range of trees produced. The analysis is re-run with the adjusted weights and this sequence is repeated until the character weightings no longer change. A strict consensus tree of the resulting 6 equally parsimonious trees produced by this method is shown in Figure 4, which is used for compiling Table 3 and for the following outline discussion.

Table 5 Groups of African species of *Viscum*. Character states for cladistic analysis. For characters see Table 3.

1. *V. triflorum*	000000000000000000000
2. *V. petiolatum*	0000000000?0???010000
3. *V. congolense*	000110100000110110000
4. *V. fischeri*	000110100010110110000
5. *V. cruciatum*	110211201121210110000
6. *V. album*	110211001101210110000
7. *V. schaeferi*	100000001111??0000011
8. *V. rotundifolium*	100000001011000000000
9. *V. pauciflorum*	100000001011000000011
10. *V. obscurum*	100000300121300110011
11. *V. oreophilum*	100000300111000110000
12. *V. crassulae*	100000300110401110000
13. *V. subserratum*	100000300121500111000
14. *V. luisingense*	000110200020110111000
15. *V. tuberculatum*	100010000010600011111
16. *V. obovatum*	110000000010400001111
17. *V. decurrens*	200010001101??1010000
18. *V. longiarticulatum*	2000003011?0??0111100
19. *V. menyharthii*	112000301110001111011
20. *V. bagshawei*	112010411111702010011
21. *V. calvinii*	112000301020??0110011
22. *V. gracile*	12000300010??0110011
23. *V. dielsianum*	111000000200??0110011
24. *V. tenue*	112000400010000110011
25. *V. loranthicola*	112010411110??1010011
26. *V. capense*	111000401000200110011
27. *V. hoolei*	111010411000200010011
28. *V. chyuluense*	112000301110001110011
29. *V. littorum*	1120003001?1???110011
30. *V. verrucosum*	112000301111002111111
31. *V. griseum*	1120003000?1002111011
32. *V. subverrucosum*	112000301111002111011
33. *V. continuum*	112000300011002110000
34. *V. hildebrandtii*	112000300011002110010
35. *V. iringense*	112000411111??2011011
36. *V. goetzei*	202000400111??1110011
37. *V. congdonii*	212000301111000110011
38. *V. combreticola*	202000301110000111011
39. *V. cylindricum*	2020004011?0000110011
40. *V. engleri*	202000300111000110000
41. *V. grandicaule*	21200040102000?011000
42. *V. schimperi*	212010410011000010011
43. *V. shirense*	202000300111000110011
44. *V. anceps*	202000300111002111100
45. *V. minimum*	112000001011001000000

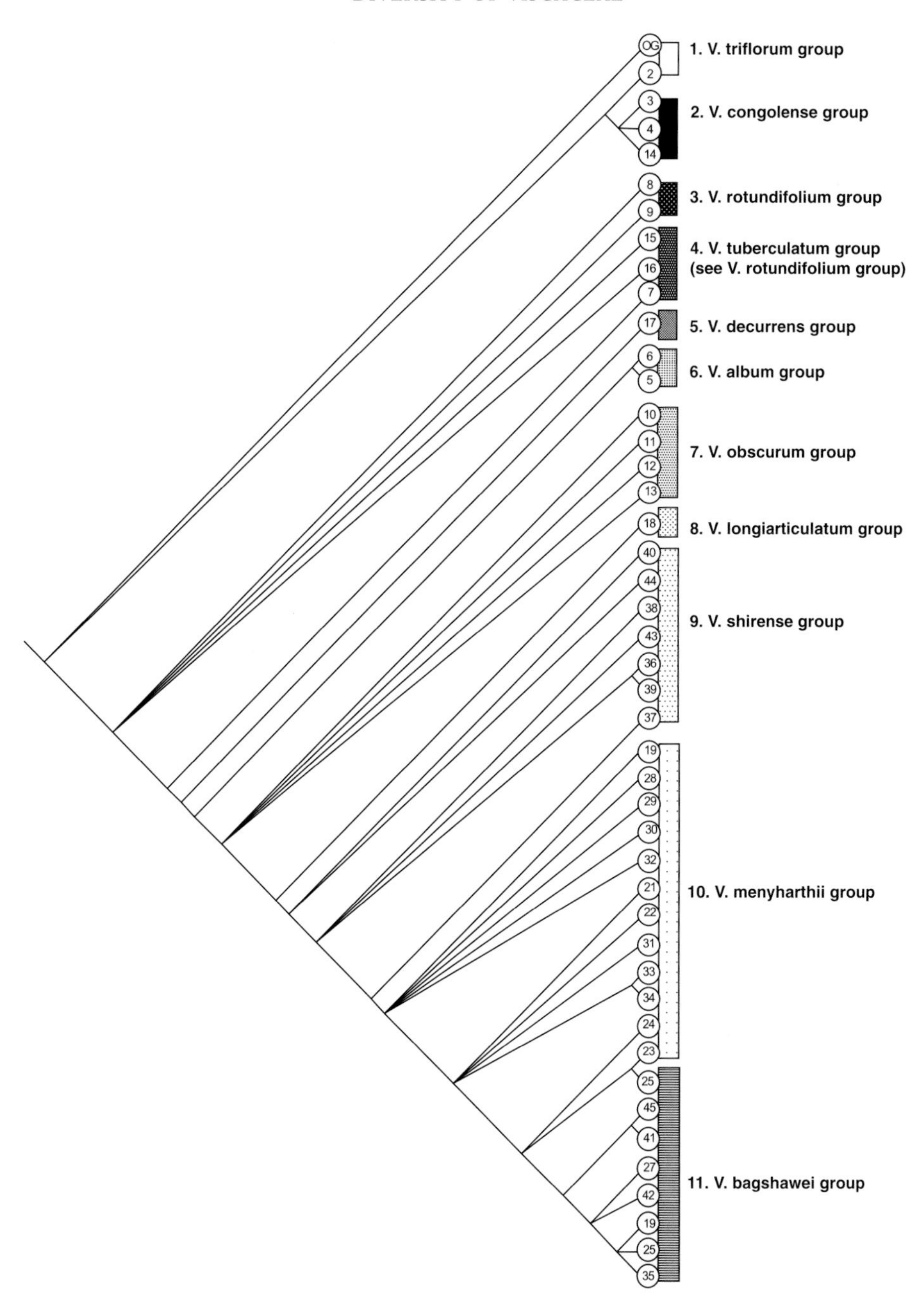

Figure 4 Groups of African Species of *Viscum*. Strict Consensus Tree from Cladistic Analysis of Characters and Character States in Tables 4 and 5 using Henig 86....**og** = outgroup (V. triflorum).

Unlike the Asian representatives, the leafless species are not so obviously related to the species with normally developed leaves. As indicated above, the dioecious *Viscum congolense* and *V. album* groups have derived features in common and may have their origins in the Sino-Himalayan region. The other groups seem to be more related to the *V. triflorum* group in Africa, either directly or indirecty.

Four of the derived groups are monoecious. The *Viscum rotundifolium* group has three species in southern Africa in mixed woodland and bushland. *V. rotundifolium* is widespread from the Cape to southern Angola and north-eastwards to central Zimbabwe, crossing several phytochoria and occurring on a wide range of hosts. *V. pauciflorum* occurs only in the mountains of the south-western Cape Province. *V. schaeferi* links with the *V. tuberculatum* and subsequent groups in the cladogram, but is closely similar to *V. rotundifolium*. It occurs in drier woodland and mixed bushland in Namibia, with isolated populations in the northern Cape and North-western Province. It commonly grows on *Boscia* and mimics that host to some extent in the clustering and shape of its small leaves. *V. rotundifolium* is closely sympatric with *V. pauciflorum* in the mountains east of Worcester and with *V. schaeferi* in the northern Cape and southern Namibia. They all have n = 14 and inflorescences like the *V. triflorum* group, but with ribbed, subterete internodes and relatively small, somewhat thickened leaves.

The *Viscum tuberculatum* group has just two closely related species. *V. tuberculatum* is widespread in drier forests and associated bushland the length of the Eastern Rift Valley in a broad zone from Ethiopia to the Northern Province of South Africa and Angola. *V. obovatum* is closely related and extends the range south-eastwards along the coast from southern Mozambique to the eastern Cape and Swaziland. Both occur on a wide variety of hosts. They are vegetatively similar to the *V. rotundifloum* group, but the inflorescences and berries tend to be sessile rather than stalked and the fruits are tuberculate. More importantly they are evolutionarily separated from that group and from each other by their chromosome numbers, n = 23 in *V. tuberculatum* and n = 12 in *V. obovatum*. The group would seem to be independently derived from the *V. triflorum* group.

Viscum decurrens (Figure 5.A) is an isolated species with unusually flattened internodes for a leafy species. It occurs only on *Symphonia globulifera* L.f. and, although often overlooked, has quite a wide range in the Congolian forests from southern Nigeria to Angola and western Uganda. It is monoecious like *V. triflorum*, but the dichasia are usually unisexual, sometimes with more than 3 flowers and the male inflorescences much sparser than the females, suggesting links to the *V. orientale* and *V. congolense* groups. Two other rarely collected and apparently local western forest species may belong to this affinity. *V. petiolatum*, known only from the type gathering made in montane forests on the Western Rift of eatern Zaire, is provisionally included in the *V. triflorum* group. It is monoecious, with n = 14, but usually has unisexual dichasia. *V. grandicaule* (Figure 6.A & 6.B) is another isolated species from Pagalu (Annobon) Island of the west coast of Gabon. It is also monoecious, with unisexual and 1-flowered dichasia, the internodes are strongly flattened and the leaves are reduced to scales. It might belong in this affinity or with the *V. bagshawei* group, a provisional assemblage of dioecious and anomalously monoecious species discussed at the end.

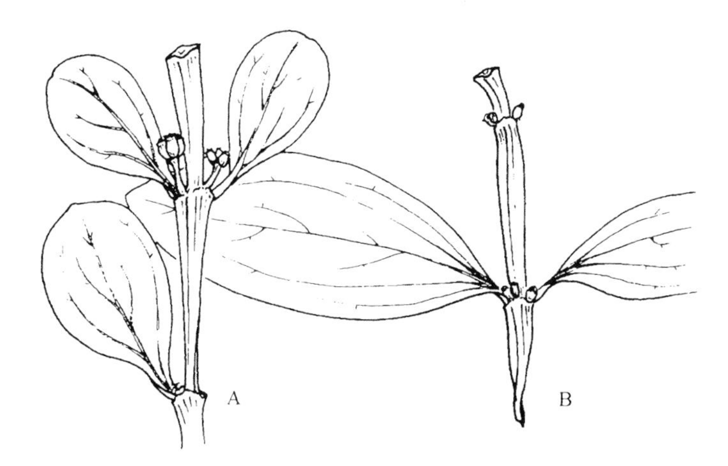

Figure 5 Fruiting branches, × 1, of A, *Viscum decurrens;* B, *V. longiarticulatum.* Drawn by Christine Grey-Wilson and reproduced from Polhill and Wiens (1998).

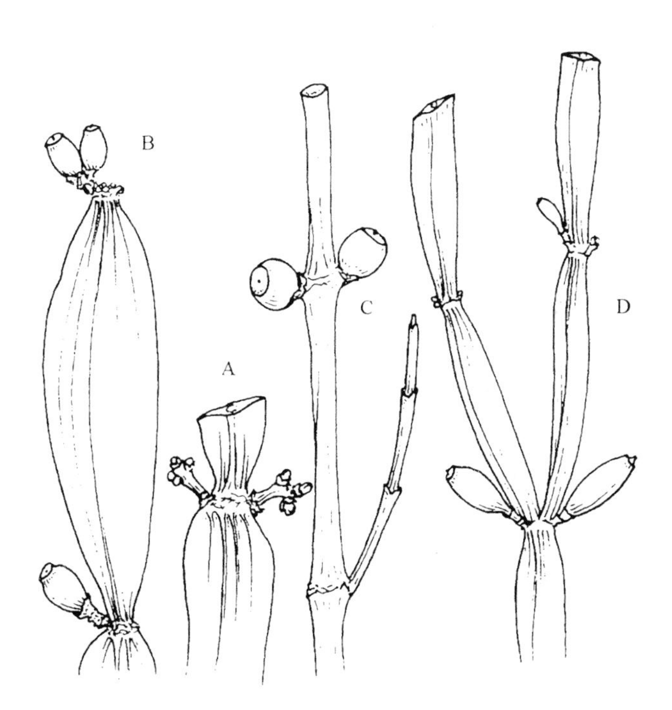

Figure 6 *Viscum grandicaule.* A, staminate flowering node × 1; B, fruiting branchlet × 1. *V. calvinii.* C, fruiting branchlet × 1. *V. engleri.* D, Fruiting branchlet × 1. Drawn by Christine Grey-Wilson and reproduced from Polhill and Wiens (1998).

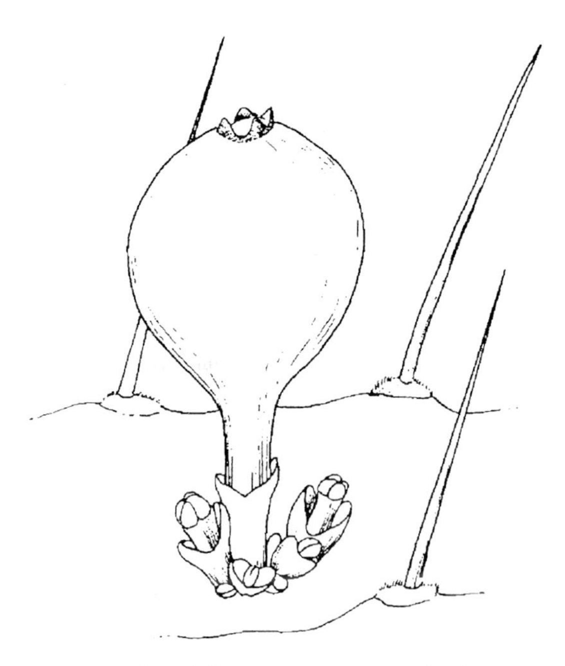

Figure 7 *Viscum minimum*. Habit of plant growing on *Euphorbia* sp. Drawn by Marguerite Scott and reproduced from Polhill and Wiens (1998).

Viscum minimum (Fig. 7) is also of unknown affinity. It is largely endophytic, growing within the stems of two species of succulent *Euphorbia* in the eastern Cape Province of South Africa. The visible parts are minute, about 1 mm high including the inflorescence. The fruit is at least twice the height of the plant and many times its volume. Despite the extreme physiological specialisations and reduction in size, the species retains many reproductive features characteristic of the least specialised species in Africa. It is monoecious, with n = 14, bearing terminal and lateral 3-flowered dichasia, the central flower male, the laterals female. This places it closest to the *V. rotundifolium* group, but it is also fairly close to the least specialised representatives of the *V. bagshawei* group, where it is placed in the cladogram.

There are four reasonably discrete groups of dioecious species, two with normal leaves, two with reduced leaves, each subdivided into groups with the internodes rounded or strongly flattened. Most species examined have n = 14, but in the *V. obscurum* group n = 15, 14, 12 or 11. Twelve species examined by Wiens and Barlow (1979) had translocations that were smaller and simpler than in the *V. album* and *V. congolense* groups. The sex-associated complex consistently appeared in males as rings of 4 or 6. In addition most had a floating ring of 4.

The *Viscum obscurum* group, with rounded internodes and well-developed leaves, has four species in South Africa. Each has a different chromosome number. *V. oreophilum,* with n = 14, occurs in bushland associated with forest edges in the Drakensburg Mountains of Swaziland and the north-eastern provinces of South Africa. *V. obscurum*, with n = 15, occurs in forest associations nearer the coast from near the Cape to the Northern Province, while *V. subserratum*, with n = 11, occurs in drier woodland from northern Kwa Zulu to the Northern and North-West Provinces. *V. crassulae*, with n = 12, is in the Eastern Cape Province and almost exclusively on *Portulacaria afra*, rarely on *Euphorbia*.

Viscum longiarticulatum (Figure 5.B) has not been examined cytologically, but is a seemingly isolated species without clear affinities. It has flattened internodes, the small leaves are deciduous, the bracteal cups are pedunculate, but, like other dioecious groups, the male flowers are in threes and the females are single. It could be envisaged as link between the *V. triflorum* group and the *V. shirense* group.

The two groups with leaves reduced to scales, the *V. shirense* group with flattened stems, and the *V. menyharthii* group with rounded stems, have eighteen species between them and nearly the same number of species in each. They all have n = 14 and inflorescences similar to the *V. obscurum* and *V. longiarticulatum* groups. The *V. shirense* group occurs in the montane and riverine forests of eastern and

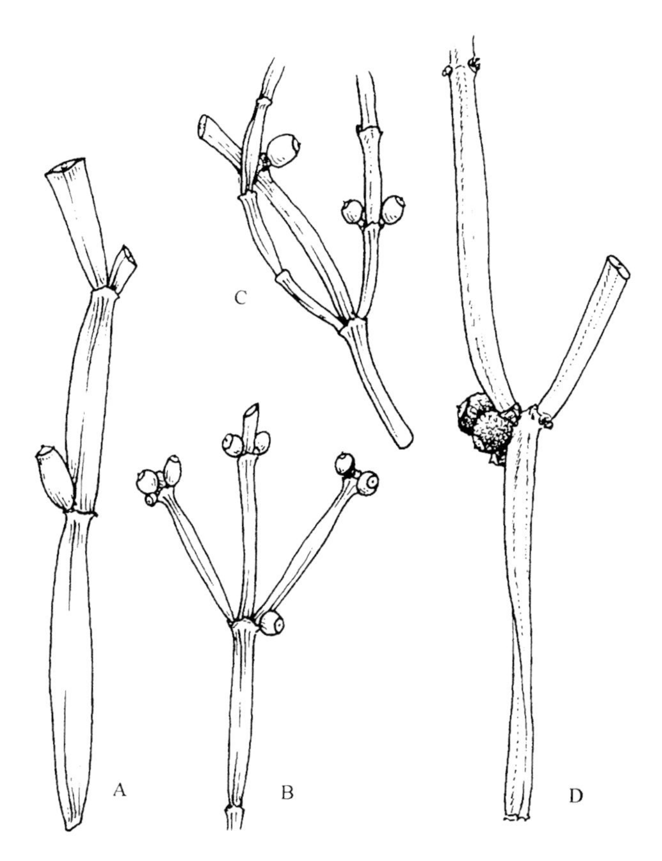

Figure 8 Fruiting branches, × 1, of A, *Viscum cylindricum;* B, *V. congdonii*; C, *V.* goetzei; D, *V. combreticola*. Drawn by Christine Grey-Wilson and reproduced from Polhill and Wiens (1998).

south-eastern Africa, with *V. combreticola* (Figure 8.D) extending further into the drier wodlands, generally on *Combretum* or legumes. *V. engleri*, the species with the most robust habit, is restricted to the forests of the Usambara Mountains of NE. Tanzania. *V. congdonii* (Figure 8.2), *V. cylindricum* (Figure 8.A) and *V. goetzei* (Figure 8.C) occur in the montane forests between southern Tanzania and eastern Zimbabwe. *V. goetzei* is generally, at least, an epiparasite of Loranthaceae. *V. shirense* and *V. anceps* occur in riverine forest and associated bushland, *V. shirense* in tropical Africa from Zaire to Mozambique and *V. anceps* in South Africa from Kwa Zulu to the eastern Cape Province.

The *V. menyharthii* group occurs principally in the woodlands of the Zambezian region, with outliers in the *Acacia* woodlands of the Somali-Masai region and the Karoo. *V. tenue* is an exception, occurring in the montane forests of Tanzania, but could be linked with the *V. shirense* group. The main species are separated eco-geographically, either allopatric or on different hosts. Five of them occur only on species of *Acacia*. *V. chyuluense*, on the Kenya-Tanzania border region, is apparently an obligate parasite of Loranthaceae. *V. menyharthii* generally occurs on *Ficus* and always so in some areas. When on other hosts physiological races may have been established.

There is a residual group of leafless species that seem to have reverted in part to the monoecious condition. The species fall into two subgroups, one tropical, the other in South Africa. The inflorescences are produced with the dichasia one-flowered and the male flowers sparse. The monoecious species tend to have the anthers fused into a central synandrium (Kuijt *et al.* 1979). The tropical species of the *V. bagshawei* group are similar to the *V. shirense* and *V. menyharthii* groups and often confused with them. *V. bagshawei* is a polyploid, with n = 28, and superficially similar to *V. hildebrandtii* in the *V. menyharthii* group, and like it is a parasite of *Acacia*. It occurs in mesic *Acacia* woodland at higher elevations along the Eastern Rift from Ethiopia to the Kenya highlands and westwards to the northern part of the Western Rift in eastern Zaire, Rwanda, Burundi and Uganda. It is replaced by *V. iringense* in the drier *Acacia* woodlands of central and southern Tanzania. *V. loranthicola* occurs a little further west and south in the Zambezian region and, as the epithet implies, is an epiparasite of Loranthaceae. *V. schimperi* has n = 28, like *V. bagshawei*, but the anthers show no more than a tendency to fuse and the stems are flattened more like the *V. shirense* group. It occurs in drier forests along the Eastern Rift in Ethiopia and Kenya. *V. grandicaule*, from Pagalu Island, is much more robust with very flattened stems and has not been examined cytologically. It might belong here or in the affinity of the *V. decurrens* group, as mentioned above.

In South Africa there is another series showing similar modifications. *V. capense* is a dioecious species with n = 10. It forms small shrublets principally in the winter rainfall areas of South Africa from the Cape to central Namibia, but with outlying populations in the Northern Province of South Africa. The scale-leaves are conspicuous and distinctive. *V. hoolei* replaces *V. capense* from the eastern Cape to the southwestern part of the Free State and Lesotho. It is so similar to *V. capense* that it has not been recognised as distinct until recently, but is monoecious with the anthers forming synandria. The male flowers are very sparse, so that populations often seem to be wholly female. *V. dielsianum* has also been confused with *V. capense* in the past. It has similar distinctive scale-leaf rudiments, but the dichasia tend to be 3-flowered. It occurs in the Northern Cape Province to SW. Namibia. It is odd that the secondarily monoecious species all tend to have developed synandria, but the phenomenon does seem to have arisen twice from fairly closely related stocks.

Overall *Viscum* seems to have a substantial genepool in Africa. The endemic African species may have arisen from two major immigrant stocks, the *V. triflorum* stock, which has radiated mostly down the Eastern Rift Valley to South Africa, and the *V. congolense* complex, mostly in the forests of equatorial Africa. The *V. album* group comes closest to the latter but may have affinities with the groups in tropical Asia also.

MEDICAL VALUES IN AFRICA

The medicinal values of Viscaceae in Africa have not been investigated to any significant extent. As in other parts of the world, mistletoes in general are considered to have magical properties and uses tend to be primarily for illnesses thought to be of mystic origin. Burkill (1995) gives a general account of the uses of Loranthaceae in West Africa and his remarks can be extrapolated to *Viscum*, which is relatively much less conspicuous in the region. They are used alone or in prescription. They are not used with the host plant, but the mistletoe has to come from the appropriate host and collection has to follow established ritual. The West African names for mistletoes generally include that of the host and group terms for mistletoes as a whole are often used.

Kokwaro (1993) notes that *Viscum fischeri* and *V. tuberculatum* have been used in Kenya as a poultice on the chest for pneumonia. The latter has also been used for liver troubles. Watt & Breyer-Brandwijk (1962) note minor uses in southern Africa by both African and European cultures. *V. capense* and *V. rotundifolium* have been used by Europeans to remove warts. The former species has also been used for bronchial problems and as an astringent, also as a blood coagulant. *V. rotundifolium* is used as a herbal tea, known as Teemohlware. It is believed to cure heart ailments and purify the blood (Oliver 1987). There is no doubt considerable scope for a more rigorous analysis of the putative values of Viscacae in Africa, where Loranthaceae are the more conspicuous group of mistletoes.

REFERENCES

Aparicio, A. (1993) Sex-determining and floating translocation complexes in *Viscum cruciatum* Sieber ex Boiss. (Viscaceae) in southern Spain. Some evolutionary and ecological comments. *Botanical Journal of the Linnean Society*, **111**, 359–369.

Balle, S. (1964a) Les Loranthacées de Madagascar et des archipels voisins. *Adansonia,* nouvelle sér. 8, **4**, 105–141.

Balle, S. (1964b) Loranthacées. In H. Humbert, (ed.), *Flore de Madagascar,* **60**, Muséum national d'Histoire naturelle, Paris, pp. 124.

Barlow, B.A. (1964) Classification of the Loranthaceae and Viscaceae. *Proceedings of the Linnean Society of New South Wales,* **89**, 268–272.
Barlow, B.A. (1981) *Viscum album* in Japan: Chromosomal translocations, maintenance of heterozygosity and the evolution of dioecy. *Botanical Magazine, Tokyo,* **94**, 21–34.
Barlow, B.A. (1983a) Biogeography of Loranthaceae and Viscaceae. In M. Calder and P. Bernhardt, (eds.), *The Biology of Mistletoes*, Academic Press, Sydney, pp. 19–46.
Barlow, B.A. (1983b) A revision of *Notothixos* (Viscaceae). *Brunonia,* **6**, 1–24.
Barlow, B.A. (1983c) A revision of the Viscaceae of Australia. *Brunonia,* **6**, 25–57.
Barlow, B.A. (1984) Loranthaceae and Viscaceae. In A.S. George, (ed.), *Flora of Australia*, **22**, Australian Government Publishing Service, Canberra, pp. 68–145.
Barlow, B.A. (1990) Biogeographical relationships of Australia and Malesia: Loranthaceae as a model. In P. Baas, K. Kalkman and R. Geesink, (eds.), *The Plant Diversity of Malesia*, Kluwer Academic Publishers, Dordrecht, pp. 273–292.
Barlow, B.A. (1997) Viscaceae. In C. Kalkman, D.W. Kirkup, H.P. Noteboom, P.F. Stevens and W.J.J.O. de Wilde (eds.), *Flora Malesiana*, ser. 1, **13**, 403–42.
Barlow, B.A. and Wiens, D. (1971) The cytogeography of loranthaceous mistletoes. *Taxon,* **20**, 291–312.
Barlow, B.A. and Wiens, D. (1975) Permanent translocation heterozygosity in *Viscum hildebrandtii* Engl. and *V. engleri* Tiegh. (Viscaceae). *Chromosoma,* **53**, 265–272.
Barlow, B.A. and Wiens, D. (1976) Translocation heterozygosity and sex ratio in *Viscum fischeri. Heredity,* **37**, 27–40.
Barlow, B.A., Wiens, D., Wiens, C., Busby, W.H. and Brighton, C. (1978) Permanent translocation heterozygosity in *Viscum album* and *V. cruciatum*: sex association, balanced lethals, sex ratios. *Heredity,* **40**, 33–38.
Burkill, H.M. (1995) *The Useful Plants of West Tropical Africa,* ed. 2, 3. Royal Botanic Gardens, Kew.
Calder, D.M. (1983) Mistletoes in focus. In M. Calder and P. Bernhard, (eds.), *The Biology of Mistletoes*, Academic Press, Sydney, pp. 1–18.
Danser, B.H. (1941) The British-Indian species of *Viscum* revised. *Blumea,* **4**, 261–319.
Dixit, S.N. (1962) Rank of the subfamilies Loranthoideae and Viscoideae. *Bulletin of the Botanical Society of India*, **4**, 49–55.
Engler, A. and Krause, K. (1935) Loranthaceae. In A. Engler and K. Prantl, (eds.), *Die natürlichen Pflanzenfamilien*, ed. 2, W. Engelmann, Leipzig, **16B**, 98–203.
Farris, J.S. (1989) Hennig86. An interactive program for phylogenetic analysis. Port Jefferson Station, New York.
Grierson, A.J.C. and Long, D.G. (1983) Loranthaceae. *Flora of Bhutan,* **1**, 143–151.
Grove, A.T. (1983) Evolution of the physical geography of the East African Rift Valley region. In R.W. Sims, J.H. Price and P.E.S. Whalley, (eds), *Evolution, Time and Space: The Emergence of the Biosphere*, Academic Press, London, pp. 115–155.
Hawksworth, F.G. and Wiens, D. (1996). Dwarf Mistletoes: Biology, Pathology, and Systematics. *Agriculture Handbook*, USDA Forest Service, Washington DC, **709**, 1–410.
Kokwaro, J.O. (1993) *Medicinal Plants of East Africa*, ed. 2. Kenya Literature Bureau, Nairobi. pp. 401.
Kuijt, J. (1959) A study of heterophylly and inflorescence structure in *Dendrophthora* and *Phoradendron* (Loranthaceae). *Acta Botanica Neerlandica,* **8**, 506–546.
Kuijt, J. (1961) A revision of *Dendrophthora* (Loranthaceae). *Wentia,* **6**, 1–145.
Kuijt, J. (1968) Mutual affinities of Santalalean families. *Brittonia,* **20**, 136–147.
Kuijt, J. (1969) *The Biology of Parasitic Flowering Plants.* University of California Press, Berkeley and Los Angeles, p. 246.

Kuijt, J., Wiens, D. and Coxson, D. (1979) A new androecial type in African *Viscum*. *Acta Botanica Neerlandica,* **28**, 349–355.

Lavin, M. and Luckow, M. (1993) Origins and relationships of tropical North America in the context of the boreotropics hypothesis. *American Journal of Botany,* **80**, 1–14.

Nickrent, D.L. (1996) Molecular systematics. In F.G. Hawksworth and D. Wiens, (eds.), *Dwarf Mistletoes: Biology, Pathology, and Systematics.* USDA Agriculture Handbook, **709**, 155–172.

Nickrent, D.L. and Franchina, C.R. (1990) Phylogenetic relationships of the Santalales and relatives. *Journal of Molecular Evolution,* **31**, 294–301.

Nickrent, D.L. and Soltis, D.E. (1995) A comparison of angiosperm phylogenies from nuclear 18S rDNA and *rbc*L sequences. *Annals of the Missouri Botanical Garden,* **82**, 208–234.

Oliver, I. (1987) Teemohlware – a refreshing bush tea. *Veld and Flora,* **73**, 16.

Polhill, R.M. and Wiens, D. (1998). *Mistletoes of Africa.* Royal Botanic Gardens, Kew, pp. 370.

Watt, J.M. and Breyer-Brandwijk, M.G. *The Medicinal and Poisonous Plants of Southern and eastern Africa*, ed. 2. Livingstone, Edinburgh, pp. 1457.

White, F. (1983). The Vegetation of Africa. UNESCO, Paris.

Wiens, D. (1975) Chromosome numbers in African and Madagascan Loranthaceae and Viscaceae. *Botanical Journal of the Linnean Society*, **71**, 295–310.

Wiens, D. and Barlow, B.A. (1979) Translocation heterozygosity and the origin of dioecy in *Viscum*. *Heredity,* **42**, 201–222.

3. EUROPEAN MISTLETOE: TAXONOMY, HOST TREES, PARTS USED, PHYSIOLOGY

HANS BECKER

Institute of Pharmacognosy and Analytical Phytochemistry, University of the Saarland, P.O. Box 151150, 66041 Saarbrücken, Germany

TAXONOMY OF VISCACEAE

Parasitic plants are found in different plant families that are not related to each other. However, within the subclass of Rosidae and the order of Santalales, there is a group of plant families that contains a large number of species with a parasitic or hemiparasitic life style. The genus *Viscum* has been grouped earlier in the family Loranthaceae but it is now widely accepted to establish two different families, the Loranthaceae and the Viscaceae (Calder 1983). The major features distinguishing the two families are listed in a table by Calder (1983). According to this, the main macroscopic difference is that Loranthaceae have larger flowers that are bright coloured and represent both sexes, whereas the latter is monoecious and highly reduced. The family Viscaceae sens. str. comprises seven genera with approximately 400 species (Barlow 1983): *Arceuthobium*, *Ginalloa*, *Notothixus*, *Korthalsella*, *Phoradendron*, *Dendrophthora*, *Viscum*.

The genus *Arceuthobium* has its widest distribution with twenty-four species in North America, and the greatest concentration in the western United States and central Mexico. *Arceuthobium oxycedri* has a distribution between the 30th and 40th northern parallel from the Azores through northern Algeria and Southern Europe to the Himalaya. The other four species are found in restricted areas in East Asia and temperate Africa. *Ginalloa* is the smallest genus of the Viscaceae with five species in South East Asia. *Notothixos* is found mostly in tropical forests from Sri Lanka to eastern Australia and the Salomon Islands. *Korthalsella* has a distribution from eastern and southern Asia through the Malesian region to Australia and New Zealand and to many islands in the Indian and Pacific Ocean. *Phoradendron* is found in South America and in the southern parts of North America. *Dendrophthora* has a similar distribution as *Phoradendron* but goes less south and less north than the latter. The genus *Viscum* comprises about ninety species, among which two thirds are native to Africa (see Kirkup *et al.*, this book), the rest is native to Eurasia and Australia. In Europe we have two *Viscum*-species: *Viscum cruciatum* ex Boiss. and *Viscum album* L.

VISCUM CRUCIATUM

Viscum crucitatum is native to South Spain and East Portugal. The stem of V. *cruciatum* is up to 60 cm long, yellowish green. Leaves are 2–4 cm long and 1 to 2 cm broad, often whorled, obovate-oblong, obtuse, yellowish green. Cymes shortly predunculate. Inflorescence cymose, the flowers crowded. 4-toothed in female flowers, absent in male flowers. Petals usually 4. Stamens almost completely connate with the petals; anthers opening by pores. Berry 6–10 mm red (Ball, 1993).

VISCUM ALBUM

The area of distribution of *Viscum album* L. European mistletoe is Central Europe (from North Africa to Southern England and Southern Scandinavia), Southwest- and East Asia to Japan.

Common Names

German: Mistel, Vogelmistel, Leimmistel, Affolter, Bocksfutter, Drudenfuá, Elfklatte, Geiákrut, Guomol, Hexenbesen, Hexennest, Immergrüne, Kluster, Marenklatte, Marentaken, Mischgle, Mischgelt, Misple, Nistle, Uomol, Vogelchrut, Vogelkläb, Vogellim, Wespe, Wintergrün, Wispen, Wäsp.
English: mistletoe, all-heal, masslin.
French: gui, gui commun, gui de druides.
Italian: vischio, visco, vescovaggine, guatrice, pania, scoaggine.
Spanish: muerdago.

Host Specifity

The European population is devided into three sub-species, that have different hosts.
– V. *album* L. ssp. *platyspermum* Kell. (ssp. *album*) growing on hardwood trees;
– V. *album* L. ssp. *abietis* Beck. growing on *Abies* sp.;
– V. *album* L. ssp. *laxum* Fick (= ssp. *austriacum* Wiesb. Vollmann) growing on pine trees and very rarely on spruce.
In East Asia we find V. *album* L. var. *colaratum* Ohwi (see Park, this book).

Leaf area may vary for the subspecies (Ball 1993, Singer 1958). Also the amount of embryos per berry varies. While ssp. *platyspermen* tends to diembryonal, the percentage of monoembryonal berries is higher in ssp. *abietes* and ssp. *laxum*. Tubeuf (1923) already pointed out that any criteria used to distinguish the subspecies can be found in the other two subspecies as well, e.g. the size of the leaves depends very much from the nutrition of the host tree and the position of the parasite within the host tree.

Nagl and Stein (1989) characterised DNA from mistletoes grown on various hosts. Small but not significant differences were found between the 2 C DNA contents of the three subspecies and the base composition. Significant differences were detected in the patterns of sequence organisation.

Earlier reports (Freudenberg 1968) that lignins from mistletoes grown on softwood trees (ssp. *abietis* and ssp. *laxum*) contain coniferous lignin, whereas lignin from mistletoes grown on hardwood trees (ssp. *album*) produce hardwood lignin could not been proven (Becker and Nimz 1974).

The pattern of flavonoid aglycones may be different for certain samples. But like morphological features there is no clear-cut distinction between the three different subspecies (Becker and Exner 1980).

The only criteria to distinguish the three subspecies by morphological criteria was described by Grazi and Urech (1981). If the ripe berries are squeezed, the seed of the ssp. *abietes* and ssp. laxum are easily freed from the skin and from the viscous layer. When the berries of ssp. *platyspermum* are squeezed, the viscous layer remains attached to the skin and the seed. The two subspecies on softwood trees can be distinguished from the appearance of their embryos. The embryo of ssp. *abietes* has a swollen end at the hypocotyl pole, whereas the respective embryo from ssp. *laxum* is cylindrical. These differences may be of advantage for the attachment of the seed to the bark of the respective host trees.

There are only very rare cases, where a natural or artificial infection was successful, but not conform to the above-mentioned host restriction. The only case in nature where a species hosted V. *album* ssp. *laxum* as well as V. *album* ssp. *platyspermum* is *Genista cinerea* observed by Grazi and Zemp (1985).

Some hardwood trees (e.g. beech, boxtree) are resistant to mistletoe infections; other species e.g. European elm (*Ulmus campestris* and *U. montana*) and European oak (*Quercus robur* and *Qu. petraea*) are rarely infected by mistletoe. However, if an infection is successful for one tree this tree can host several mistletoe bushes (Ramm *et al.*, this book). A detailed list of host trees of *Viscum album* ssp. *platyspermum* is listed in Luther and Becker (1987). Hariri *et al.* (1991, 1992) studied different trees that were known to have different resistances to mistletoe infections. A histocytological observation demonstrated that two barriers were involved in the resistance: a mechanical one and a chemical one. The mechanical barrier concerns the thickness of the phellem (the outer part of the bark) and the location and quantity of lignified fibers. The bark of resistant trees has a thicker phellem and more superficial fibers. The chemical barrier is related to an important secretion of polyphenols, which is stimulated by the attack of the parasite. The effect of mistletoe infection on young apple trees has been studied by Preston (1977, see below).

APPEARANCE OF *VISCUM ALBUM*

The primary haustorium grows vertically through the host tissue. In addition, cortical strands develop which are orientated parallel to the surface. In contrast to the primary haustorium, the cortical strands retain their mitotic activity throughout the year (Sallé 1978). The cortical strands can give rise to secondary haustoria or to aerial shoots. They have no cell-to-cell connection as seen for the haustoria.

Figure 1 Image of *Viscum album* growing on poplar.

The shoot forms usually one node every year so that the age of a mistletoe bush can be approximately determined by counting the nodes. During the first three to four years, the shoot grows without branching. Afterwards a dichotomous shoot system or systems of higher order develops. Mistletoe bushes tend to form a globular shape which may reach over 1 m in diameter (Dorka 1996). (Fig. 1)

The coriaceous leaves have a life span of usually two years. Their size varies from about 3.5 to 6 cm in length and 1–2 cm in breadth. The length to breadth index has been used to classify subspecies but the variation is too big to make a clear cut

Figure 2 Male flower of *Viscum album* (from Becker, 1986; with kind permission of Karger Verlag, Basel) which consists of four perianth members (for better insight, the front segment was cut off). The anthers are implanted on the perianth members; they dehisce by numerous pores, shedding the spherical tricolpate pollen, as shown in Figure 3. Photo: Erbar (Heidelberg)

classification. They are aequifacial as seen by cross section and bear numerous oxalate crystals.

Viscum album is dioecious, i.e. part of the plants is female, the other part is male. The expression of sex is determined by a translocation heterozygoty (Mechelke 1976). Flowers are highly reduced and inconspicuous. According to weather conditions, they open between the end of February and April. Despite their insignificant appearance, they are insect pollinated. Insects are attracted by a sweet fruit-like smell and by floral nectar. Male flowers have no anthers of the usual structure, four tepals bear around 50 pollen compartments that open by pores (see Fig. 2). The pollen grains are oval (35×60 μm), tricolpate with numerous spinulae. Female

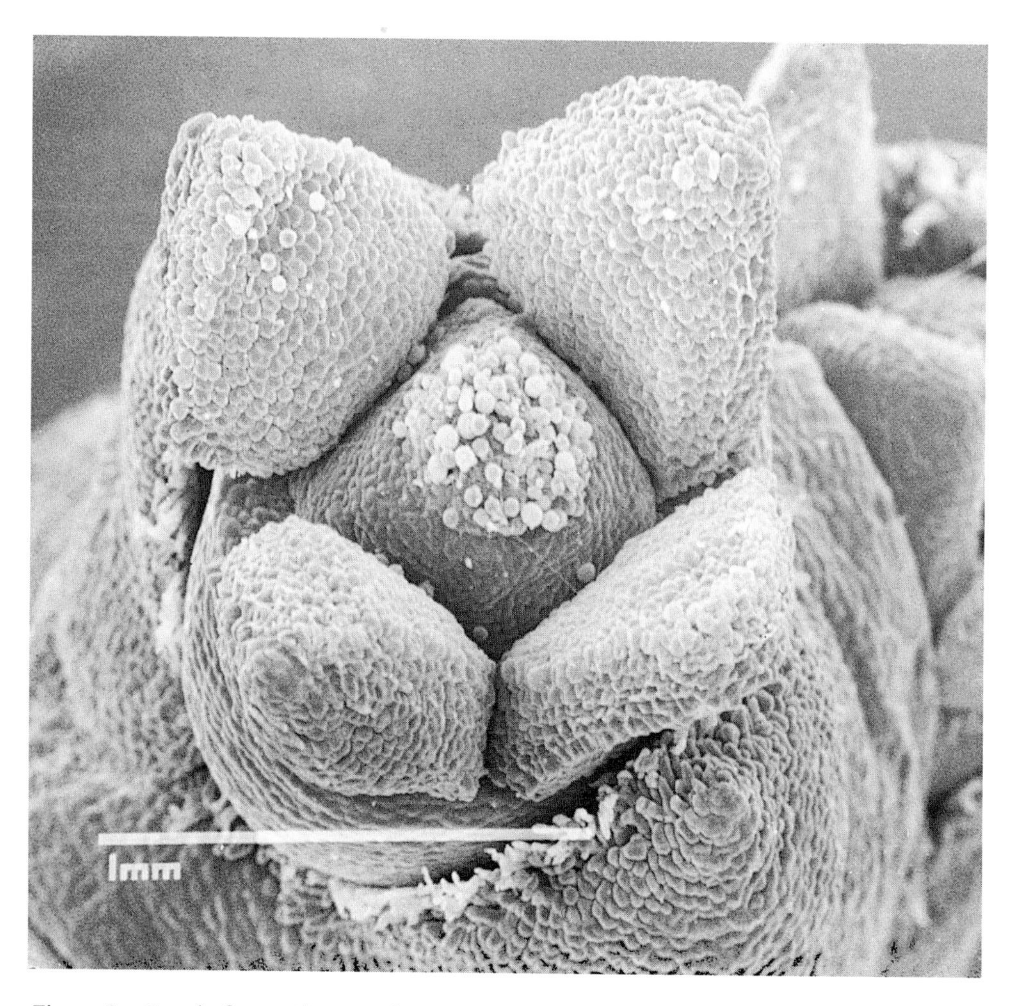

Figure 3 Female flower (from Becker, 1986; with kind permission of Karger Verlag, Basel) which consists of four perianth members with a simple style in the centre. Photo: Erbar (Heidelberg)

flowers (see Fig. 3) are even smaller than male ones. True ovula are not formed (Bhandari and Vohra 1983); the embryo develops at the base of the ovary.

Macroscopic Description of the Drug

Viscum album is also used in homeopathic practice. For this purpose, the shoot bearing leaves and fruits are harvested in autumn. The German Homoeopathic Pharmacopoiea (HAB) gives a detailed description of the drug. The following comments are mainly based on that source.

The drug is derived from plants that grow mostly on hardwoods (ssp. *album*), in rare cases also on fir (ssp. *abietis*) and on pine (ssp. *austriacum*). The shoots have no significant odour and a slightly tart taste. The viscous fruits are of a sweetish, slightly bitter taste. The bushes easily break at the nodes. The shoots repeatedly ramified in form of a dichasium are round or slightly flattened. The internodes reach a length of up to 80 mm and are up to 4 mm in diameter in the middle. The surface is smooth, bare and of dark green colour; twigs of ssp. *austriacum* are often yellow green. They emerge in the axil of a pair of two oppositely inserted scale leaves with a narrow, reddish border and whitish, fringed leaf margin. The terminal shoots culminate in an inflorescence head of up to 7 mm length usually with three male or of female flower buds, which are surrounded by thick scaled leaves.

The leaves are inserted in twos, rarely in threes or more. They are oblong, obovate, maximally three times as long than broad (ssp. *album*, ssp. *abietes*) or linear lanceolate and four to five times as long than broad (ssp. *austriacum*). They are entire, dark green (ssp. *album*, ssp. *abietis*) or slightly yellow-green (ssp. *austriacum*) and bare. There are three distinct veins and two less distinct veins running parallel to the leaf margin. In young leaves, the veins are visible on both sides, while in older leaves they are mainly seen on the underside.

The pseudofruits derived from the terminal shoots of female plants are solitary, or are borne in twos or threes or in rare cases multiples. Initially, they are globose or pyriform and green, the ripe fruits are shiny-white (ssp. *album*) and longer then broad or milky white (ssp. *abietis*) or yellow-white (ssp. *austriacum*) with fine, bright, longitudinal stripes. As epigynous fruits they show the remain of stigma and the four petals at the tip of the fruit. The fruit pulp is tough and mucilaginous, and surrounds one, in rare cases two pseudo-seeds. The latter are peltate shield-shaped to triangular and with one to four embryos. These are dark-green and consist of a haustorium which emerges from the endosperm, a hypocotyl and two rudimentary cotyledons.

Microscopic Description

The German Pharmacopoiea (DAB) includes a microscopic as well as a macroscopic description of *Viscum album*. Epidermial cells of young shoots are papilliform, those of aged shoots tangentially elongated and the cells are divided by thin cross walls. Stomata of the paracytic type are very large. The guard cells are located in pits and partly covered by papilliform auxiliary cells. The cleft is rectangular to the shoot axis. In cross section there are several, usually eight vascular bundles which are accompanied by two bundles of fibers each. The pith is partly hollow. The cell walls of wood parenchyma, pith parenchyma and pith rays are greatly thickened and pitted. They contain druses of calcium oxalate (see Fernandez, this book) with a diameter of about 30 to 40 μm, with a grey well marked centre in the middle. On the leaves the cells of the epidermal layer are covered by a thick cuticle. Stomatas of the paracytic type are found on both sides of the leaves. Guard cells on the leaves are less exposed as these on the shoots. The mesophyll cells contain the same druses of calcium oxalate as mentioned for the shoots. In addition there are starch granules of 8 and 15 μm length or small refractive droplets.

The pulverised drug is yellow-green. It is examined after heating in a solution of chloral hydrate (80 g in 20 ml water). The powder shows the following characteristics: druses of calcium oxalate of 30 to 40 μm diameter, stomata of the paracytic type whose guard cells are located in pits and partly covered by the papillate auxiliary cells; cells of the epidermal layer with greatly thickened cell walls; papillate cells of the epidermal layer of young shoots; cells of wood parenchyma, pith parenchyma, and of pith rays highly thickened and pitted.

NUTRITION OF *VISCUM ALBUM*

Viscum album is hemiparasitic, i.e. it depends for water and mineral nutrition on its respective host but is able to produce carbohydrates by photosynthesis. It contains all pigments, chlorophyll a and b as well as carotinoids that are necessary for photosynthesis (Luther and Becker 1987). Through radiotracer experiments it has been shown that mistletoes have the same photosynthetic activity per unit leaf surface as the respective host plants. As there is no phloem connection between mistletoe and host, organic substances from the host are only transported *via* the xylem. This transport includes amino acids, cyclohexols (Richter and Popp 1992) and thiols (Renneberg *et al.* 1994). Mistletoe have a much higher transpiration rate than their respective hosts. Schulze *et al.* (1984) found that mistletoes on pine trees have a more than 3-fold transpiration rate, calculated for leaf surface, as compared to the pine tree. In a broad study on three continents, Ehleringer *et al.* (1985) investigated transpiration and mineral nutrition of different mistletoes. They conclude that nitrogen supply is the limiting factor and that mistletoe transpiration is higher on hosts with low nitrogen content than on hosts with higher nitrogen content in their xylem sap. All mistletoes including *V. album* usually have a higher mineral content than the respective host and especially the infected host branch (Lamont 1983).

All mistletoes affect their respective host trees in many ways (Hawksworth 1983): They adversely affect height and diameter growth, lower the vigor of the host, induce premature mortality, adversely affect the quality and quantity of wood produced, reduce fruiting of infected trees, and predispose trees to attack by other agents, such as insects or decay fungi.

All these aspects also apply to *V. album*. Cervera and Villaescusa (1977) estimate that fir trees infected by *V. album* heave a reduced secondary growth of about 40%.

Table 1 Secondary compounds from *Viscum album* (for lectins and vicotoxins, see Pfüller, this book).

Class of compounds	*Individual compounds*	*References*
Flavonoids	*V. album ssp. platyspermum:* *(2R)-5,7-Dimethoxyflavonon-4′-O-glucoside, (2S)-Homoeriodictyol, (2S)-Homoeriodictyol-7-O-glucoside, 2-Hydroxy-4′,6′-dimethoxychalcon-4-O-glucoside, 2′-Hydroxy-4′,6′-dimethoxychalcon-4-O[apiosyl-(1→2)]glucoside, 2′-Hydroxy-3,4′,6′-trimethoxychalcon-4-O-glucoside, Isorhamnetin-3-O[apiosyl(1→6)]-glucosyl-7-O-rhamnoside, Isorhamnetin-3-O-rutinoside, (2S)-3′,5,7-Trimethoxyflavanon-4′-O-glucoside, after hydrolysis of mistletoe extracts: Quercetin, 6 different Quercetinmonomethyl-, Quercetin-dimethyl resp. Quercetintrimethylether, Homoeriodictyol and Sakuranetin (4,5′-Dihydroxy-7-methoxyflavanon).*	Fukunaga *et al.* (1987) Fukunaga *et al.* (1988) Becker *et al.* (1978) Becker and Exner (1980) Lorch (1993)
	V. album ssp. *laxum:* 2′-Hydroxy-4′,6′-dimethoxychalcon-4-O-glucoside; after hydrolyisis: 5 Quercetinmethylether, 5,7-Dimethoxy-4′-hydroxyflavon, Sakuranetin.	Becker *et al.* (1978) Becker and Exner (1980) Lorch (1993)
	V. album ssp. *abietis:* after hydrolysis: Homoeriodictyol, Sakuranetin and 8 different Quercetinmethylether	Becker and Exner (1980) Lorch (1993)
	V. album ssp.*coloratum:* 7,3′-O-Dimethylquercetin-3-O-glucoside (Flavoyardorinin A), 7,3′-O-Dimethylluteolin-4′-O-glucoside (Flavoyardorinin B), 7,3′-O-Dimethylluteolin-4′-O-[apiosyl(1→2)]glucoside (Homoflavoyardiorinin B), (2S)-Homoeriodictyol-7-O-[apiosyl(1→2)]glucoside and Rhamnazin-3,4′-di-O-glucoside.	Ohta and Yagashita (1970) Fukunaga *et al.* (1989)

Table 1 *continued.*

Class of compounds	*Individual compounds*	*References*
Phenylpro-panoids	Caffeic acid, Sinapic acid, Ferulic acid, Protecatechic acid, Syringic acid, Vanillic acid, Anisic acid and - Gentisic acid. Coniferyl-4′-O[apiosyl(1→2)]glucoside, Syringin (= Syringoside, Syringenin-4′-O-glucoside), Syrin-genin-4′-O-[apiosyl(1→2)]glucoside; the lignans syringaresinol and its glucoside Syringaresinol-4,4′-O-glucoside (= Eleutheroide E)	Fukunaga *et al.* (1987) Wagner *et al.* (1984, 1986)
Triterpenes	β-Amyrin (α-Amyrin), β-Amyrin-acetate, Betulinic acid, Oleanolic acid, Ursolic acid	Fukunaga *et al.* (1987) Wagner *et al.* (1986)
Phytosterol	β-Sitosterol, Stigmasterol and their respective glucosides (1970)	Ohta and Yagashita
Alkaloids	Alkaloids have been repeatedly recorded, but no definite structure has yet been given	Khwaja *et al.* (1981)
Cyclic peptide	Viscumamide (cyclo(Leu-Ile-Leu-Ile-Leu) from *V. album* ssp. *coloratum*	Okumara and Sakurai (1973)
Cyclitols	Mannitol, myo-Inositol, Quebrachitol, Pinitol, Viscumitol (myo-Inositol-dimethylether)	Richter (1992) Richter and Popp (1992)
Polysaccha-rides	Stems and leaves contain mainly methylester of 1→α4 Galacturonic acid. Berries contain Rhamnogalacturonanes as basic structures to which individual branched (1β→6)-D-galactan chains are linked via O-4 of rhamnose while arabinosyl residues or complex arabinan side chains are linked via 0–3 of galactose units.	Wagner and Jordan (1988)

In some cases, when the mistletoe occupies a large surface of the canapy the host trees die, especially if there is a period of extreme dryness (Brossier 1969).

The influence of mistletoe on artificially infected apple trees was studied in detail by Preston (1977). Trees of different growth characteristics were included in these studies: dwarfing (M.9), semi-dwarfing (M.26), vigorous (M.M.111), and very vigorous (M.25). Nine years after inoculation, host tree size and cropping were reduced on trees with mistletoe on all rootstocks except M.25. Fresh weights of mistletoe vegetation, when related to host tree size, showed an inverse relationship to the known rootstock effects on scion size, trees on the dwarfing M.9 bearing a heavier weight of mistletoe to their size than trees on the more vigorous rootstocks. The crop weight of the apple trees 11 years from planting was reduced between 7 and 56%.

SECONDARY COMPOUNDS FROM *VISCUM ALBUM*

Secondary compounds from *Viscum album* are listed in Table 1. Lectins and viscotoxins will be described in detail by Pfüller, this book.

REFERENCES

Ball, P.W. (1993) Viscum. In T.G. Tuttin, N.A. Burges, A.O. Chater, J.R. Edmondson, V.H. Heywood, D.M. Morre, and D.H. Valentine, (eds.), *Flora Europaea*, Cambridge University Press, Cambridge, pp. 86.

Barlow, B.A. (1983) Biogeography of Loranthaceae and Viscaceae. In A. Keast, (ed.), *The Biology of Mistletoes,* Academic press, Paris, San Diego, San Francisco, Sao Paulo, Tokyo, Toronto, pp. 19–46.

Becker, H. and Exner, J. (1980) Vergleichende Untersuchungen von Misteln verschiedener Wirtsbäume anhand der Flavonoide und Phenolcarbonsäuren. *Z. Pflanzenphysiol.*, **97**, 417–428.

Becker, H., Exner, G., and Schilling, G. (1978) Isolierung und Strukturaufklärung von 2′-Hydroxy-4′,6′-dimethoxychalkon-4-glukosid aus *Viscum album* L. *Z Naturforsch.*, **33c**, 771–773.

Becker, H. and Nimz, H. (1974) Untersuchungen des Lignins der Mistel (*Viscum album* L.) in Abhängigkeit von der jeweiligen Wirtspflanze. *Z. Pflanzenphysiol.*, **72**, 52–63

Becker, H. (1986) Botany of European mistletoe (*Viscum album* L.) *Oncology*, **43** (suppl. 1), 2–7.

Bhandari, N.N. and Vohra, S., C., A. (1983) Embryology and affinities of Viscaceae. In M. Calder, and P. Bernhardt, (eds), *The Biology of Mistletoes,* Academic Press, Sydney, New York, London, Paris, San Diego, San Francisco, Sao Paulo, Tokyo, Toronto, pp. 69–86.

Brossier, J. (1969) Réflexions sur le qui du sapin. *Revue forestière francaise XXI,* **6**, 558–561.

Calder, M. (1983) Mistletoes in focus: An introduction. In M.Calder and P. Bernhard, (eds.), *The Biology of Mistletoes,* Academic Press, Sydney, New York, London, pp. 1–17.

Cervera, J.M. and Villaescusa, R. (1977) Inventario de los abetares del Valle de Aran affectados por el Muerdago, *Rapport du Service de l'I.C.O.N.A.*, Lerida, pp. 41–47.

Dorka, R. (1996) Zur Chronobiologie der Mistel. In R. Scheer, H. Becker, and P.A. Berg, (eds), *Grundlagen der Misteltherapie*, Hippokrates, Stuttgart, pp. 28–45.

Ehleringer, J., R., Schulze, E., D., Ziegler, H., Lange, O., L., Faguhar, G., D. and Cowar, I., R. (1985) Xylem-tapping mistletoes: Water or nutrient parasites. *Science,* **227**, 1479–1481.

Freudenberg K. (1968). In K. Freudenberg and A.C. Neish, (eds.), Constitution and biosynthesis of lignin, Springer, Berlin, Heidelberg, New York, pp. 113–114.

Fukunaga, T., Kajikawa, I., Nishiya, K., Watanabe, Y., Takeya, K., and Itokawa, H. (1987) Studies on the constituents of the European mistletoe, *Viscum album* L. *Chem. Pharm. Bull.*, **35**, 3292–3297.

Fukunaga, T., Kajikawa, I., Nishiya, K., Watanabe, Y., Suzuki, N., Takeya, K., and Itokawa, H. (1988) Studies on the constituents of the European mistletoe, *Viscum album* L. II, *Chem. Pharm. Bull.*, **36**, 1185–1189.

Fukunaga, T., Kajikawa, I., Nishiya, K., Watanabe, Y., Takeya, K., and Itokawa, H. (1989) Studies on the constituents of the Japanese mistletoe, *Viscum album* L. var. *coloratum* Ohwi grown on different host tress. *Chem. Pharm. Bull.*, **37**, 1300–1303.

German Pharmacopoeia, "*Deutsches Arzneibuch*" DAB 1999 Deutscher Apotheker Verlag, Stuttgart.

German Homeopathic Pharmacopoeia, "*Homöopathisches Arzneibuch*" (HAB) (1998), Deutscher Apotheker Verlag, Stuttgart.

Grazi, G. and Urech K. (1983) Einige morphologische Merkmale der Mistelbeere (*Viscum album* L.) und deren taxonomische Bedeutung. Beitr. Biol. Pflanzen, **56**, 293–306.

Grazi, G. and Zemp, M. (1985) Über den Ginster, *Genista einerea*, als Sammelwirt für Laubholzmistel (*Viscum album* ssp. *album*) und Kiefernmistel (*Viscum album* ssp.*austriacum*). 8. *Symposium Morphologie, Anatomie und Systematik*, Hamburg 3–7 März 1985.

Hariri, E.B., Sallé, G., and Andary, C. (1991) Involvement of flavonoids in the resistance of two poplar cultivars to mistletoe ((*Viscum album* L.) *Protoplasma*, **162**, 20–26.

Hariri, E.B., Jeune, B., Bandino, S. Urech, K., and Sallé, G. (1992) Elaboration d'un coefficient de résistance au gui chez le chêne. *Can. J. Bot.*, **70**, 1239–1246.

Hawksworth, F.G. (1983) Mistletoes as Forest Parasites. In M. Calder and P. Bernhardt, (eds.), *The Biology of Mistletoes*, Academic Press, Sydney, New York, London, pp. 317–333.

Khwaja, T.A., Dias, C.B., Papoian T., and Pentecost, S. (1981) Studies on cyctotoxic and immunologic effects of *Viscum album* (Mistletoe). *Proc. Am. Assoc. Cancer Res.*, **22**, 253.

Lamont, B. (1983) Mineral nutrition of mistletoes. In M. Calder and P. Bernhardt, (eds.), *The Biology of Mistletoes*, The Academic Press, Sydney, New York, London, San Diego, San Francisco, Sao Paulo, Tokyo, Toronto, pp. 185–204.

Lorch, E. (1993) Neue Untersuchungen über Flavonoide in *Viscum album* L. *abietis, album* and *austriacum. Z. Naturforsch.*, **48c**, 105–107.

Luther, P. and Becker, H. (1987) *Die Mistel. Botanik, Lektine, medizinische Anwendung*. Springer-Verlag, Berlin.

Mechelke, F. (1976) Sex-correlated complex heterozygoty in *Viscum album* L. *Naturwissenschaften*, **63**, 390–391.

Nagl, W. and Stein, B. (1989) DNA characterization in host-specific *Viscum album* subspecies (Viscaceae). *Pl. Syst. Evol.*, **116**, 243–248.

Ohta, N. and Yagashita, K. (1970) Isolation and structure of new flavonoids, flavoyardorinin A, flavoyadorimin B and homoflavoyadorinin B in the leaves of *Viscum album* var. *coloratum* epiphysic on *Pyrus communis. Agric. Biol. Chem.*, **34**, 900–907.

Okumura, Y. and Sakurai A. (1973) Chemical studies on the mistletoe. II. The structure of viscumamid, a new cyclic peptide isolated from *Viscum album* Linn. var. *coloratum Ohwi. Bull. Chem. Soc. Jpn.*, **46**, 2190–2193.

Preston A.P. (1977) Effects of mistletoe (*Viscum album*) on young apple trees. *Hort Res.*, **17**, 33–38.

Renneberg, H., Schupp, R., and Schneider, A. (1994) Thiol composition of a xylon-tapping mistletoe and the xylem sap of its hosts. *Phytochemistry*, **37**, 975–977

Richter, A. (1992) Viscumitol, a dimethylether of *muco*-inositol from *Viscum album. Phytochemistry*, **31**, 3925–3927.

Richter, A., and Popp, M. (1992) The physiological importance of accumulation of cyclitols in *Viscum album* L. *New Phytol.*, **121**, 431–438

Sallé, G. (1978) Nuclear DNA metabolism in the tip of the cortical strands of *Viscum album*. L. *Ann. Bot.*, **42**, 171–176.

Singer, O. (1958) Ein Beitrag zur Kenntnis der Mistel. *Pharmazie*, **13**, 781–783.

Schulze, E.D., Turner, N.,C., and Glatzel, G. (1984) Carbon, water and nutrient relations of two mistletoes and their hosts: A hypothesis. *Plant Cell Environm.*, **7**, 293–299.

Tubeuf Frh. von K. (1923) Monographie der Mistel. R. Oldenburg, München, Berlin.

Wagner, H., Feil, B., and Bladt, S. (1984) *Viscum album* – die Mistel. Analyse und Standardisierung von Arzneidrogen durch Hochleistungsflüssigkeitschromatographie (HPLC) und andere chromatographische Verfahren (III). *Deutsche Apotheker Zeitung*, **124**, 1429–1432.

Wagner, H., Feil., B., Seligmann, O., Petricic, J., and Kologjera, Z. (1986) Phenylpropanes and lignans of *Viscum album* L. Cardioactive drugs V. *Planta Medica*, **52**, 102–104.

Wagner, H., and Jordan, E. (1988) An immunologically active arabinogalactan from *Viscum album* berries. *Phytochemistry*, **27**, 2511–2517.

4. KOREAN MISTLETOES AND OTHER EAST-ASIAN POPULATIONS

WON-BONG PARK

College of Natural Science, Seoul Women's University, Seoul, 139–774, Korea

INTRODUCTION

Semiparasitic plants, mistletoes distributed in Korea and other East-Asian countries (mainly China and Japan) have long been recognised as therapeutic herbs (Li, 1975). Mistletoes are traditionally used as sedative, analgesic, spasmolytic, cardiac and anticancer agent; the herbs are also used to tone the liver and kidneys, strengthen tendons and bones, expel pathogens associated with rheumatism, stabilise the fetus and cause lactogenesis [Hsu Hong-yen, 1972; Zee-Cheng, 1997].

The use of mistletoes as a remedy for circulatory and nervous disorders dates from antiquity. Modern pharmacological studies have demonstrated their cardiotonic, antihypertensive (Wagner *et al.*, 1986; Fukunaga *et al.*, 1989b) and antiplatelet aggregatory activity *in vitro* (Zhu *et al.*, 1985; Samal *et al.*, 1995). In Korea, mistletoes were also used to treat tachyarrhythmia (Wu *et al.*, 1994a,b), acute myocardial infarction (Zhu *et al.*, 1984, 1985, Cheng *et al.*, 1985) and schizophrenia (Okuda *et al.*, 1987). The herbs contain substances such as choline and acetylcholine which are known for their regulatory effect on blood pressure and circulation for toning the heart muscle (Pora *et al.*, 1957; Paskov *et al.*, 1958a,b; Samuelsson *et al.*, 1959; Samal *et al.*, 1995; Wagner *et al.*, 1986). In animals, the herbs also have diuretic (Li *et al.*, 1959), antibacterial (Xu, 1947) and antihepatoxic effects (Yang *et al.*, 1987).

Since the 1920s, extracts from European mistletoe (*Viscum album* L., VA-E) are popular in Europe as an unconventional cancer treatment. These extracts have been used in adjuvant cancer therapy with immunostimulatory, cytostatic/cytotoxic and DNA stabilising properties (Hajto *et al.*, 1989; Stein and Berg, 1997; Beuth, 1997; Büssing *et al.*, 1995, 1996). VA-E contains various toxic and nontoxic proteins, alkaloids, flavonoids and polysaccharides (Pfüller *et al.*, 1993; Ribéreau-Gayon, *et al.*, 1986, 1993).

While the European mistletoe has been studied intensively, we know less about Korean mistletoes and other East-Asian populations as therapeutic herbs, especially as a suggested anticancer drug. This chapter will present an overview on the current knowledge of the botany and recent investigations on the chemical components and their biological activities of Korean mistletoes and other East-Asian populations.

BOTANY, TAXONOMY, HOST TREES AND GEOGRAPHY

The Korean name of mistletoe is "gyo-uh-sari" which means "the plant that survives in cold winter". Three genera with four species of mistletoes are distributed in Korea: *Viscum album*. var. *coloratum* Ohwi. (Figure 1a), *Korthalsella japonica* Engl., *Loranthus yadoriki* Sieb. and *Loranthus tanakae* Fr. et Sav. (Figure 1b).

Table 1 Korean mistletoes and other East-Asian populations.

Name	*Characteristics*	*Host trees (Genus)*	*Distribution*
Viscum album L. var. *coloratum*	yellow berry, lanceolate leaf	*Quercus* *Ulmus* *Salix* *Castanea crenata* S. at Z. *Celtis sinensis* Pers *Alnus hirsuta* (Spach) Rupr. *Acer palmatum* Thunb	Korean, Japan, China
Viscum album L. var. *coloratum* for. *Rubroauratiacum*	Red berry		Cheju island in Korea
Korthalsella japonica Engler	Degenerated leaf	*Camellia japoanica* L. *Euonymus japonica* Thunb. *Eurya japonica* Thunb. *Vaccinium bracteatum* Thunb.	Korea, Japan
Loranthus yadoriki Sieb.	Broad leaf, yellow berry	*Camellia japoanica* L. *Machilus thunbergii* S. et Z. *Castanopsis cuspidata* var. *sieboldii* Nakai *Actinodaphne lancifolia* (S. et Z.) Neisn	Cheju island in Korea
Loranthus tanakae Fr. et Sav.	Deciduous, broad leaf, berries forming tail-shape	*Quercus genus* *Castanea crenata* S. et Z	Korea
Loranthus parasiticus (L.) Merr	Broad leaf	*Morus alba* L., etc	China
Taxillus kaempferi Danser		*Quercus genus* Thunb. *Prunus yedoensis* Matsum. *Camellia japoanica* L. etc.	Japan
Taxillus yadoriki Danser		*Catanopsis cuspidata* var. *sieboldi* Nakai. etc.	Japan

(A)

(B)

Figure 1 *Viscum album*. var. *coloratum* (A) and *Loranthus tanakae* (B), two yellow-berry mistletoes distributed in Korea.

Loranthus parasiticus (L.) Merr is distributed in China, and *Taxillus kaempferi* Danser and *Taxillus yadoriki* Danser are distributed in Japan (Table 1).

The yellow-berry mistletoe, *Viscum album* L. var. *coloratum* is a half-parasitic plant on oak (*Quercus genus*), chestnut tree (*Castanea crenata* S. et Z.), hackberry (*Celtis sinensis* Pers), birch (*Betula platyphylla* var. *japonica* Hara), *Alnus hirsuta* (Spach) Rupr., elm (*Ulmus* genus), willow (*Salix* genus) and maple (*Acer palmatum* Thunb). The plant is distributed widely over the Korean peninsula, the Northern part of China and Japan. Plants often develop a roundish form up to 30 to 60 cm or more in diameter. The thick, glossy and lanceolate leaves are arranged oppositely. The plant is dioecious with inconspicuous and light yellow male and female flowers. They bloom as clusters in the forks of the branches in spring. The round light-yellow berries are sticky and ripen during the mid-winter. The red-berry mistletoe, *Viscum album* var. *coloratum* for. *rubroaurantiacum* Ohwi is distributed in Cheju island, the southernmost island of Korea.

In the 1980s, the Korean mistletoe (*Viscum album* L. var. *coloratum*) was introduced and cultivated in Switzerland, Lucas garden, Arlesheim. There are at present 87 Korean mistletoes successfully growing on *Quercus* genus, *Betula* alba, *Cytisus, Nerium oleander, Populus trichocarpa, Prunus avium, Robinia pseudoacacia*, and *Salix caprea*. Recently, hyperparasitic white-berry European mistletoe is growing on the twigs of yellow-berry Korean mistletoe (personal communications from F. Grazi in Arlesheim).

Korthalsella japonica Engler is distributed in southern islands of Korea and Japan. The plant is parasitic on camellia (*Camellia japonica* L.), spindle trees (*Euonymus japonica* Thunb), *Eurya japonica* Thunb and *Vaccinium bracteatum* Thunb. The small degenerated leaves are protruded from the lots of joints of the green stem.

There are three species of broad leaf mistletoes in Korea and China. *Loranthus yadoriki* Sieb. and *Loranthus tanakae* Fr. et Saw. distributed in Korea and *Loranthus parasiticus* (L.) Merr distributed in China.

Loranthus yadoriki Sieb. is distributed in Cheju island, the southernmost island of Korea. The evergreen parasite shrub grows on camellia (*Camellia japonica* L.), silver magnolia (*Machilus thunbergii* S. et Z.), *Castanopsis cuspidata* var. *sieboldii* Nakai and *Actinodaphne lancifolia* (S. et Z.) Neisn. The broad-elliptical leaves are 1 to 3 cm long and arranged oppositely or alternately. The flowers are bisexual and the sticky-yellow berries are attached to the stem continuously forming tail-shape.

An extraordinarily deciduous plant, *Loranthus tanakae* Fr. et Sav. is distributed in Cheju island and in the middle of Korean peninsula. The plant grows on oak (*Quercus* genus) and chestnut trees (*Castanea crenata* S. et Z.). The tongue-shaped leaves are 2 to 4 cm long and arranged oppositely. The dark purplish brown stem is freely forked and the grey part of the bark is peeled after winter.

Loranthus parasiticus (L.) Merr grows on mulberry (*Morus alba* L.) and many other trees in Southern China. The broad leaves are ovate or oblong and are arranged oppositely or alternately. The red brown flowers are bisexual and umbrella-shaped. The elliptical berries are ripen in autumn.

CHEMICAL COMPONENTS

Viscum album var. *coloratum* Ohwi (= *V. album* var. C.)

Various chemical components have been isolated and identified from the extracts of *V. album* var. C., such as lectins, steroids, triterpenes, sesquiterpene lactones, carbohydrates, flavonoids, organic acids and amines, alkaloids, amino acids and peptides.

Lectins

An agglutinin from *Viscum album* var. C. harvested in Korea (VCA) was isolated by gel filtration using Sephadex-G 75 (Park *et al.*, 1994a), affinity chromatography using acid-treated Sepharose 4B (Park *et al.*, 1997a). VCA was also isolated using Sepharose 4B modified by lactose-bovine serum albumin (BSA) conjugate synthesised by reductive amination of ligand (lactose) to ϵ-amino groups of lysine residues of spacer (BSA) after reduction by $NaCNBH_3$ (Park *et al.*, 1998-a). Recently the VCA was efficiently isolated from Korean mistletoe by affinity chromatography using asialofetuin immobilised Sepharose 4B. Using this optimised isolation procedure the Korean mistletoe lectin (VCA) binds to galactose and lactose, similar to ML I isolated from *V. album* L. The molecular weight of the VCA determined by SDS-PAGE was 60 kDa, with a 31.5 kDa A-chain and a 34.5 kDa B-chain, and has an isoelectric point of 8.0–8.7 (Pfüller and Park, in preparation).

The investigation of binding kinetics and sugar inhibition of VCA by BIACore (Pharmacia) showed different association and dissociation steps and sugar specificity as compared to those of European mistletoe lectins (ML I, ML II, ML III). Both, association and dissociation rate of VCA to asialofetuin immobilised on sensorchip CM5 (Pharmacia) was lower than those of European ML, and the recognition of VCA by asialofetuin was inhibited by D-galactose and lactose, which is similar to ML I. Using an ELLA system (asialofetuin-lectin-polyclonal antibody-streptoavidin/peroxidase-OPD/H_2O_2), VCA reacts with polyclonal anti-ML antibodies. Further investigations using sandwich ELISA with monoclonal anti-ML antibodies revealed

Table 2 Lymphocyte stimulating activity of VCA.

Mitogen	*MTT assay*
Concanavalin A	0.201
VCA from unfermented sample	0.236
VCA from fermented crude extract (1 d)	0.181
VCA from fermented crude extract (2 d)	0.159
VCA isolated from fraction 1* of sample fermented for 3 d	0.157
VCA isolated from fraction 2* of sample fermented for 3 d	0.131

Metabolic activity of lymphocytes stimulated with VCA (2 HU) for 24 h was measured by MTT assay as described (Park *et al.*, 1995).
* fraction from ion exchange chromatography.

that VCA was recognised by an antibody detecting ML I (MNA9-TA5b), but not by a monoclonal antibody to ML II/III (C12-H11b) (Pfüller and Park, in preparation), indicating that VCA shares distinct epitopes with ML from European mistletoe.

The VCA was recognised to increase the metabolic activity of lymphocytes, as measured by MTT assay, more strongly than the mitogenic lectin concanavalin A (Park *et al.*, 1994a). In contrast, an additional lectin of 18.5 kDa MW, identified by fermentation with *Lactobacillus plantarum*, decreased the metabolic activity of lymphocytes (Park *et al.*, 1995) (Table 2). An apoptosis-inducing four chain lectin from Korean mistletoe (termed KML-C) with 27.5, 30, 31 and 32.5 kDa was described by Yoon *et al.* (1999).

The effects of pH, temperature and guanidine chloride on the activities of VCA were investigated by measuring its intrinsic fluorescence and compared with its hemagglutinating activities. There are significant relationships between activities and conformations of the lectin (Park *et al.*, 1998a): The hemagglutinating activity of the lectin was stable at the pH range of 4.0 to 8.0, decreased up to 50% at pH 9.0, and disappeared completely at pH 10.0. It is noticed that the activity was enhanced to 200% at the pH range of 4.0 to 6.0. The enhancing of lectin activity induced by pH changes in fluorescence position of spectral maximum evidently indicated the significant structural transition in the environment of tryptophan residues. Blue shift was detected on the acidic pH, which suggested unfolding of the protein structure near tryptophan residues. Similar changes also occurred at the alkaline pH range of 8.0 to 10.0. These changes corresponded to the decrease of activity in the alkaline pH. From these results it is assumed that high activity of lectin at pH 4.0 to 6.0 was caused by more suitable folding structure of lectin. The hemagglutinating activity of lectin was stable at a wide range of temperature (0–45°C). Half of the activity was maintained at 55°C, but the activity disappeared over 65°C. The increase in temperature resulted in a typical denaturational shift of the protein spectra towards a position characteristic of free tryptophan (353 nm). From these results, it was assumed that there may be a relationship between activity and conformation of lectin. In denaturing conditions, such as high concentration of guanidine hydrochloride, tryptophan emission profile of lectin showed typical denaturational red shift, which also corresponded to the conformation and activity of lectin.

Organic acids

palmitic acid, lignoceric acid, cerotic acid, octacosanoic acid, succinic acid, ferulic acid, caffeic acid and protocatechuic acid (Kong *et al.*, 1989).

Steroids

β-sitosterol and daucosterol (Tseng *et al.*, 1957; Kong *et al.*, 1987a, b).

Carbohydrates

mesoinositol; the acetate has hypotensive action (Tseng *et al.*, 1957).

Triterpenes

oleanolic acid, β-amylin, β-amyrin palmitate, β-acetylamyrin, erythordiol, betulinic acid and lupeol (Tseng *et al.*, 1957); oleanolic acid, olea-12-en-3β-ol and olean-12-en-3β-ol acetate and lup-20(29)-en-3-one (Ahn, 1996).

Flavonoids

flavoyandrinin-A, flavoyandrinin-B and homo-flavoyandrinin-B (Ohta *et al.*, 1970); rhamnazin-3-O-β-D-glucoside, homoeriodictyol-7-O-β-D-glucoside; rhamnazin (Kong *et al.*, 1987a); homoeriodictyol, viscumneoside III (homoeriodictyol-7-O-β-apiosyl-l-2-β-D-glucopyranoside) (Kong *et al.*, 1988b); viscumneoside IV (rhamnazin-3-O-β-D-(6″-β-hydroxy-β-methylglutaryl)glucoside (Kong *et al.*, 1988c, 1990a); viscumneoside V (homoeriodictyol-7-O-β-apiosyl-l-5-β-D-apiosyl-l-2β-D-glucopyranoside), viscumneoside VI (homoeriodictyol-7-O-β-D-(6″-O-acyl)glucopyranoside (Kong *et al.*, 1988a, b); viscumneoside VII (rhamnazin-3-O-β-D-apiosyl-1-2 (6″-O-(3-hydroxy-3-methylglutarate)glucoside) (Kong *et al.*, 1990b); viscoside A (Li *et al.*, 1985); isorhamnazin-3-O-β-D-glucoside; isorhamnazin-7-O-β-D-glucoside; rhamnazin-3,4′-di-O-glucoside, homoeriodictyol-7-O-(apiosyl-l-2 glucoside) (Fukunaga *et al.*, 1989a, b).

Loranthus Yadoriki Sieb.

The plant contains (1→4)-linked glucan, (1→4)-linked rhamnogalacturonan and (1→4)- and (1→3)-linked galactan. (1→3)-linked galactan is the backbone with side chain of (1→5)-L-arabinofuranosyl and (1→6)-D-galactopyranosyl residues (Lee *et al.*, 1996). Triterpenoids, olean-12-en-3β-ol, olea-12-en-3β-ol acetate and lup-20(29)-en-3-one and urs-12-en-3β-ol are identified (Ahn, 1996).

Loranthus parasiticus L. Merr.

The lectins (67.5 kDa) from the plant agglutinate rabbit erythrocytes, but do not agglutinate human erythrocytes. The hemagglutination is inhibited by galactose, N-acetylgalactasamine, fructose, fucose and melecitose (Chen *et al.*, 1992). The plant also contains sesquiterpene lactones (coriamyrtin, tutin, coriatin and corianin) and flavonoids (quercetin and quercetin-3-arabinoside) (Tseng *et al.*, 1957; Wang *et al.*, 1980, 1982; Wu *et al.*, 1984; Okuda *et al.*, 1987).

Korthalsella japonica Engl.

Oleanolic acid, phytosterol, phytosterol-glucoside, flavone glycoside, chrysoeriol-4′-O-glucoside were isolated from *Korthalsella japonica* growing in Japan (Fukunaga *et al.*, 1989) and tritertpenoids (oleanolic acid, olea-12-en-3β-ol acetate and ursolic acid derivatives) from the plant growing in Korea (Ahn, 1996). The chloroform fraction inhibits the growth of lymphoma cell line $P388D_1$ and transformed mouse embryo fibroblasts. The fraction contains three kinds of alkaloids, the suggested main components for the anticancer activities of Korean mistletoe (Kim *et al.*, 1996).

Taxillus yadoriki Danser and *Taxillus kaempferi* Danser

From *Taxillus yadoriki* Danser growing in Japan, flavonoid glycosides (hyperin and quercitrin) were identified, and quercetinin, avicularin, taxillusin, quercitrin and hyperin from *Taxillus kaempferi* Danser growing in Japan (Fukunaga *et al.*, 1989).

BIOLOGICAL ACTIVITIES

Aqueous VA-E has been used in adjuvant cancer therapy with both immunostimulatory and cytostatic/cytotoxic properties (Hajto *et al.*, 1989; Stein and Berg, 1997; Beuth, 1997; Büssing *et al.*, 1996). VA-E stimulate the immune system nonspecifically, as it increases the number and activity of natural killer cells and neutrophils, induces cytokines such as tumour necrosis factor alpha (TNF-α), interferon-gamma (IFN-γ), interleukin-1 (IL-1) and IL-6 (Hajto *et al.*, 1990; Müller and Anderer, 1990; Stein *et al.*, 1998a and 1998b). While the aqueous VA-E has been studied intensively, the biological studies mainly with *V. album* L. var. C. have been carried out recently.

The extract of *V. album* L. var. C. did not exhibit mutagenic properties as measured by *Salmonella typhimurium* TA98 and Ames test. Further, the extract inhibited mutagenesis induced by mitomycin C and N-methyl-N′-nitro-N-nitrosoguanidine by spore rec-assay (Ham *et al.*, 1998). The ethanol and heated aqueous extracts inhibited mutagenesis induced by benzo-α-pyrene [B(α)P] and Trp-P-1 by SOS chromotest (Ham *et al.*, 1998).

The aqueous extract of the plant was cytotoxic (ID_{50}: 8 μg/ml) to both, non-tumorigenic A31 and tumorigenic MSV cells, while the heat-treated extract was less cytotoxic (ID_{50}, 300 μg/ml) (Park *et al.*, 1997b and 1998b). A fresh extract of Korean mistletoe was reported to be more active (ID_{50}: 0.1 μg/ml) than fermented one (ID_{50}: 9.6 μg/ml) (Khwaja *et al.*, 1986), but the activity of the heat-treated extract increased by fermentation. In fermented Korean mistletoe, the lectins decreased significantly, while the presence of a new lectin was recognised (Park *et al.*, 1994b, 1995). It is supposed that the heat-denatured lectins in the heat-treated extract were digested by fermentation process and some fragments of soluble cytotoxic peptide were produced. The extract was also cytotoxic against 6 human tumour xenografts (i.e. ovarian cancer, small cell and large cell lung carcinoma, colon, renal and melanoma xenografts) (Choi *et al.*, 1996).

Similar effects of VA-E have been reported (Ribereau-Gayon *et al.*, 1986): The unfermented VA-E was approximately 10 times more cytotoxic to leukemic Molt 4 cells than fermented one; and the fermented one contained lower amount of lectin, while unfermented preparation contained about 10 times more.

The aqueous extract of the Korean mistletoe also inhibited tumour metastasis and angiogenesis caused by hematogenous and non-hematogenous tumour cells in mice

[Yoon *et al.*, 1995]. The extract significantly inhibited lung metastasis produced by highly metastatic murine tumour cells (Yoon *et al.*, 1995). *In vivo* analysis for tumour-induced angiogenesis revealed that the extract suppressed tumour growth and inhibited the number of blood vessels oriented towards the tumour mass (Yoon *et al.*, 1995). The culture supernatants of murine peritoneal macrophages treated with the aqueous mistletoe extract showed tumour necrosis factor-α (TNF-α) activity, and the supernatant used inhibited the growth of rat lung endothelial cells *in vitro*. The antimetastatic effect may result from the suppression of tumour growth and the inhibition of tumour-induced angiogenesis by inducing TNF-α (Yoon *et al.*, 1994, 1995). Further, the extract induced the secretion of IL-1, IL-6, and INF-γ from macrophage; these cytokines were found to be cytotoxic against carcinoma cells (Yoon *et al.*, 1997).

The prophylactic effect of the extract on tumour metastasis produced by highly metastatic tumour cells, colon 26-M3.1 carcinoma, B16-BL6 melanoma and L5178Y-ML25 lymphoma cells, was recognised using experimental models in mice. Intravenous administration of the extract 2 days before tumour cell inoculation significantly inhibited lung metastasis of B16-BL6 and colon 26-M3.1 cells, and liver and spleen metastasis of L5178Y-ML25 cells (Yoon *et al.*, 1994, 1995). Furthermore, mice administered the extract 2 days before tumour cell inoculation showed significantly prolonged survival rates compared with the untreated mice. The administration of the extract significantly augmented NK cytotoxicity to Yac-1 tumour cells after the treatment (Yoon *et al.*, 1994, 1995). Furthermore, depletion of NK cells by injection of rabbit anti-asialo GM1 serum completely abolished the inhibitory effect of the extract on lung metastasis of colon 26-M3.1 cells. These results suggest that the extract possesses immunopotentiating activity which enhances the hose defence system against tumours, and that its prophylactic effect on tumour metastasis is mediated by NK cell activation.

Kwaja *et al.* (1980) reported the presence of alkaloids in Korean mistletoe, and addressed the anticancer activity of *V. album* L. var. C. to the alkaloids. Experiments with extracts from Californian, European and Korean mistletoe (*Phoradendron villosum*, *V. album* and *V. album* var. C.) showed that Korean mistletoe was more active in inhibiting the growth of leukemia L1210 cells as compared to the other extracts. Similarly, alkaloidal fraction II isolated from Korean mistletoe extract was the most active (ID_{50}, 0.17 μg/ml) of all other fractions. The data suggest that like viscotoxins and mistletoe lectins, biologically active mistletoe alkaloids and other compounds may have a contribution to the mechanism of cytotoxicity of mistletoe extracts, and that all biologically active compounds of mistletoe may exert synergistically therapeutic activity (Khwaja *et al.*, 1986). However, the structures of these compounds have not been elucidated as they are extremely labile (Khwaja *et al.*, 1980); others were unable to clearly show the presence of alkaloids in European mistletoe (Becker and Pfüller, personal communications).

Cytotoxicity of an aqueous extract from Korean mistletoe was strongly related to the activity of VCA. Using human lymphocytes and leukaemic Molt-4 T cells treated for 24 h with the whole plant extract and the purified VCA, a dose-dependent induction of mitochondrial Apo2.7 molecules (Figure 2), active caspase-3 and

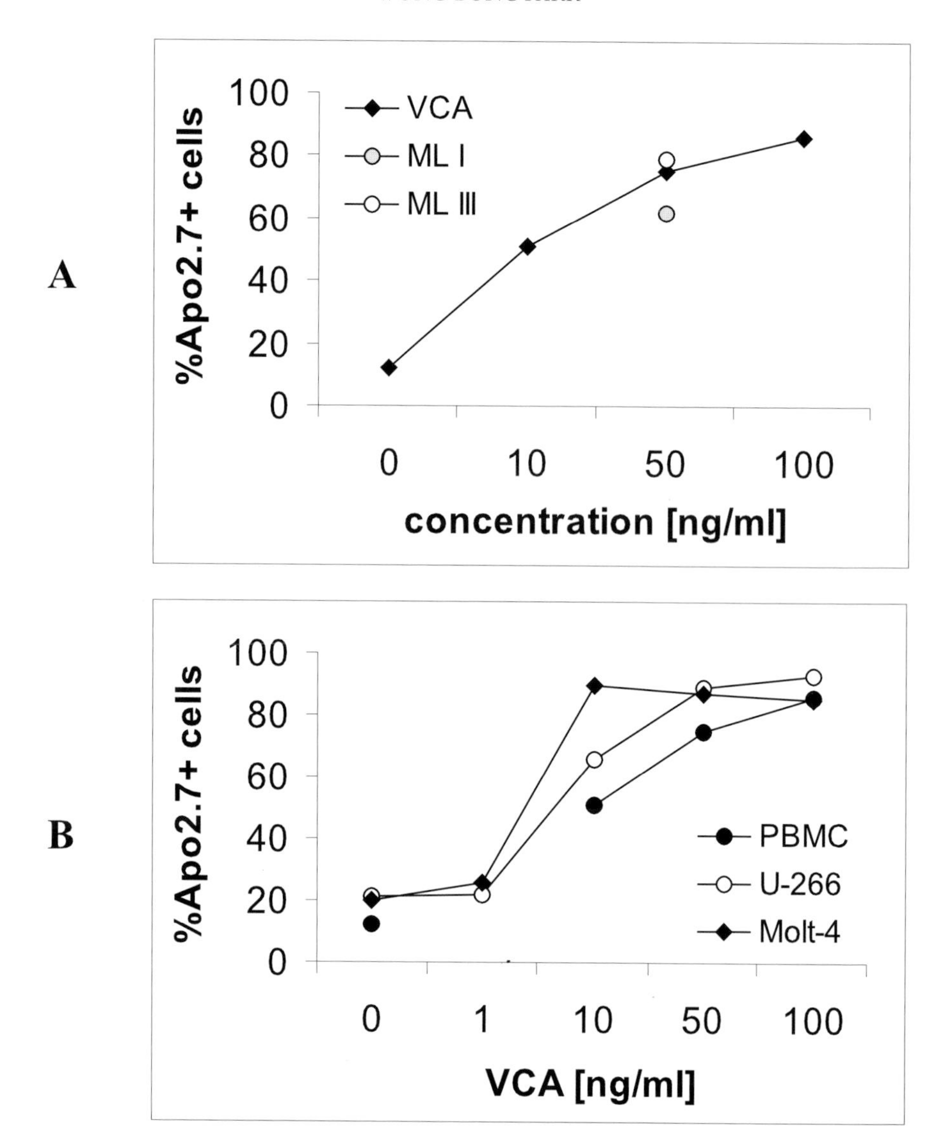

Figure 2 Induction of apoptosis by by VCA. A. Peripheral blood mononuclear cells (PBMC) from a healthy individual were treated for 24 with VCA at 0, 10, 50 and 100 ng/ml, and the toxic lectins from European mistletoe (ML I and ML III) at 50 ng/ml. Expression of mitochondrial apoptosis marker Apo2.7 was measured by flow cytometry as described [Büssing *et al.*, 1999]. VCA was the same effective to induce apoptosis as compared to the galNAc-binding ML III. B. Apo2.7 expression in peripheral blood mononuclear cells (PBMC; ●), IgE-producing myeloma cell line U-266 (○) and the leukaemic Molt-4 T cells (◆) treated for 24 h with VCA at 0, 1, 10, 50 and 100 ng/ml. As observed also for the toxic lectins from *Viscum album* L. (ML I, ML II, ML III), Molt-4 blast cells were more sensitive towards the toxin than normal PBMC. Results with kind permission of A. Büssing, Herdecke.

Table 3 The biological activities of Korean mistletoe.

Cell line	*Effects of processed plant material*	*References*
Leukemia L1210	Fresh extract (ID_{50} 0.1 μg/ml)	Khwaja *et al.*, 1986
Leukemia L1210	Fermented extract (ID_{50} 9.6 μg/ml)	Khwaja *et al.*, 1986
Leukemia L1210	Alkaloid (IC_{50} 0.17 μg/ml)	Khwaja *et al.*, 1986
B16-BL6 melanoma, colon 26-M3.1 carcinoma, lymphoma	inhibition of tumour metastasis and angiogenesis, induction of IL-1, IL-6, and IFN-γ, TNF-α and NK cell activity	Yoon *et al.*, 1995
Ovarian and lung cancer; colon, renal and melanoma xenografts	Fresh extract (IC_{70} 1.2 μg/ml)	Choi *et al.*, 1996
Non-tumourigenic A31	Non-tumourigenic A31	Park *et al.*, 1997-b
Tumourigenic MSV	Heat-treated extract (IC_{50} 300 μg/ml)	Park *et al.*, 1997-b
Murine cell lines: Colon 26, B16-BL6, 3LL, Meth-A, L1210, Yac-1, L5178Y, L929, 3T3	Aqueous extract (IC_{50}, 0.4–307 μg/ml) Lectin (IC_{50}, 1–210 ng/ml)	Yoon *et al.*, 1999
Human cell lines; U937, HL60, K562, THP-1, Jurkat, Raji, Hs578T	Aqueous extract (IC_{50}, 0.5–42 μg/ml) Lectin (IC_{50}, 0.1–62 ng/ml)	Yoon *et al.*, 1999

degradation of Bcl-2 proteins was observed (Büssing and Park, in preparation), indicating the onset of the apoptotic cell death. The biological effects of Korean mistletoe are summarised in Table 3.

CONCLUSIONS

The present knowledge has provided some fundamental understanding of these products, which will be beneficial for further studies. There is increasing evidence that not only lectins, but also other substances present in European mistletoe, show cytotoxic and immune modulating activity. Mistletoe extracts intrinsically consist of a large number of substances. Among them, the lectins, viscotoxins, and polysaccharides are the best described. According to the recent studies, *V. album* L. var. C. growing in Korea shows similar cytotoxic and immunological activities as compared to those of European mistletoe which has been studied intensively. Extracts from *V. album* L. var. C. have inhibitory effects on tumour angiogenesis and metastasis. It will be necessary to further investigate the chemical components and biological activities of the plants and the possible mechanism of actions, such as the synergistic interactions between the individual components for the cancer therapy. Those results will shed light on the cancer therapy in the future.

Although Korean mistletoes and other East-Asian populations have been used traditionally as therapeutical herbs for a long time, scientific investigations of those plants are very limited. It is also necessary to investigate scientifically the mistletoe distributed in Korea and other East-Asian countries as traditional therapeutic herbs.

ACKNOWLEDGEMENTS

The author wishes to thank authors of the articles cited in this review, especially Prof. J.B. Kim for valuable discussions.

REFERENCES

Ahn, W.Y. (1995) Analysis of chemical constiuents of saccharides and triterpenoids in the Korean native mistletoes. *Mokchae Konghak*, **24**, 27–33.

Beuth, J. (1997) Clinical relevance of immunoactive mistletoe lectin-1. *Anticancer Drugs*, **8** (suppl 1), S53-S55.

Büssing, A., Regnery, A., and Schweizer, K. (1995) Effects of *Viscum album* L. on cyclophosphamide-treated peripheral blood mononuclear cells in vitro: sister chromatid exchanges and activation/proliferation maker expression. *Cancer Letters*, **94**, 199–205.

Büssing, A., Suzart, K., Bergmann, J., Pfüller, U., Schietzel, M., and Schweizer, K. (1996) Induction of apoptosis in human lymphocytes treated with *Viscum album* L. is mediated by the mistletoe lectins. *Cancer Letters*, **99**, 59–72.

Büssing, A., Vervecken, W., Wagner, M., Wagner, B., Pfüller, U., and Schietzel, M (1999) Expression of mitochondrial Apo2.7 molecules and caspase-3 activation in human lymphocytes treated with the ribosome-inhibiting mistletoe lectins and the cell membrane permeabilizing viscotoxins. *Cytometry*, **37**, 131–139.

Choi, O.B., Yoo, T.J., Drees, M., Scheer, R., and Kim, J.B. (1996) Inhaltsstoffe und in-vitro-Zytotoxizität eines Extraktes aus *Viscum album* L. ssp. *coloratum* (Koreanische Mistel)-Konsequenzen für die Standardisierung von Mistelpräparaten. *Zeitschrift für Onkologie*, **28**, 77–81.

Chen, X.H., Zeng, Z.K., and Liu, R.H. (1992) Purification and characterization of lectin from *Loranthus parasiticus* (L) Merr. *Shengwu Huayue Zazhi*, **8**, 150–156.

Cheng, B.H., and Zhu, S.H. (1985) An experimental observation on the preventive and curative effects on acute myocardial infarction through improvement of the myocardial oxygen consumption. *Chung Hsi i Chieh Ho Tsa Chih*, **5**, 565–566.

Fukunaga, T., Kajikawa, I., Nishiya, K., Takeya, K, and Itokawa, H. (1989a) Studies on the constituents of the Japanese mistletoe, *Viscum album* L. var. *coloratum* Ohwi grown on different host trees. *Chem. Pharm. Bull.*, **37**, 1300–1303.

Fukunaga, T., Kajikawa, I., Nishiya, K., Takeya, K., and Itokawa, H. (1989b) Studies on the constituents of the Japanase misteltoe *Viscum album* L. var. *coloratum* Ohwi grown on different trees and their antimicrobial and hypotensive properties. *Chem. Pharm. Bull.*, **37**, 1543–1546.

Hajto, T., Hostanska, K., and Gabius, H.J. (1989) Modulatory potency of the β-galactoside-specific lectin from mistletoe extract (Iscador) on the host defense system in vivo in rabbits and patients. *Cancer Res.,* **49**, 4803–4808.

Hajto, T., Hostanska, K., Frey, K., Rordorf, C., and Gabius, H.J. (1990) Increased secretion of tumour necrosis factor α, interleukin 1, and interleukin 6 by human mononuclear cells exposed to β-galactoside-specific lectin from clinically applied mistletoe extract. *Cancer Res.*, **50**, 3322–3326.

Ham, S.H., Park, W.B., Kang, S.H., Choi, K.P., and Lee, D.S. (1998) Antimutagenic effects of Korean mistletoe extracts. *J. Korean Soc. Food Nurt. Sci.*, **27**, 359–365.

Kim, P.S., and Ahn, W.Y. (1996) Analysis of chemical constituents of saccharides and triterpenoids in the Korean native mistletoes (II) – Screening the extractives of Korean camellia mistletoe (*Korthalsella japonica*) for cytotoxicity. *Mokchae Konghak*, **24**, 87–94.

Khwaja, T.A., Varven, J.C., Pentecost, S., and Pande, H. (1980) Isolation of biologically active alkaloids from Korean mistletoe *Viscum album coloratum*. *Experientia*, **36**, 599–600.

Khwaja, T.A., Dias, C.B., and Pentecost, S. (1986) Recent studies on the anticancer activities of mistletoe (*Viscum album*) and its alkaloids. *Oncology*, **43** (Suppl. 1), 42–52.

Kong, D.Y., Luo, S.G., Li, H.T., and Lei, X.H. (1987a) Chemical components of *Viscum coloratum* I. *Yiyao Gongye*, **18**, 123–127.

Kong, D.Y., Luo, S.Q., Li, H.T., and Lei, X.H. (1987b) Chemical components of *Viscum coloratum* II. *Yiyao Gongye*, **18**, 445–447.

Kong, D,Y, Dong, Y.Y., Luo, S.Q., Li, H.T., and Lei, X.H. (1988a) Determination of new flavone glycosides in colored mistletoe (*Viscum coloratum*) by HPLC. *Zhong Cao Yao*, **19**, 495–496.

Kong, D.Y., Luo, S.Q., Li, H.T., and Lei, X.H. (1988b) Chemical components of *Viscum coloratum* III. Structure of viscumneoside III, V and VI. *Yaoxue Xuebao*, **23**, 593–600.

Kong, D.Y., Luo, S.Q., Li, H.T., and Lei, X.H. (1988c) Chemical components of Viscum coloratum IV. Structure of viscunmeoside IV. *Yaoxue Xuebao*, **23**, 707–710.

Kong, D.Y., Luo, S.Q., Li, H.T., and Lei, X.H. (1989) Chemical components of *Viscum coloratum* V. *Yiyao Gongye*, **203**, 108–110.

Kong, D.Y., Luo, S.Q., Li, H.T., and Lei X.H. (1990a) Chemical components of *Viscum coloratum* VI. Chirality of the acyl group of Viscumneoside IV. *Yaoxue Xuebao*, **25**, 349–352.

Kong, D.Y., Li, H.T., and Luo, S.Q. (1990b) Chemical constituents of *Viscum coloratum* VII. Isolation and structure of viscumneoside VII. *Yaoxue Xuebao*, **258**, 608–611.

Lee, S.H., and Ahn,W.Y. (1996) Chemical constiuents of saccharides and triterpenoids in the Korean native mistletoes (III) – Structual features of water-soluble polysaccharides from Korean oak mistletoe (*Loranthus yadoriki* SIEB). *Mokchae Konghak*, **24**, 28–36.

Li, Y.S., Fu, S.X., Han, J., Tseng, W.F. (1959) Diuretic action of flavone arabinoside. Isolation from the Chinese drug Kwang chi-sheng (*Loranthus parasiticus*). *Yao Hsueh Hsueh Pao*, **7**, 1–5.

Li, G., Liu, Q., and Yuan, Y.M. (1995) Isolation and properties of toxic lectin from *Viscum coloratum* (Beijisheng). *Shengwu Huaxue yu Shengwu Wuli Jinzhan*, **22**, 349–352.

Li, M.H. Chemical structure of viscoside A. *Zhong Cao Yao*, **16**, 49–50.

Li, S.Z. (1975) Material Medica Principles: describes 1,892 kinds of medical material in 1578 AD. In *The Great Pharmacopoeia (Ben-Cao Gang-mu)*, Vol. 1–52, In People's Health Publication, Beijing.

Müller, E.A., and Anderer, F.A. (1990) A *Viscum album* oligosaccharide activating human natural cytotoxicity is an interferon γ inducer. *Cancer Immunol. Immunother.*, **32**, 221–227.

Ohta, N., and Yagishita, K. (1970) Isolation and structure of new flavonoids, flavoyandrinin-A, flavoyandrinin-B and homo-flavoyandrinin-B, in the leaves of *Viscum album* L. var. *coloratum* Ohwi epiphyting to *Pyrus communis* L.. *Agr. Biol. Chem.*, **34**, 900–907.

Okuda, T., Yoshida, T., Chen, X.M., Xie, J.X., and Fukushima, M. (1987) Corianin from Coriaria japonica A. Gray and sesquiterpene lactones from *Loranthus parasiticus* Merr used for treatment of schizophrenia. *Chem. Pharm. Bull.*, **35**, 182–187.

Park, W.B., and Kim, H.S. (1994a) Isolation and charaterization of lectin from *Viscum album* L, var. *coloratum. Yakhak Hoeji*, **38**, 418–424.

Park, W.B., and Kim, H.S. (1994b) Changes of lectin from *Viscum album* L, var. *coloratum*, by fermentation with *Lactobacillus plantarum* – Isolation and purification. *Yakhak Hoeji*, **38**, 687–695.

Park, W.B., Kim, H.S., Na, H.B., and Ham, S.S. (1995) Changes of lectin from *Viscum album* L, var. *coloratum*, by fermentation with *Lactobacillus plantarum* – Effects of pH, temperature, sugar specificity and lymphocyte stimulating activity. *Yakhak Hoeji*, **39**, 24–30.

Park, W.B., Han, S.K., Lee, M.H., and Han, K.H. (1997a) Isolation and characterization of lectins from stem and leaves of Korean mistletoe (*Viscum album* L, var. *coloratum*) by affinity chromatography. *Arch. Pharm. Res.*, **20**, 306–312.

Park, W.B., Ju, Y.J., and Han, S.K. (1998a) Isolation and characterization of β-galactoside specific lectin from Korean mistletoe (*Viscum album* L, var. *coloratum*) with Sepharose 4B and changes of conformation. *Arch. Pharm. Res.*, **21**, 429–435.

Park, J.H., Hyun, C.K., Shin, H.K., and Yeo, I.H. (1997b) Effects of heat treatment, sugar addition and fermentation on cytotoxicy of Korean mistletoe. *Korean J. Food Sci. Technol.*, **29**, 362–368.

Park, J.H., Hyun, C.K., and Shin, H.K. (1998a) Cytotoxicity of heat-treated Korean mistletoe. *Cancer Letters*, **126**, 43–48.

Paskov, D., Rusinov, K., and Atanasova, S. (1958a) Pharmacology of Viscum album I. *Izvest Otdel Biol Med Nauki*, **2**, 53–54.

Paskov, D., Rusinov, K., and Atanasova, S. (1958b) Pharmacology of Viscum album II. *Izvest Otdel Biol Med Nauki*, **2**, 27–37.

Pfüller, U., Kopp, J., Körner, I.J., Zwanzig, M., Göckeritz, W., and Franz, H. (1993) Immunotoxins with mistletoe lectin 1 A-chain directed against interleukin-2 receptor of human lymphocytes: comparison of efficiency and specificity. In E. Driessche, H. Franz, S. Beeckmans, U. Pfüller, A. Kallikorm, T.C. Bog-Hansen, (eds.), *Lectins: Biology, Biochemistry, Clinical Biochemistry*, Vol. 8., Wiley Eastern Limited, New Delhi, pp. 34–40.

Pora, A., Pop, E., Roska, D., and Radu, A. (1957) The influence of the host plant on the content of hypotensive and cardioactive principles in mistletoe (*Viscum album*). *Pharmazie*, **12**, 528–538.

Ribéreau-Gayon, G., Jung, M.L., Scala, D.D., and Beck, J.P. (1986) Comparison of the effects of fermented and unfermented mistletoe preparations of cultured tumor cell. *Oncology*, **43** (suppl.1), 35–41.

Ribéreau-Gayon, G., Jung, M.L., Dietrich, J.B., Beck, J.P. (1993) Lectins and viscotoxins from mistletoe (*Viscum album* L.) extracts: development of a bioassay of lectins. In E. Driessche, H. Franz, S. Beeckmans, U. Pfüller, A. Kallikorm, T.C. Bog-Hansen, (eds),. *Lectins: Biology, Biochemistry, Clinical Biochemistry*.Vol. 8, Wiley Eastern Limited, New Delhi, pp. 21–28.

Samal, A.B., Gabius, H.J., and Timonshenko, A.V. (1995) Galactose-specific lectin from *Viscum album* as a mediator of aggregation and priming of human platelets. *Anticancer Research*, **15**, 361–367.

Samuelsson, G. (1959) Phytochemical and pharmacological studies on *Viscum album*. III. Isolation of a hypotensive substance, γ-aminobutyric acid. *Svensk Farmaceutisk Tidskrift*, **63**, 545–553.

Stein, G.M., and Berg, P.A. (1997) Mistletoe extract-induced effects on immunocompetent cells: *in vitro* studies. *Anticancer Drugs*, **8** (suppl 1), S39-S42.

Stein, G.M., Henn, W., von Laue, B., and Berg, P.A. (1998a) Modulation of the cellular and humoral immune responses of tumour patients during mistletoe therapy. *Eur. J. Med. Res.*, **3**, 194–202.

Stein, G.M., Schietzel, M., and Büssing, A. (1998b) Mistletoe in immunology and the clinic (short review). *Anticancer Res.*, **18**, 3247–3250.

Tseng, K.F., and Li, S.C. (1957) Chinese mistletoe I. The chemical constituents of *Viscum album* subspecies *coloratum*. *Yao Hsueh Hsueh Pao*, **5**, 169–177.

Tseng, K.F., and Chen, Z.L. (1957) Studies of the chemical compositions of *Loranthus parasiticus*. II. Isolation of quercetin and its glucoside in Kwang chi-shen. *Yaoxue Xuebao*, **5**, 317–325.

Wagner, H., Feil, B., Seligmann, O., Petricic, J., and Kalogjera, Z. (1986) Cardioactive drugs. V. phenylpropanes and lignans of Viscum album. *Planta Medica*, **2**, 102–104.

Wang, F.P., and Yuan, Y.P. (1980) Isolation of some chemical constituents from *Loranthus parasiticus* (L) Men. *Chung Tsao Yao*, **11**, 345.

Wang, Q.R., Liu, M.Y., and Wang, T.Y. (1994) Antitumor activityof total alkaloids in *Viscum coloratum*. *Zhong Kuo Chung Yao Tsa Chih*, **1**, 45–47.

Wl, J.X., Yu, G.R., Wang, B.Y., Zhong, D.S., and Huang, D.J. (1994) Effects of *Viscum coloratum* flavonoids on fast response action potential of hearts. *Chung Kuo Yao Li Hsueh Pao*, **15**, 169–172.

Wu, C.Y., and Zhang, G.D. (1984) Quantitative analysis of sesquiterpene lactones in *Loranthus parasiticus* (L) Men. parasiting on *Coriaria sinica* maxim and in seeds of *Coriaria sinica* maxim. *Yaoxue Xue bao*, **19**, 56–62.

Wu, C.Y., Zhang, G.D., and Liu, H.Y. (1982) Colorimetric method for microdetermination of tutin in *Loranthus parasiticus* (or L. yadoriki) and in injections. *Zhong Cao Yao*, **13**, 536–539.

Wu, J.X., Yu, G.R., and Wang, B.Y. (1994) Experimental study on cellular electrophysiology of *Viscum coloratum* flavonoids in treating tachyarrhythmias. *Chung Kuo Chung Hsi i Chieh Ho Tsa Chih*, **14**, 421–423.

Xu, Z. (1947) Comparative studies of pharmacological efficacy of major pathogenic bacteria by phytotherapy. *Agric. Bull.*, **1**, 17–25.

Yang, L.L., Yen, K.Y., Kiso, Y., and Hikino, H. (1987) Antihepatotoxic actions of Fonnosan plant drugs. Journal of Ethnopharmacology, **19**, 103–110.

Yoon, T.J., Yoo, Y.C., Hong, E.K., and Cho, Y.H. (1994) Effects of Korean mistletoe extracts on the induction of IL-1 and TNF-α from macrophages. Korean Journal of Pharmacognosy, **25**, 132–139.

Yoon, T.J., Yoo, Y.C., Choi, O.B., Do, M.S., Kang, T.B., Lee, S.W., *et al.* (1995) Inhibitory effect of Korean mistletoe (*Viscum album coloratum*) extract on tumour angiogenesis and metastasis of haematogenous and non-haematogenous tumour cells in mice. *Cancer Letters*, **97**, 83–91.

Yoon, T.J., Yoo, Y.C., Kang, T.B., Do, M.S., Azuma, I., and Kim, J.B. (1997) Immunological activities of Korean mistletoe extract (*Viscum album coloratum*; KM-110). *Korean Journal of Immunology*, **19**, 571–581.

Yoon, T.J., Yoo, Y.C., Kang, T.B., Baek, Y.J., Huh, C.S., Song, S.K., *et al.* (1998) Prophylactic effect of Korean mistletoe (*Viscum album coloratum*) extract on tumor metastasis is mediated by enhancement of NK cell activity. *International Journal of Immunopharmacology*, **20**, 163–172.

Yoon, T.J., Yoo, Y.C., Kang, T.B., Shimazaki, K., Song, S.K., Lee, K.H., *et al.* (1999) Lectins isolated from Korean mistletoe (*Viscum album coloratum*) induce apoptosis in tumor cells. *Cancer Letters*, **136**, 33–40.

Zhu, W.M., Xie, L., and Zheng, S.Y. (1985) Mechanism of the antiplatelet aggregative activity. *Yiyao Gongye*, **16**, 257–261.

Zhu, S.H. (1984) Effect of Viscum coloratum (Kom) Nakai on change in the cyclic nucleotides in ischemic myocardium. *Chung Hsi I Chieh Ho Tsa Chih*, **4**, 548–551.

5. MISTLETOES FROM ARGENTINA

Ligaria cuneifolia var. *cuneifolia* as a substitute for the European mistletoe (*Viscum album* L.)

TERESA B. FERNÁNDEZ[1], BEATRIZ G.VARELA[2], CARLOS A. TAIRA[4,5], RAFAEL A. RICCO[2], ALBERTO A. GURNI[2], SILVIA E. HAJOS[1,5], ELIDA M.C. ALVAREZ[1,5] and MARCELO L. WAGNER[2,3]

[1]Cátedra de Inmunología-IDEHU, [2]Cátedra de Farmacobotánica, [3]Museo de Farmacobotánica „Juan A. Domínguez", [4]Cátedra de Farmacología, [5]Member of the Research Career, CONICET. Facultad de Farmacia y Bioquímica, Universidad de Buenos Aires. Junín 956 (1113), Buenos Aires, Argentina.

INTRODUCTION

A number of hemiparasite plants belonging to the mistletoe species, which are taxonomically related to the European mistletoe (*Viscum album* L.), grow in Argentina. According to the work carried out by *Abbiatti* (1946), 23 different mistletoe species grow in diverse phytogeographic regions of the country with the exception of the Pampean savannah, the Patagonian steppes and the arid Andean regions. The Monte (shrub-like) Formation due to the abundance of leguminous trees, which are the preferred mistletoe hosts, is the most densely populated.

From the taxonomic viewpoint, native mistletoe species belong to the three different families into which Loranthaceae (*sensu latu*) has been divided. These families are Loranthaceae D. Don (*sensu strictum*), Eremolepidaceae Tiegh. and Viscaceae Miq. (Barlow, 1964; Kuijt, 1988; Subils, 1984; Barlow *et al.*, 1989) (Table 1).

Many uses have been described for native mistletoe species, which differ depending on ethnic heritage, geographic region and species. *Phoradendron pruinosum* and *Ph. liga* are both used in the Northeast of Argentina for cardiac disorders and *Ph. hieronymi* for asthma treatment (Martínez-Crovetto, 1981; Toursarkissian, 1980; Wagner *et al.*, 1986), while Tobas (Chaco Amerindians from the Northeast of Argentina) use *Ph. liga* to sedate horses (Arenas, 1982). In the provinces of Salta and Catamarca stems with white fragrant flowers of *Tripodanthus acutifolius* are used in the Corpus Christi festivity, held in June during its flowering season (Abbiatti, 1946). A viscous substance called "viscina" or "liga" obtained from the berries of *Tristerix corymbosus* and *Ligaria cuneifolia* is used to trap insects and birds (Abbiatti, 1946; Diem, 1950). *Eubrachion ambiguum*, whose use has not yet been

Table 1 Mistletoes in Argentina.

Loranthaceae D. Don		
	Tripodanthus (Eichl.) Tiegh. (= Phrygillantus Eichl.)	*Trp. flagellaris* (Cham. et Schlecht) Tiegh. *Trp. acutifolius* (Ruiz et Pav.) Tiegh.
	Struthanthus Mart	*S. angustifolius* (Griseb.) Haum.
		S. acuminatus (Ruiz et Pav.) Blume *S. uraguensis* (Hook. et Arn.) G. Don
	Ligaria Tiegh.	*L. cuneifolia* (Ruiz et Pav.) Tiegh
	Tristerix Mart.	*Trx. corymbosus* (L.) Kuijt *Trx. verticillatus* (Ruiz et Pav.)Barlow et Wiens
	Psittacanthus Mart.	*P. cordatus* (Hoffmans) Blume
Eremolepidaceae Tiegh.		
	Eubrachion Hook. f.	*E. ambiguum* (H.et A.) Engler (= *E. andalgalense* Abbiatti)
Viscaceae Miq.		
	Phoradendron Nutt.	*Ph. pruinosum* Urb. *Ph. mucronatum* (D.C.) Krug et Urb. (= *Ph. argentinum* Urb.) *Ph. piperoides* (H.B.K.) Nutt. *Ph. acinacifolium* Mart. *Ph. acinacifolium* Mart. *Ph. liga* (Gill.) Eichl. *Ph. salicifolium* (Preol.) Eichl. *Ph. hieronymi* Trel. *Ph. tucumanense* Urb. *Ph. falcifrons* (Hook. et Arn.) Eichl. *Ph. subfalcatum* Abbiatti *Ph. dipterum* Eichl. *Ph. perrottetii* (D.C.) Nutt. *Ph. quadrangulare* (Kunth) Griseb.

recorded in Argentina, is employed in the south of Brazil for the treatment of lumbar aches and pneumonia. The Wichi (native Amerindians from the Northeast of Argentina) make amulets and ritual objets with *Ph. liga*, *Ph. hieronymi* and *Trp. flagellaris* (Martinez-Crovetto, 1964).

European migrants and descendants, according to both morphologic and habitat similarities, chose *L. cuneifolia* var. *cuneifolia* as the natural substitute for the European mistletoe (*Viscum album* L.). Infusions of leaves and stems have thus been used for their putative ability to decrease high blood pressure. This species is the most widely used mistletoe in the country and is popularly known as “liga”, “liguilla” or “muérdago criollo” (Argentine mistletoe).

"MUERDAGO CRIOLLO"

Ligaria cuneifolia var. ***cuneifolia*** (= ***Psittacanthus cuneifolius*** (Ruiz et Pavon) Blume) (Figure 1)

Geographic Distribution

This South American species is found in Perú, Bolivia, Argentina, Chile, Brazil and Uruguay. It is the most widespread in Argentina and its habitat extends from Salta and Jujuy in the North to La Pampa in the South, and from Entre Ríos and Northeast Buenos Aires to the Andean foothills in the West.

Host Trees

L. cuneifolia grows mainly on leguminous trees as *Gourliea*, *Piptadenia*, *Prosopis* and *Acacia* species, but can also be found on *Celtis* (Ulmaceae), *Schinus*

Figure 1 *Ligaria cuneifolia* var. *cuneifolia*. *Ligaria cuneifolia* with a red flower found in central and western Argentina (upper right – *Ligaria cuneifolia* var. *cuneifolia*) and with a more orange/yellow colour in the east of the country (lower right – *Ligaria cuneifolia* var. *flava*).

(Anacardiaceae), *Bulnesia*, *Schinopsis* (Anacardiaceae) and *Ephedra* (Ephedraceae). Among cultivated plants it parasites *Pyrus malus*, *P. communis*, *Prunus* sp. (Rosaceae) and *Robinia* (Leguminosae).

Morphology

L. cuneïfolia var. *cuneifolia* is a shrubby, glabrous plant, devoid of aerial roots. Adult branches are thick and cylindrical, while young branches are almost flattened and both are striated-wrinkled. Arranged alternately in stems, fleshy wrinkled leaves measuring 1.5–6.5 cm long and 0.4 to 1.5 cm wide, may be sub-sessile, linear, oblong, lanceolate or linear-spatulate, while midribs are unconspicuous. Flowers present in solitary racemes and have oval basal scales (prophylls) 0.1 cm long. The calyx is scyathiform, triangular and tridentated. Hexamerous flowers have an intense red colour in specimens from central and western Argentina, while they are orange/yellow in those from the east of the country (Figure 1). Tepals are linear-spatulate and on occasion joined in a tube from their lower middle portion. Stamens are unequal and inserted into the tepals up to the middle of their length. The versatile anterae are oblong and apiculated. The stigma is capitated, the stylus filiform and the ovary inferous. The fruit is a globular, dark reddish berry, crowned by the tubiform calycle. Seeds are endospermous (unlike *Psittacanthus*) and germinate during early November (Abbiatti, 1946). The flowering season starts in spring and persists till autumn.

Anatomical Characterisation

Leaves

Epidermis: Epidermic cells are square and regular with thick walls. They are covered by a thin cuticle. The lower epidermis presents numerous paracytic stomata, whereas the upper epidermis has very few.

Mesophyl: The mesophyl is isobilateral and formed by two layers of radially elongated cells bordering both epidermis while central cells are shorter. Scattered irregular, ramified, branched stone cells with calcium oxalate crystals can be observed. The conductive tissue is made up by a large vascular bundle in the centre and smaller lateral bundles all surrounded by cells with thick walls. Vessels frequently end on enlarged tracheids (Figure 2a).

Stems

Epidermis: It presents square cells covered by a thick papillous yellow cuticle and a moderate amount of paracytic stomata.

Bark: The heterogeneous cortical parenchyma consists of two or three layers of radially enlarged chlorenchymatic cells and various layers of inner cells tangentially enlarged (Figure 2b). In the first zone branched crystalliferous stone cells are present, while in the second area groups of fibres with thick walls and narrow lumen appear.

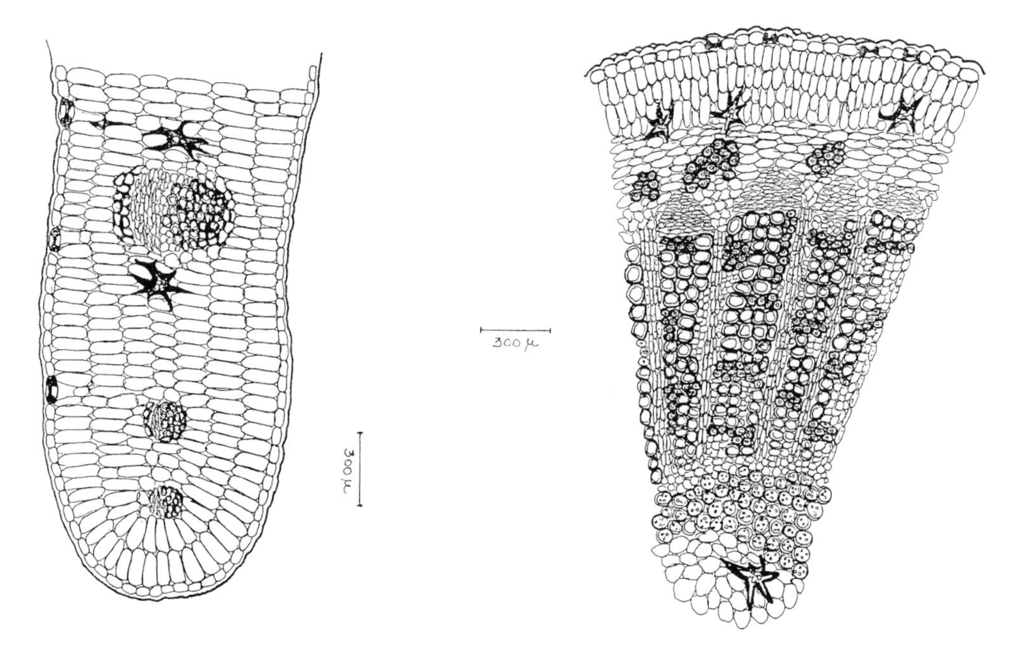

Figure 2 Anatomical structure of the leaf (left) and stem (right) of *L. cuneifolia* (× 40).

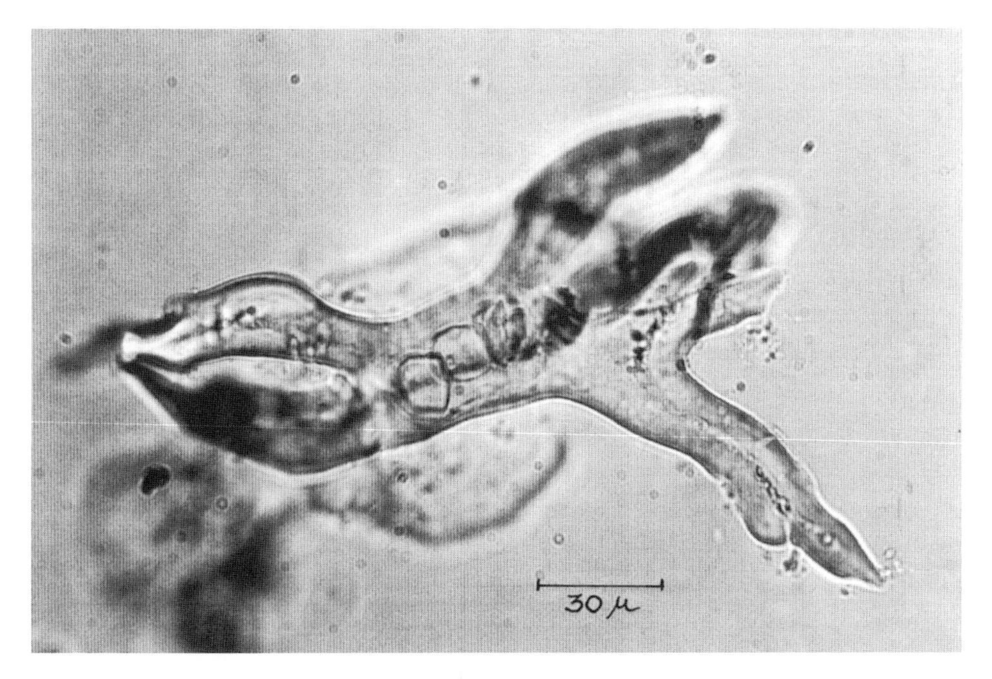

Figure 3 Crytalliferous branched stone cell of *L. cuneifolia (R. et P.) Tiegh*. (× 400).

Central cylinder: The central cylinder is made up by opened collateral conducting bundles. The xylem forms a continuous ring presenting a great number of thick-walled lignified fibres in variable groups per bundle. Medullar rays consist of two to five layers of parenchymatous radially enlarged cells. A perimedullar zone is formed by round cells with thick sclerosed walls having simple pits. Around the pith, round cells with thick cellulose walls are found, accompanied by crystalliferous branched stone cells (Varela and Gurni, 1995) (Figure 3).

CONSTITUENTS

Micromolecular Compounds

Aminated compounds

In studies carried out on samples growing on diverse host trees, collected from diverse geographic locations, tyramine is present in leaves and stems (Vazquez y Novo *et al.*, 1989). The detected concentration in most samples remains below 10 mg% but roughly 10% yield over 100 mg%. Plants parasiting *Geoffroea decorticans* (H. et Arn.) Burkart (Fabaceae) are the ones presenting the greatest amounts of tyramine, ranging from 120 to 360 mg per 100 g of dried material.

Flavonoids

In all plant samples, regardless of the host tree, the phytochemical study of the flavonoids has disclosed the presence of quercetin as the only flavonol (Graziano *et al.*, 1967; Wagner, 1993). Quercetin occurs free and monoglycosylated with xylose, rhamnose and arabinose at the hydroxyl group in position 3 of the flavonol skeleton (Wagner, 1993; Fernández *et al.*, 1998).

Leucoanthocyanidins, catechin-4-β-ol and proanthocyanidins with variable degrees of polymerization (dimers, oligomers and polymers of catechin and epicatechin) which yield cyanidin after acid treatment are also present (Wagner, 1993; Fernández *et al.*, 1998) (Figure 4).

In *L. cuneifolia* the precursor dihydroquercetin may follow one of two metabolic pathways: (1) The enzyme flavonol synthase oxidises dihydroquercetin leading to quercetin, part of which accumulates, while most is glycosilated in the hydroxyl group in the C3 by the uridine diphosphate-sugar-flavonoid-3-O-glycosyl-transferase. (2) The alternative pathway is activated by NADPH-dependent 3-hydroxyflavanone-4-reductase, which reduces the carbonyl group rendering leucocyanidin. This latter compound may be transformed into flavan-3-ol (catechin or epicatechin) by 3,4 cys-diol-reductase. Both leucocyanidin and flavan-3-ol may be condensed originating dimers, oligomers and polymers by the proanthocyanidin synthase enzymatic complex (Stafford, 1990).

Flavonoid synthesis in *L. cuneifolia* is more simple than in *V. album* where biosynthesis is more diversified since in the latter S-adenosyl-L methionine-X-O-methylase (SAM) gives rise to the methylated flavonoids not detected in the Argentine mistletoe (Figure 5).

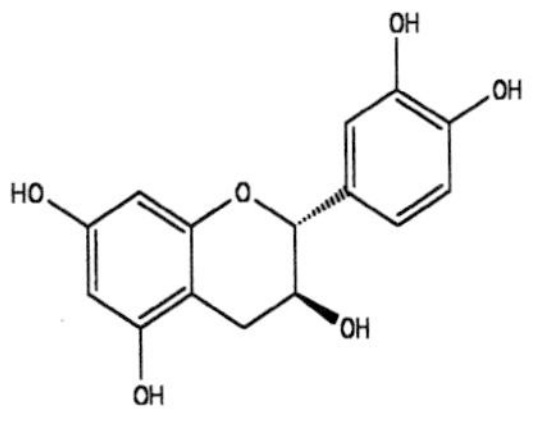

Catechin 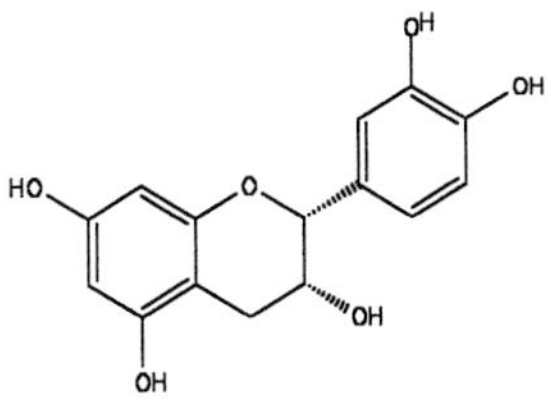 Epicathechin

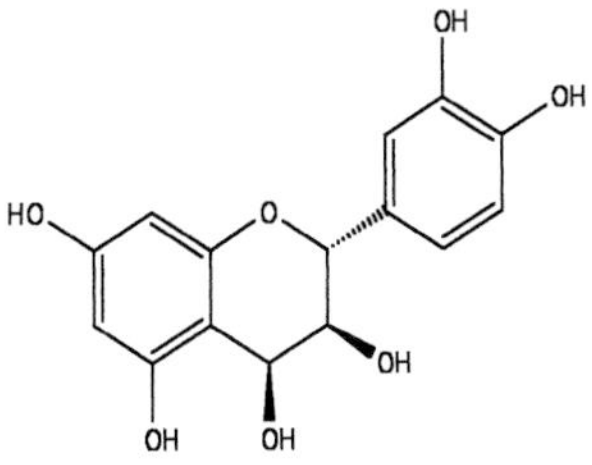

Catechin- 4 β ol

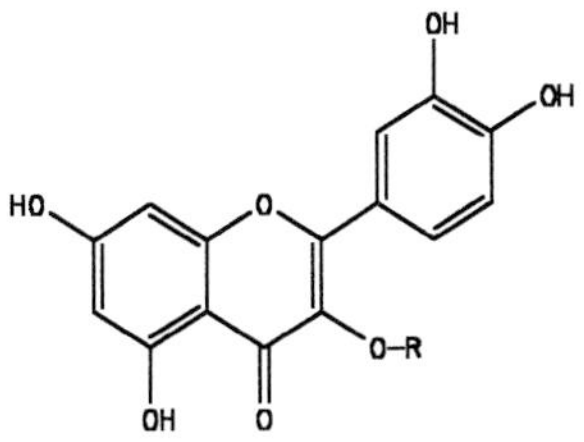

R = H Quercetin

R = Sugar Quercetin-3-O-monoglycoside

Figure 4 Chemical structure of the flavonoids.

Macromolecular Compounds

Macromolecular components of acellular extracts analysed by electrophoresis show a complex protein pattern ranging from 14 to 90 kDa in molecular weight, quite distinct from the one obtained with *V. album* (Fernández *et al.*, 1998). When extracts are analysed under denaturing and reducing conditions and transferred onto nitro-cellulose, mouse anti-Ligaria antiserum reacts not only with acellular extracts of *L. cuneifolia*, but also with those of *V. album* and other Viscaceae species such as *Ph. liga*, thus demonstrating that the proteins present in the above species exhibit antigenically related epitopes (Wagner *et al.*, 1998).

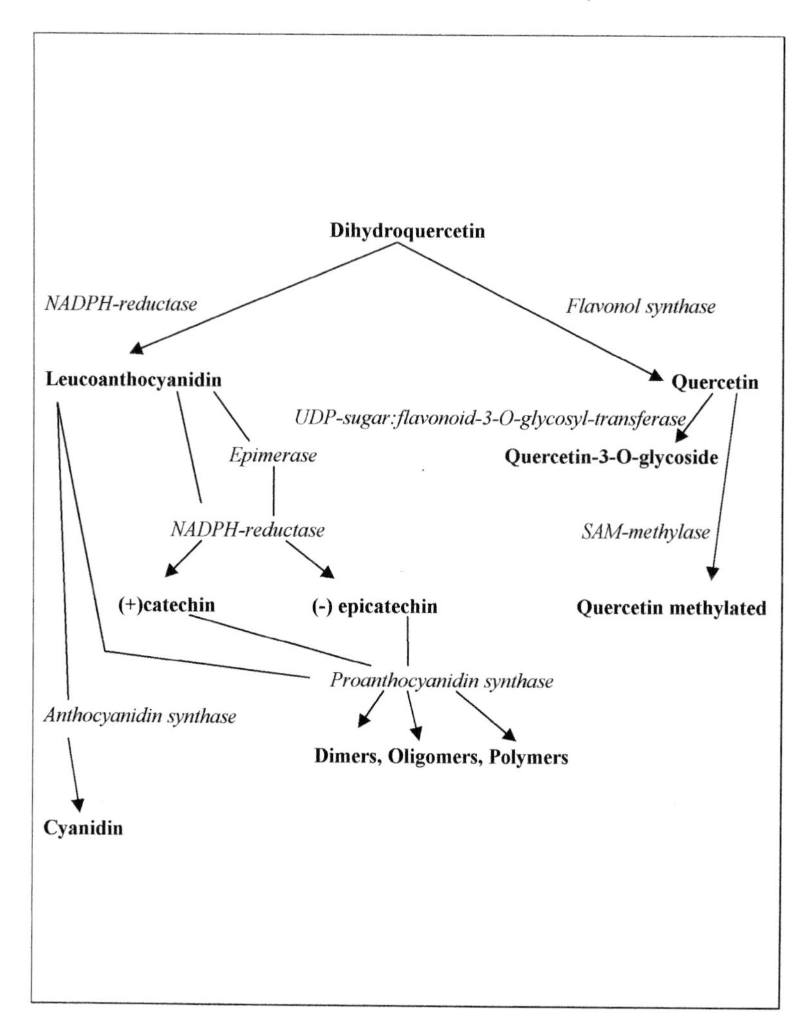

Figure 5 Metabolic pathways.

BIOLOGICAL AND PHARMACOLOGICAL PROPERTIES

Vasoactive Activity

Given its outward similarity to *V. album*, Argentine mistletoe (*L. cuneifolia*) is widely employed in infusions as an alternative medicine to treat high blood pressure (Domínguez, 1928; Ratera and Ratera, 1980). The first pharmacological and phytochemical studies performed on *L. cuneifolia* samples disclosed a degree of hypotensive action (Domínguez, 1928). Since extracts administered parenterally have been reported to exert experimental hypertension, indiscriminate use should be guarded against (Domínguez, 1928; Ratera and Ratera, 1980).

It was Izquierdo and co-workers who originally demonstrated the presence of tyramine, a compound exerting sympathicomimetic activity (Izquierdo *et al.* 1955).

Recent studies compare the cardiovascular effects of samples of *L. cuneifolia* growing on three host trees: *Schinus polygamus* (Cav.) Cabr. (Anacardiaceae), *Acacia caven* (Mol.) Molina (Mimosaceae) and *G. decorticans* (Taira *et al.*, 1994). The specimen growing on *S. polygamus* exerts a dose-dependent pressor effect (Figure 6) accompanied by a secondary hypotensor phase, whereas high doses produce a drop in heart rate. The fact that the α-adrenergic antagonist phenoxy-

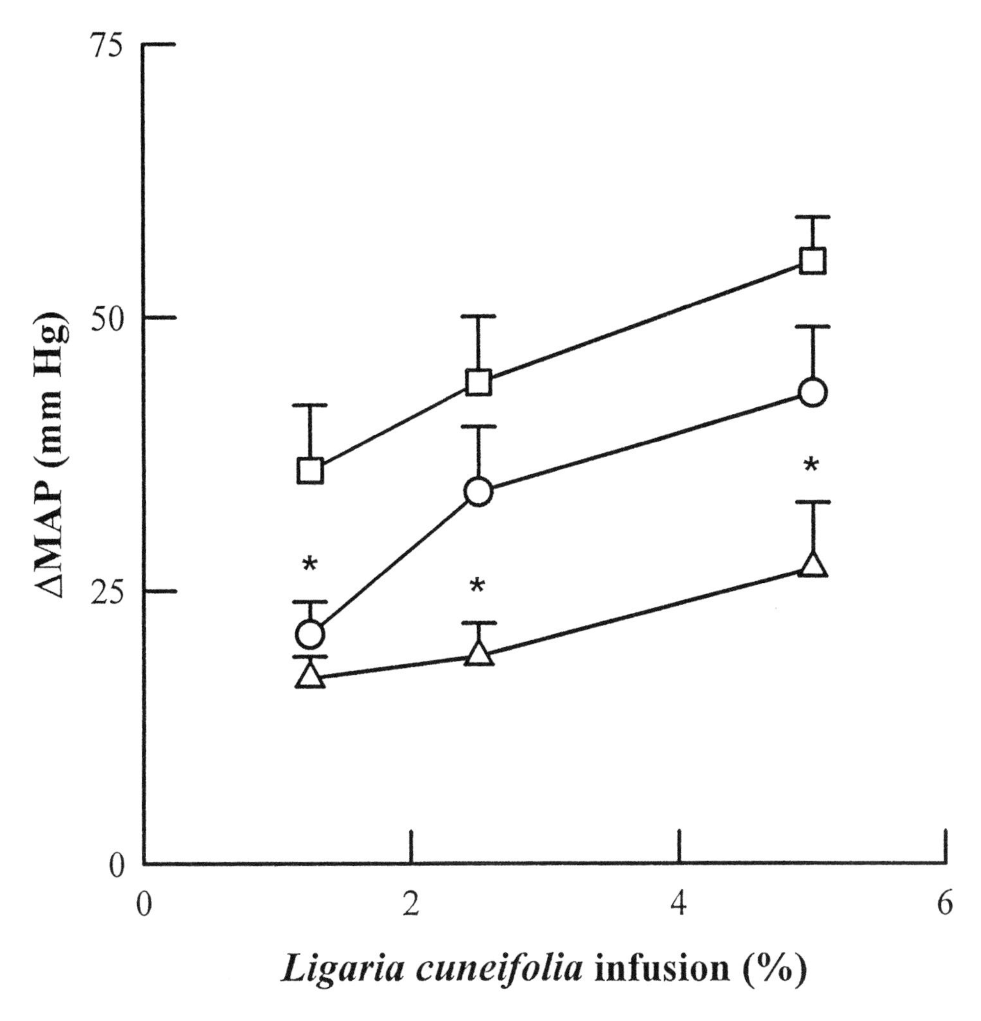

Figure 6 Pressor effect of infusions of *Ligaria cuneifolia*. Change of mean arterial pressure (MAP) induced by infusions of *L. cuneifolia* (1.25, 2.5 and 5.0% P/V) by intravenous injection in Wistar rats (each point is the mean ± SEM of four experiments). Host tree: □ *A. cavens*, O *S. polygamus*, Δ *G. decorticans* (*$p < 0.05$).

benzamine inhibits the pressor action of the extract suggests that the effect is of adrenergic origin (Lefkowitz *et al.*, 1991). Cholinergic blockade brought about by atropin evidence that both hypotension and bradicardia are related to muscarinic receptors (Lefkowitz *et al.*, 1991).

In the Argentine mistletoe specimen parasiting *A. caven*, an α-adrenergic compound with pressor effect is also present. Unlike the *S. polygamus* sample, the hypotensive effect may be observed after adrenergic blockade, but the nature of the hypotensive agent is unknown. High concentrations of the extract also produce a drop in heart rate. *L. cuneifolia* collected from *G. decorticans* only exerts a minor pressor effect and no changes in heart rate are observed.

On the basis of these results, it may be posited that the host tree is capable of modulating the profile of vasoactive components produced by the mistletoe.

Immunomodulating Activity

A wide range of biological activities has been reported for various mistletoes including antiviral, antitumoural, immunomodulating and inflammation modifying effects. *V. album* extracts specifically have been popular in Europe for seven decades as an unconventional approach to cancer treatment (Bloksma *et al.*, 1982, Hajto, 1986, Jurin *et al.*, 1993). Tumour-reducing or modulating components have been identified as lectins, viscotoxins, proteins, peptides, oligosaccharides, alkaloids, polyphenolic compounds and flavonoids (Khwaja *et al.*, 1986; Hostanska *et al.*, 1995; Kuttan *et al.*, 1997; Stein and Berg, 1997; Beuth *et al.*, 1996; Büssing *et al.*, 1996; Gabius *et al.*, 1992; Zee Cheng, 1997). Anti-cancer activity may not only be due to inhibition of cellular proliferation but also to cytokine induction and immunoadjuvant effects (Hostanska *et al.*, 1995; Männel *et al.*, 1991; Müller and Anderer, 1990).

To study the possible immunomodulating effect exerted by *L. cuneifolia* extracts, parameters of cell proliferation and function have been evaluated. When variable concentrations of plant extract are added to normal murine splenocytes, their

Table 2 Effect of *L. cuneifolia* on the proliferation of murine splenocytes.

	L. cuneifolia extract [µg/ml]			
	0.1	*1*	*10*	*100*
splenocytes	88.19 ± 11.54	105.45 ± 10.27	115.53 ± 42.43	131.97 ± 29.85
splenocytes + ConA	72.99 ± 6.90	51.23 ± 1.18	19.70 ± 2.15	23.73 ± 2.61
splenocytes + LPS	99.68 ± 1.84	93.93 ± 5.41	82.03 ± 3.01	66.79 ± 9.88
LB cells	68.00 ± 21.52	56.88 ± 13.45	46.73 ± 8.23	8.07 ± 1.06

Proliferation of murine splenocytes (10^6 cell/ml) and murine leukaemic LB cells (5.10^5 cell/ml) was meausred by (^{3}H)-thymidine uptake. The cells were incubated for 24 h at 37°C, pulsed with 1 µCi (^{3}H)-thymidine for further 24 h. The cells were stimulated with Concanavalin A (ConA; 7 µg/ml) or with LPS (20 µg/ml). Control cellular growth in absence of *L. cuneifolia* extract was taken as 100%. Proliferation was calculated as $(cpm_{experiment} \times 100)/cpm_{control})$.

growth is slightly stimulated (Table 2), while mitogen-activated splenocytes were inhibited, an effect more pronounced for the mitogenic lectin Concanavalin A than for lipopolysaccharide. In contrast to normal murine splenocytes, leukaemic LB cells were strongly inhibited by *L. cuneifolia* extracts (Table 2) (Fernández *et al.*, 1998), indicating that leukemic cells are more sensitive towards the drug than normal cells. These cells were killed by apoptosis (Figure 7).

Since the inducible form of the arginine-dependent enzyme nitric oxide synthase generates toxic amounts of nitric oxide, which enables the activated macrophage to destroy tumour cells and microorganisms, *L. cuneifolia* effect on nitric oxide production has been evaluated (Cui *et al.*, 1994). Murine macrophages from normal mice collected after thyoglycolate and Concanavalin A stimulation were cultured alone or in the presence of lipopolysaccharide or variable doses of recombinant murine interferon-γ. In all cases, *L. cuneifolia* acellular extracts are capable of enhancing nitric oxide production (Fernández *et al.*, 1998) (Table 3).

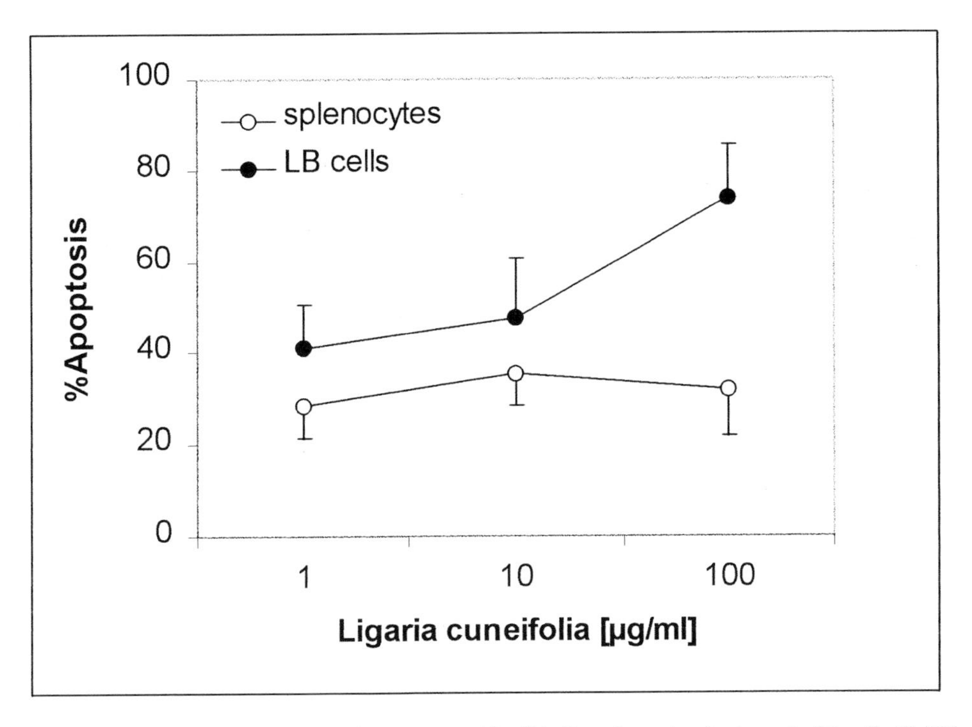

Figure 7 Apoptosis in murine splenocytes (10^6 cell/ml) and murine leukaemic LB cells (5.10^5 cell/ml) treated with *L. cuneifolia* extract for 48 h. Apoptotic nuclei were measured by UV microscopy (epillumination) using acridine orange and ethidium bromide. Results are means ± SD from 5 different experiments and are given as % of cells (total number of cells with apoptotic nuclei per total number of cells × 100 from 5 different experiments). Apoptosis was verified by the characteristic DNA labelling in agarose gel (data not shown).

Table 3 Nitric oxide production by murine macrophages.

L. cuneifolia	*alone*	*+ LPS [100 μg/ml]*	*+ IFNγ [0.1 U/ml]*	*+ IFNγ [1 U/ml]*	*+ IFNγ [10 U/ml]*
0 μg/ml	11.98 ± 2.32	10.23 ± 4.18	13.72 ± 2.63	15.38 ± 1.64	22.84 ± 3.52
100 μg/ml	18.74 ± 6.03	18.22 ± 4.72	24.55 ± 1.65	28.18 ± 3.85	41.58 ± 2.41

Peritoneal murine macrophages (10^6 cell/ml) were cultured in absence or presence of *L. cuneifolia* extract [100 μg/ml] and either alone or in the presence of LPS or recombinant murine IFNγ. Nitric oxide production (nM/10^5 cells) was measured based on the Griess reaction, as described elsewhere (Fernández *et al.*, 1998).

CONCLUSIONS

L. cuneifolia var. *cuneifolia* which is the natural substitute for the European mistletoe (*Viscum album*) in Argentina exerts distinct effects: hypotension, immunomodulation and induction of apoptosis. Incubation of murine cells with acellular extracts of the plant resulted in an antiproliferative effect on both, activated splenocytes and leukaemic cells, while normal splenocytes were stimulated. Apart from the induction of an apoptotic cell death in leukemic cells, the plant enhances the production of macrophage nitric oxide.

Anatomical study shows that the main microscopical features to identify this species are the presence of crystalliferous branched stone cells in leaves and stems, the absence of other crystals and the lack of cork in stems. No lipids are detected. Along with leucoanthocyanidins and proanthocyanidins, the only detected flavonol is quercetin glycosylated with three sugars. The precursor dihydroquercetin follows two metabolic pathways: one leading to quercetin by means of a flavonol synthase and a second one leading to leucocyanidin by means of a 3-hydroxyflavanone-4-reductase (Stafford, 1990). The simultaneous presence of both flavonoids is useful to characterise this species when compared to other Loranthaceae and Viscaceae. Macromolecular protein components of the extracts analysed by electrophoresis present a pattern quite dissimilar to *V. album* but proteins present exhibit related antigenic epitopes. In most samples tyramine concentration fails to exceed 10 mg%, but may reach 360 mg% in samples parasiting *G. decorticans*.

ACKNOWLEDGMENTS

This study was supported by grants FA 093 and FA 127 from the University of Buenos Aires.

REFERENCES

Abbiatti, D. (1946) *Las Lorantáceas Argentinas*. Revista del Museo de La Plata (nueva serie) 7 (sección botánica)

Arenas, P. (1982) Recolección y agricultura entre los indíginas Maká del Chaco Boreal. *Parodiana*, **1**, 171–243.
Barlow, B.A. (1964) Classification of the Loranthaceae and Viscaceae. *Proceedings of The Linneaen Society of New South Wales*, **89**, 268–272.
Barlow, B.A., Hawsksworth, F.G., Kuijt, J., Polhill, R.M., and Wiens, D. (1989) Genera of Mistletoes. *The Golden Bough*, **11**, 1–3.
Beuth, J., Stoffel, B., Samtleben, R., Stoak, O., Ko, H.L., Pulverer, G., *et al.* (1996). Modulating activity of mistletoe lectins 1 and 2 on lymphatic system in BALB/c mice. *Phytomedicine*, **2**, 269–273.
Bloksma, N., Schmiermann, P., de Reuver, M., van Dijk, H., and Willers, J. (1982) Stimulation of humoral and cellular immunity by *Viscum* preparations. *Planta Medica*, **46**, 221–227.
Büssing, A., Suzart, K., Bergmann, J., Pfüller, U., Schietzel, M., and Schweizer, K. (1996) Induction of apoptosis in human lymphocytes treated with *Viscum album* L. is mediated by the mistletoe lectins. *Cancer Letters*, **99**, 59–72.
Cui, S.J., Reichner, J.S., Mateo, R.B., and Albina, J.E. (1994) Activated murine macrophages induce apoptosis in tumor cells through nitric oxide- dependent or independent mechamism. *Cancer Research*, **54**, 2462–2467.
Diem, J. (1950) Las plantas huéspedes de la lorantácea *Phrygillantus tetrandrus* (Ruiz et Pavon) Eichl. *Boletín de la Sociedad Argentina de Botánica*, **3**, 177–179.
Domínguez, J.A. (1928) *Contribuciones a la Materia Médica Argentina*. Peuser Ed., Buenos Aires.
Fernández, T., Wagner, M.L., Varela, B.G., Ricco, R.A., Hajos, S.E., Gurni, A.A. (1998) Study of an Argentine Mistletoe, the hemiparasite *Ligaria cuneifolia* (R.et P.) Tiegh. (Loranthaceae). *Journal of Ethnopharmacology*, **62**, 25–34.
Gabius, S., Joshi, S., Kayser, K., and Gabius, H. (1992) The galactoside-specific lectin from mistletoe as biological response modifier. *International Journal of Oncology*; **1**, 705–708.
Graziano, M.N., Widmer, G.A., Juliani, R., and Coussio J.D. (1967) Flavonoids from the argentine mistletoe *Psittacanthus cuneifolius*. *Phytochemistry* **6**, 1709–1711.
Hajto, T. (1986) Immunomodulatory effects of Iscador: A *Viscum album* preparation. *Oncology*, **43** (suppl. 1), 51–65.
Hostanska, K., Hajto, T., Spagnoli, G., Fischer, J., Lentzen, H., and Herrmann, R. (1995) A Plant Lectin derived from *Viscum album* induces cytokine gene expression and protein production in cultures of human peripheral blood mononuclear cells. *Natural Immunity*, **14**, 295–304.
Izquierdo, J.A., Izgellaris and Starita, J.A. (1955) Acciones vasculares de *Phrygilantus* y del *Psittacanthus cuneifolius*. *Revista Farmacéutica*, **97**, 177–181.
Jurin, M., Zarkovic, N., Hrzenjak, M., and Ilic, Z. (1993) Antitumorous and immunomodulatory effects of the *Viscum album* L. Preparation Isorel. *Oncology*, **50**, 1–6.
Khwaja, T., Dias, C.B. and Pentecost, S. (1986). Recent studies on the anticancer activity of Mistletoe (Viscum album) and its alkaloids. *Oncology*, **43** (suppl. 1), 42–50.
Kuijt, J. (1988a) Revision of *Tristerix* (Loranthaceae). *Systematic Botany Monographs*, **19**, 1–61.
Kuijt, J. (1988b) Monographs of the Eremolepidaceae. *Systematics Botany Monographs*, **18**, 1–60.
Kuttan, G., Menon, L.G., Antony S., and Kuttan, R. (1997) Anticarcinogenic and antimetastasic activity of Iscador. *Anticancer Drugs*, **8** (suppl 1), 15–16.
Lefkowitz, R.J., Hoffman, B.B., and Taylor P. (1981) Transmisión neurohumoral: los sistemas nerviosos autónomos y motor somático. In A. Goodman Gilman, T.W. Rall, A.S. Nies,

P. Taylor, (eds.), *Las Bases farmacológicas de la terapéutica. 8° ed.* (Spanish transduction), Ed. Médica Panamericana, Buenos Aires, pp. 97–113.

Martínez Crovetto, R. (1981) *Las plantas utilizadas en Medicina Popular en el Noroeste de Corrientes (República Argentina). Miscelanea N° 69*, Fundación Miguel Lillo ed., S.M. Tucumán.

Martínez Crovetto, R. (1964) Estudios etnobotánicos I. Nombres de plantas su utilidad, según los indios Tobas del Este del Chaco. *Bonplandia*, **1**, 279–333.

Männel, D., Becker, H., Gundt, A., Kist, A., and Franz, H. (1991) Induction of tumor necrosis factor expression by a lectin from Viscum album. *Cancer Immunology Immunotherapy*, **33**, 177–82.

Müller, E. and Anderer, F. (1990) A Viscum album oligosaccharide activating human natural cytotoxicity is an interferon gamma inducer. *Cancer Immunology Immunotherapy*, **32**, 221–227.

Ratera, E.L. and Ratera, M.O. (1980) *Plantas de la flora Argentina empleadas en Medicina Popular*, Hemisferio Sur ed., Buenos Aires, p. 82.

Stafford, H.A.: *Flavonoid Metabolism*. Boca Raton, CRC Press. Inc., Florida.

Stein, G.M. and Berg, P.A. (1997). Mistletoe extract-induced effects on immunocompetent cells: in vitro studies. *Anticancer Drugs*, **8** (suppl.1), 39–42.

Subils, R. (1984) Eremolepidaceae, Loranthaceae, Viscaceae. In *Boletín de la Sociedad Argentina de Botánica*; **23**, pp. 121, 176, 264.

Taira, C.A., Wagner, M.L, Adrados, H.M., Pino, R., and Gurni, A.A. (1994) Estudio Farmacológico de un Agente Vasoactivo presente en *Ligaria cuneifolia* var. cuneifolia. *Acta Farmacéutica Bonaerense*, **13**, 91–95.

Thunberg, E., and Samuelsson, G. (1982). Isolation and properties of ligatoxin A, a toxic protein from the mistltoe Phoradendron liga. *Acta Phram. Suecica*, **19**, 285–292.

Toursarkissian, M. (1980) *Plantas Medicinales de la Argentina sus nombres botánicos, vulgares, usos y distribución geográfica*, Hemisferio Sur ed., Buenos Aires.

Varela, B.G., and Gurni, A.A. (1995) Anatomia foliar y caulinar comparativa del muérdago criollo y del muérdago europeo. *Acta Farmaceútica Bonaerense*, **14**, 21–29.

Vazquez y Novo, S.P., Wagner, M.L., Gurni, A.A., and Rondina, R.V.D. (1989) Importancia Toxicológica de la Presencia de Sustancias Aminadas en Ejemplares de *Ligaria cuneifolia* var. *cuneifolia* Colectados en Diferentes Areas de la República Argentina. *Acta Farmacéutica Bonaerense*, **8**, 23–29.

Wagner, M.L., Vaccaro, M.C., Gurni, A.A., Coussio, J.D., and Rondina, R.V.D. (1986) Estudio de la variabilidad en compuestos aminados de diferentes ejemplares del género Phoradendron que crecen en las zonas Centro-oeste y misionera argentinas. *Acta Farmacéutica Bonaerense*, **5**, 139–148.

Wagner, M.L. (1993) *Estudios Fitoquímicos Comparativos de los Flavonoides de Loranthaceae de la Flora Argentina. Relación con el Muérdago Europeo*. Doctoral Thesis, Universidad de Buenos Aires.

Wagner, M.L., Fernández, T., Varela, B., Alvarez, E., Ricco, R., Hajos, S., *et al.* (1998) Anatomical, Phytochemical and Immunochemical Studies on *Ligaria Cuneifolia* (R ET P.) Tiegh (Loranthaceae) *Pharmaceutical Biology*, **36**, 1–9.

Zee Cheng, R.K.Y. (1997) Anticancer research on Loranthaceae plants. *Drugs of the Future* **22**, 519–530.

6. CULTIVATION AND DEVELOPMENT OF *VISCUM ALBUM* L.

HARTMUT RAMM, KONRAD URECH, MARKUS SCHEIBLER AND GIANFRANCO GRAZI

Institut Hiscia, Verein für Krebsforschung, Kirschweg 9, 4144 Arlesheim, Switzerland

INTRODUCTION

In his *Historia naturalis* (Liber XVI, 95), Pliny the Elder (AD 23–79) not only wrote of the way in which mistletoe was specially venerated by the Druids of Gaul, but also referred to the rarity of mistletoe-bearing oaks. To this day, *Viscum album* rarely grows on oaks (*Quercus robur/petraea*). Oak-grown mistletoe is however important in the range of anthroposophical mistletoe preparations, and special efforts have therefore been made in recent years to cultivate mistletoe on oaks. The aim was to ensure availability and quality of the pharmaceutical raw material in the long term. Problems that arose made it necessary to go deeply into mistletoe biology. They mainly concerned host tree resistance, soil conditions, interaction with the animal world, and mistletoe development.

NATURAL OCCURRENCE OF MISTLETOE-BEARING OAKS

The pioneers of oak mistletoe cultivation experienced difficulties in transferring *Viscum album* on indigenous oaks (Tubeuf, 1923; Bellmann, 1963; Grazi, 1987). Experimental sowing of mistletoe seed did not prove a promising method for the detection of mistletoe-receptive specimens of *Quercus robur* and *Q. petraea*, but the success rate considerably increased when use was made of mistletoe-bearing oaks and their progeny. French oaks naturally bearing *V. album* plants provide the basis for oak mistletoe cultivation today. The natural occurrence of mistletoe-bearing oaks will therefore be considered in more detail below.

Systematic searches in France allowed to identify more than 200 mistletoe-bearing indigenous oaks (*Q. robur* and *Q. petraea*). These were isolated individual specimens well removed one from the other. Mistletoe-bearing American oaks (*Q. rubra, Q. palustris, Q. coccinea*), on the other hand, which were also registered, were often found to grow in groups. They are distinctly more common (more than 450 mistletoe-bearing American oaks recorded) than mistletoe-bearing European oaks. The available data do not permit to calculate the percentage of mistletoe-bearing oaks among the total population. The low number of registered trees among the huge oak populations in the main areas where mistletoe is found does show, however, that indigenous oaks are largely resistant to *V. album*, and that American oaks also display a restricted receptivity.

Apart from being relatively rare, mistletoe-bearing oaks also tend to bear only a small number of mistletoe bushes. It is interesting to compare the number of mistletoe plants found on the following 4 groups of rare host trees (Figure 1):

- indigenous oaks (*Q. robur* and *Q. petraea*)
- *Q. rubra*
- *Q. palustris/coccinea*
- *Ulmus* sp.

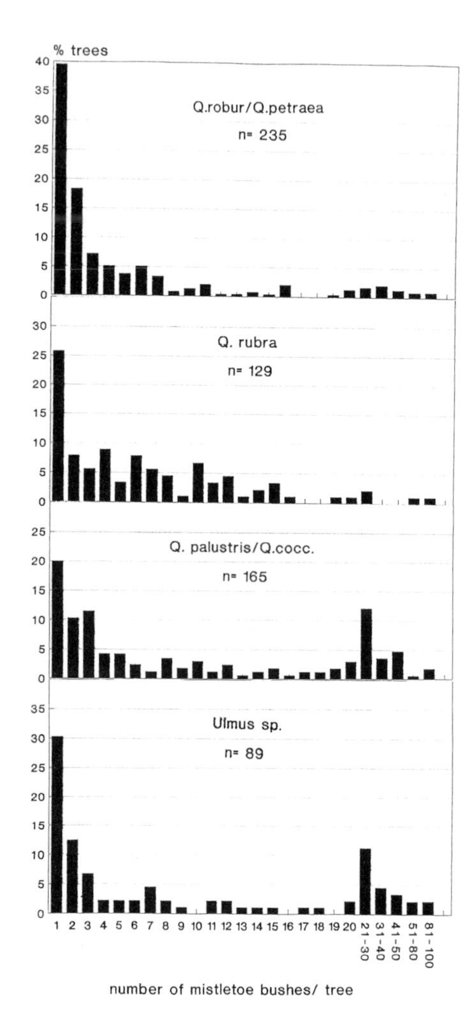

Figure 1 Incidence of different *Viscum album* colonization densities on different host trees (*Q. robur/petraea*, *Q. rubra*, *Q. palustris/coccinea* and *Ulmus* sp.) in France plotted against percentage of host trees (% of total number investigated). n = no. of trees investigated.

Three quarters (74.0%) of all mistletoe-bearing indigenous oaks (*Q. robur* and *Q. petraea*) bear only 1–5 mistletoe bushes. The corresponding percentages are also high for *Q. rubra*, *Q. palustris/coccinea* and *Ulmus* sp. at about 50%. The histograms in Figure 1 show that the trees become all the rarer the higher the number of mistletoe plants they bear. *Q. robur/petraea* and *Q. rubra* differ clearly from *Q. palustris/coccinea* and *Ulmus* sp. in the frequency of individuals carrying large numbers of mistletoe plants. Mistletoe numbers of 21 to 100 bushes are four times less common with the former (5.8%) than the latter (22.1%).

Experimental sowing of mistletoe seed on wild-growing indigenous mistletoe-bearing oaks showed that about 40% of the trees would not accept new mistletoe plants, even if repeated sowings were made (Table 1). The mistletoe seed used was investigated and did not show reduced vitality. It is probable, therefore, that these mistletoe-bearing oaks had an inherent resistance. This indicates that receptivity might be expressed only during a limited phase of the trees' life.

Mistletoe oaks on which it proved possible to establish new mistletoe plants showed a correlation between receptivities and the number of preestablished native mistletoe bushes (Table 1). Oaks bearing only few native bushes would as a rule only accept a few additional mistletoe bushes, whereas the proportion of trees accepting many (>10) new mistletoe plants was high within the group of oaks with many native bushes (Table 1).

Differences in mistletoe frequencies on oaks, therefore, are unlikely to be due only to differences in colonisation pressure from mistletoe. Differential expression of receptivity might be the major cause for the observed differences. A number of structural elements in oak and poplar bark have been identified as definite resistance parameters. A correlation was in fact established between a resistance coefficient

Table 1 Success rate of artificial *V. album* sowings on native oaks (*Q. robur* and *Q. petraea*).

No. of mistletoe bushes per tree from artificial sowing	*% proportion (n = 100%)*		
	oaks with 1–3 native bushes per tree (n = 73)	*oaks with 4–10 native bushes per tree (n = 28)*	*oaks with >10 native bushes per tree (n = 12)*
0	45.2%	42.8%	41.70%
1–3	21.9%	14.3%	8.30%
4–9	19.2%	17.9%	16.60%
> 10	13.7%	21.4%	33.30%

Mistletoe was sown on oaks growing in the wild with different levels of native mistletoe colonisation (1–3, 4–10, > 10 bushes per tree). At least 2,000 mistletoe seeds were carefully applied in April to the bark of younger branches of the mistletoe-bearing oaks, putting them in groups of 10. Sowing was repeated at least once in subsequent years, with the count made at least 7 years after the last sowing.

taking account of these parameters and receptivity of the oaks under investigation (Hariri *et al.*, 1991; Hariri *et al.*, 1992).

There was only a small difference in the average number of mistletoe bushes on naturally occurring mistletoe-bearing *Q. robur* and *Q. petraea*. The mean was 6.4 and 7.1 per tree, respectively. *Q. rubra* also had relatively few mistletoe plants, the average being 8.3, compared to *Q. palustris/coccinea* which appear to be 1.7 times as receptive (Table 2).

About 78% of the indigenous mistletoe-bearing oaks detected in France we studied were *Q. robur* and only 22% *Q. petraea*. With 24 random samples, we sought to establish if mistletoe shows a preference for hybrids of the two species *Q. robur* and *Q. petraea* (Table 3). The degree of hybridisation was detected by a high-resolution analysis according to Kissling (1980). 14 of the 17 mistletoe oaks tested and identified as *Q. robur* were pure species, whilst the rest had a minor component of *Q. petraea*. Mistletoe-bearing *Q. petraea* on the other hand all showed minor hybridisation with *Q. robur*. These results, however, don't make it possible to establish any preferential hospitality of one of the two indigenous oak species, nor of particular hybrids unless the composition of the oak populations in the mistletoe distribution area is taken into account.

Table 2 Mean number of *V. album* bushes on mistletoe-bearing oaks (*Q. robur, Q. petraea, Q. rubra, Q. palustris/coccinea*) and elms growing in the wild in France.

Host species	*n*	*Mean no. of mistletoe bushes per mistletoe-bearing tree*
Q. robur	192	6.4
Q. petraea	55	7.1
Q. rubra	149	8.3
Q. palustris/coccinea	314	13.8
Ulmus sp.	89	12.9

Table 3 Estimated degree of hybridization for *V. album*-bearing native oaks in France.

Hybridisation	*No. of trees*	
	Q. robur	*Q. petraea*
pure species	14	0
grade 1	3	6
grade 2	0	1
grade 3	0	0

Leaf samples (≥ 10 leaves) were taken from the crown region in June and assessed for a large number of biometric parameters, using the method described by Kissling (1980). Grade 1 hybridization means up to 12.5%, grade 2 12.5–37.5% and grade 3 37.5–50% of characteristics from the other species.

SELECTION OF SUITABLE HOST TREES

The resistance phenomenon is of critical importance for an effective mistletoe cultivation. The hypothesis that resistance of oaks could be overcome only by the virulence of selected mistletoe seed could be ruled out. Seeds from any hardwood mistletoe (*V. album* ssp. *album)* could become established on mistletoe-receptive oaks. It was found that mistletoe resistance appears to be genetically fixed by the host tree (Grazi and Urech 1983). The work of Frochot *et al.* (1978) had already suggested that the resistance might be bound up with the trees' genetic constitution.

The oak mistletoe cultivation methods given below thus all depend on mistletoe receptive parent trees.

1. Sowing mistletoe on mistletoe-bearing oaks in natural sites: A naturally weak mistletoe distribution potential is made up for by applying vital mistletoe seeds to young branches of mistletoe oaks growing in the wild (Table 1).
2. Grafting: Scions of mistletoe-bearing oaks are grafted on to any oak material of local origin. Depending on scion quality and care taken, the proportion of mistletoe-bearing progeny may be up to 100%. The growing trees will, however, assume the "physiological age" of the scions, which may limit growth and the sustainability of mistletoe production.
3. Rooting mistletoe-bearing oak cuttings: This requires considerable experience in biology and technical equipment. Very much as with grafting, growth and mistletoe production are likely to be limited if cuttings are of a greater physiological age.
4. Sowing acorns from mistletoe-bearing oaks: This yields a relatively high percentage of mistletoe-receptive progeny (12–19%, Table 4). The method of choice is lining out in nursery style, as selecting suitable specimens will require

Table 4 Percentage of *V. album*-receptive oak seedlings among progeny of mistletoe-bearing parent trees.

Seed provenance	*Parent trees*	*n*	*Mistletoe receptive seedlings* No.	%
Escy	*Q. petraea*	42	8	19.0
Pala	*Q. robur*	34	5	14.7
Langra	*Q. robur*	25	3	12.0
Corgi	*Q. robur*	8	1	12.5
Total		109	17	15.6

Ripe acorns gathered off the trees or germinating acorns which had dropped off (only where no interference from neighbouring oaks existed) were collected in France and sown in our nursery. After *c.* 4 years, mistletoe seed was sown for at least 4 years on all oak seedlings, always in April, and observed for at least another 4 years. Mistletoe receptivity was considered established if one or more mistletoe plants had reached the first foliage leaf stage that follows the primary leaves.

little space. Oak seedlings can be tested for mistletoe receptivity after three to five years and then transferred to final growing sites. The young trees are lined out two or three times during the five to seven years of early growth and adequately protected against damage from wild animals. Sowing acorns directly in their definitive growth site, with mistletoe-receptive specimens selected afterwards, has the advantage of avoiding loss of oak-specific root potential, with the tap root being able to penetrate deep soil layers without impairment due to lining out. Connection with ground water is ensured, mineral resources in deep soil layers are made available, and conditions created for sustainable mistletoe production. To avoid damage from wild animals it will, however, be necessary to expend more on root and trunk protection.

Practical experience has also been gained in cultivating *V. album* on elms. The natural occurrence of mistletoe-bearing elms is limited, and pathogens are a major threat to the host. *Ulmus campestris* receptive to *V. album* was propagated by isolating root suckers. Young plants were dug up in winter when the trees were leafless and cool temperatures reduced the risk of infection. It was possible to obtain up to 100% of mistletoe-receptive progeny by this vegetative method. The figure for mistletoe-bearing *Ulmus glabra* grown from seed was between 20 and 40%.

With other pharmaceutically relevant mistletoe hosts, long-term practical experience yields estimated figures of 40–50% receptive trees for seedlings of fir (*Abies alba*) and pine (*Pinus silvestris*), and 90–100% for grafted apple trees (*Malus domestica*) and cuttings of poplar (*Populus trichocarpa*) (Grazi and Scheibler, unpublished results).

CHOOSING A SITE

Mistletoe receptivity among oaks has a definite genetic origin, but site conditions also proved vital in maintaining optimum receptivity and long-term mistletoe cultivation. One important factor is an adequate water supply to the host roots to compensate for the relatively high transpiration rate of mistletoe (Schulze *et al.*, 1984).

In 1923, Tubeuf wondered whether levels of specific minerals such as calcium in the site may have an effect on mistletoe occurrence, but no definite conclusions could be drawn from his findings. With oaks, however, we could make relevant observations. Chlorosis was repeatedly seen in mistletoe-bearing oaks grown in plantations in the Jura mountains where the lime content is high, resulting in quite considerable losses as regards both mistletoe and oaks. Leaf and soil analysis showed lime-induced iron and above all manganese deficiency. Furthermore extensive soil analysis in France showed that almost 80% of mistletoe-bearing oaks grow in moderately to highly acid soils for which relatively low lime and high iron and manganese levels are typical (Ramm 1994; Figure 2). In experiments with oak seedlings grown in pots, it was possible not only to deal with the chlorosis but also to increase the proportion of mistletoe-receptive specimens by changing the root environment from basic lime to an acid mixture of clay and sand (Ramm, unpublished results).

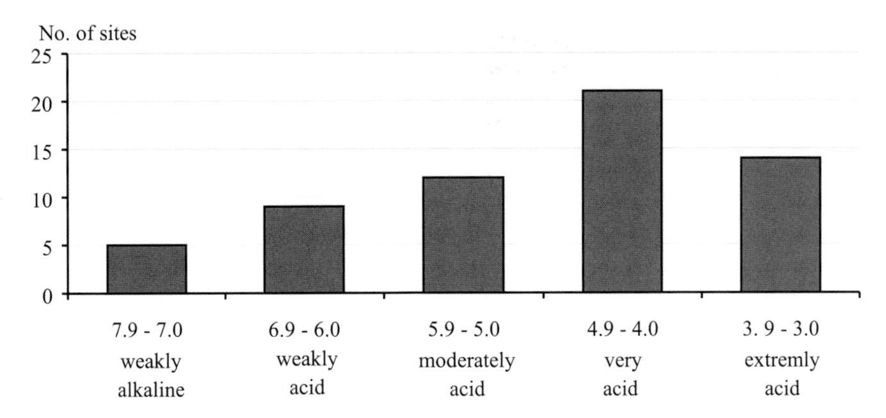

Figure 2 Distribution of pH readings in the soils of French mistletoe oaks (n = 61). Soil samples from horizon A, pH classes according to Scheffer and Schachtschabel (1979).

SOWING MISTLETOE

Mistletoe fruits with viable embryos are available from October to May. The seed can, however, only be sown to good effect during a relatively short time that is limited by a number of factors. Mistletoe seeds applied to the host in late autumn or winter are frequently eaten by tits (Grazi, 1986; Weber, 1993; Grazi and Urech, 1996). On the other hand, availability of mistletoe fruit may be greatly reduced by the beginning of the new vegetative period, as some species of birds show a preference for eating mistletoe berries in winter. Germination of the remaining seeds may also be advanced so far that the elongated hypocotyls are no longer able to connect with the host branch. The optimal time for applying mistletoe seed is from mid-March to mid-May, depending mainly on the geographical situation.

It is therefore advisable to build up a good store of mistletoe seed that may be sown later. In practice, the following method has proved effective. We harvest the fruits of healthy, robust mistletoe bushes from January onwards. The exocarp is removed, leaving the inner mesocarp around the seeds (Figure 3A) which then are placed in Petri dishes. The slimy, sticky pulp is carefully dehydrated over a period of 24–48 hours. The stabilised seeds can then be stored outside until spring. As mistletoe embryos quickly lose vitality if kept in the dark (Tubeuf, 1923), it is necessary to make sure that the seeds get enough light all the time. Care must be taken, on the other hand, to prevent the live tissues getting overheated.

Before sowing, the mistletoe seeds are covered with water so that the dried-up mesocarp swells up again and becomes slimy. Using tweezers, single or groups of seeds are transferred to the bark of host branches. They will firmly adhere within a few days because of the glue-like substances in the mesocarp. To prevent seeds from getting washed off, it is advisable to choose a time when precipitation levels are low. Seeds are applied to young branches in the host periphery. This avoids damage to the central trunk caused by the sinker of the mistletoe. In damp growth sites with

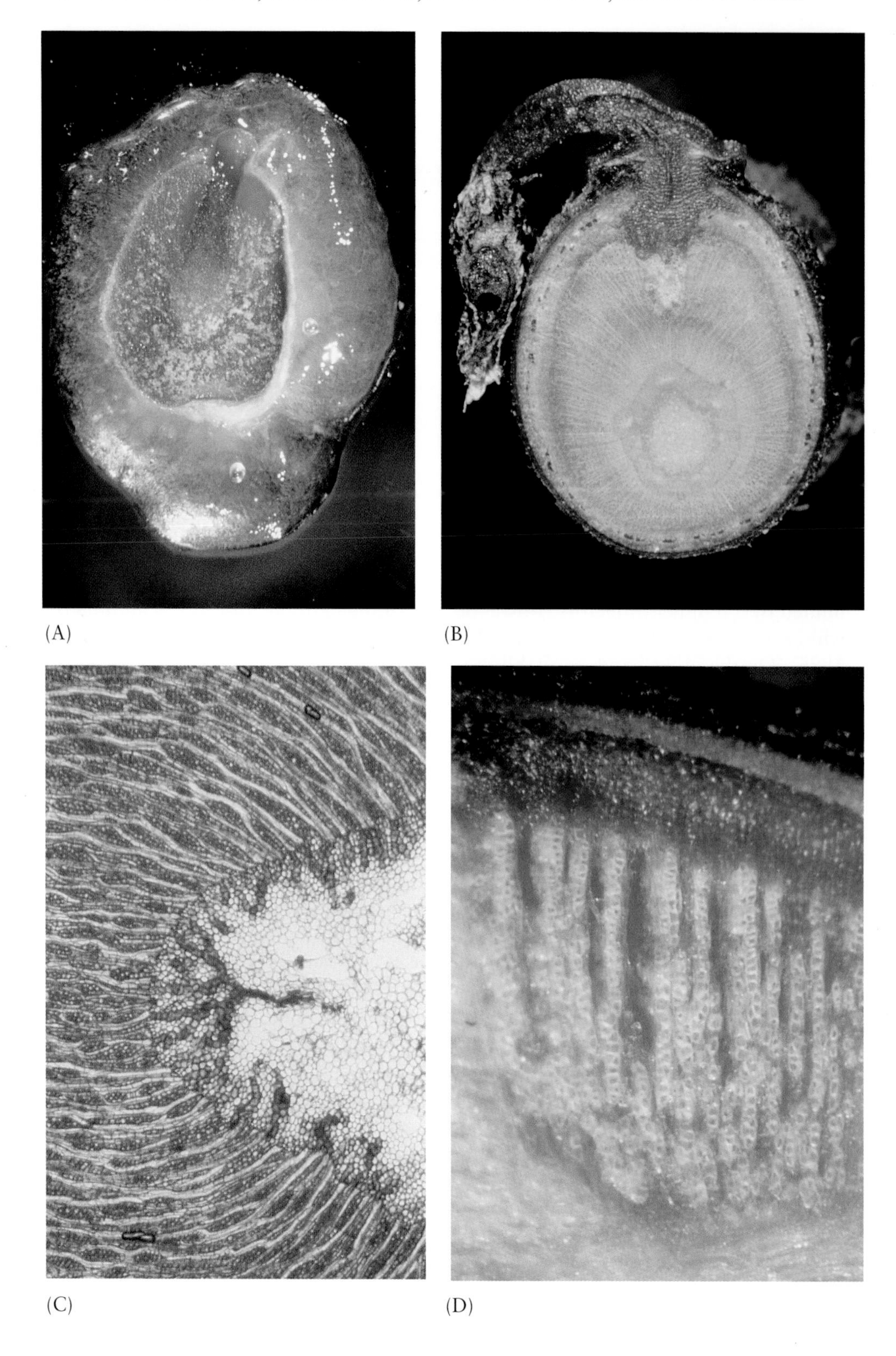

(A) (B)

(C) (D)

dense herbage or grass cover under the trees, the seeds should be protected from snails and slugs with light-permeable gauze strips.

Successful sowing is evident from the following:

1. Hypocotyl and holdfast stay green.
2. Host branch shows distinct swelling under the holdfast.
3. Hypocotyl comes upright.
4. Two elongated, pale green primary leaves grow from the bud on the shoot.

The last of these four steps is the most definite visible sign that the mistletoe haustorium has connected with the host's water conduit system inside the branch. The tree is mistletoe-receptive and the young mistletoe plant usually should be able to grow into a typical mistletoe bush in the years ahead.

DEVELOPMENT OF HAUSTORIAL SYSTEM

Mistletoe embryos germinate whilst still in the fruit, but are unable to break through the tough exocarp without help. In nature, birds such as blackcap (*Sylvia atricapilla*) and mistletoe thrush (*Turdus viscivorus*) remove the seed from the fruit, more or less effectively getting it in contact with the branch of a host (Grazi, 1986). Apart from mucous polysaccharides which in time are washed out, the mesocarp attached to the endocarp contains glutinous substances which firmly attach the seed to the host bark as they dry. After a period of winter rest induced by cold temperatures, the embryos begin to grow in April by elongating the hypocotyl. Mainly negatively phototropic and if necessary also negatively geotropic growth (Tubeuf, 1923) directs the tip of the hypocotyl towards the host bark. The epidermal cells of the hypocotyl's tip secrete a viscous fluid which enables close contact with the bark and helps to affix the embryo directly to the host (Löffler, 1923).

Thoday (1951) describes the way papillae subsequently grow from the epidermis, connecting the base of the hypocotyl even more firmly with the host bark. The tip of the hypocotyl broadens to a flat disk, the so-called holdfast, and the papillae connected to the host are drawn to the periphery, opening up the host periderm layer by

◀ **Figure 3** A. Transverse section of slime-enveloped "seed" of hardwood-grown mistletoe (*Viscum album* ssp. *album*). Green endosperm enveloped in whitish, translucent mesocarp, with one embryo embedded in it, its hypocotyl pointing to the periphery (× 9). B. Transverse section of an apple tree branch with a growing embryo of *Visucm album* on it. In the autumn, the sinker has become embedded in the wood of the host which responds with hypertrophic growth (× 10). C. Part of mistletoe sinker in an apple tree branch, tangential section, stained with astral blue/safranin red. Host wood brownish, with light-coloured vascular strands coming in from the left. On the right the light-coloured sinker parenchyma, with dark xylem structures growing into it from the periphery, where host xylem and sinker tissue meet. These structures join to form a central vascular strand (× 33). D. Longitudinal section of a secondary sinker of *V. album* ssp. *album* growing from a cortical strand (top, dark green). The sinker tissue consists of stratified xylem strands with green haustorial parenchyma intercalated between (× 34). Photos: Raman layer. The slightly oval holdfast shows bilateral symmetry, including a meristematic zone along the major axis and adjacent to the host bark. With cell divisions starting from here, the meristematic tissue is penetrating into the opened-up host periderm. A typical apical meristem develops, driving a wedge through the bark. Apart from mechanical forces, enzymatic processes coming from the mistletoe haustorium probably also help to open up the host tissues (Sallé, 1983). On young apple tree branches the apical meristem, pushing its way in centripetally, will reach the host's cambium about two months after holdfast attachment, which is towards the end of June.

Cell division activity in the mistletoe haustorium is going on then in an intercalary meristem which is established at the level of the host cambium. Like the cambium which produces cells centripetally that differentiate out into woody tissue, the intercalary mistletoe meristem produces new cells towards the central part of the branch. Enclosed by the young wood of the host, this new tissue forms the primary sinker (Figure 3B). *Viscum album* does not grow actively through the host cambium, but is passively embedded in the host's secondary xylem (Thoday, 1951; Sallé, 1979, 1983).

Host tissues close to the sinker are stimulated into hypertrophic growth. The branch swells, an important signal that the sinker is connected with the host xylem (Figure 3B). Secondary thickening will widen the intercalary mistletoe meristem in the following years, removing it from the centre of the branch in line with the host cambium. The primary sinker gradually assumes the form of a wedge brought in alignment with the host xylem's direction of flow. The oldest tissue of the sinker is resting deep in the wood and dies off only after some years.

For as long as the surrounding host tissue is still young and little differentiated, the sinker consists of pale green, undifferentiated parenchyma. But as soon as the vascular tissues of the host begin to differentiate out and solidify, corresponding differentiation begins also in the adjacent periphery of the sinker where this comes in contact with the host vessels. Parenchyma cells that follow one another are transformed into vessels by dissolving the organelles and hollowing out the cells, with cell walls transverse to the direction of flow reabsorbed, whilst cell walls that lie in the flow direction are reinforced with spiral or reticular thickening. As Melchior (1921) noted, this also involves secondary cell division in the sinker parenchyma. The xylem strands of the mistletoe's sinker sprout at right angles to the original flow direction of the host xylem, from the periphery to the central axis of the haustorium, opening out into the central vessels (Figure 3C).

At the same time as the primary sinker develops, so-called cortical strands begin to grow in longitudinal and also circular direction from the haustorial stem through the host's secondary phloem. Contact between the growing tip of cortical strands and actively dividing host cambium triggers the development of secondary sinkers. As in primary sinker development, their further development is taken over by intercalary meristems embedded in the host cambium. In contrast to the intercalary sinker meristems, which are physiologically adapted to the host cambium's growth rhythms and come to rest in winter, the apical meristems of the cortical strands do not have seasonal growth rhythms. The rate of secondary sinker development, on the other hand, is dependent on the host's cambial growth rhythms and reaches a maximum during the first half of summer (Sallé, 1978).

Cortical strand morphology is fundamentally different from sinker morphology. Growth starts from an apical meristem protected by an anterior zone of elongated cells (Melchior, 1921; Thoday, 1951). Apart from xylem which conducts sap taken up in the secondary sinkers, they also have phloem (Sallé, 1979; Sallé, 1983) which supports the growing tip with organic matter. Primary and secondary sinkers, on the other hand, have no phloem at all. And whilst open connections exist between the host's and the mistletoe's xylem in the sinker region (Melchior 1921; Sallé 1983; Becker 1986), there are no connections between host and mistletoe in the region of the cortical strands. Finally it is worth noting that adventitious shoots may grow from the cortical strands. These break through the cortex at a greater or lesser distance from the primary shoot.

A remarkable feature in the haustorial system of *Viscum album* is the green colour of the sinker parenchyma (Figure 3D), which persists for several years even deep down in the wood where no light should penetrate. In the sinker of *Korthalsella*, Fineran (1995) identified green pigments as chlorophyll. He speculated that this mistletoe species might be able to utilise even extremely low radiation potential by photosynthetic activity. Another aspect of interest is the way xylem vessels in the sinker of *V. album* are piled up (Figure 3D). In studies on *Phoradendron*, Calvin and Wilson (1995) established that this North American member of the Viscaceae only realises a rather small proportion of theoretically possible connections with adjacent host xylem by differentiating out xylem structures in its sinker. As the relative proportion of xylem structures also shows seasonal variations, they assume that *Phoradendron* is reacting to different pressures and vascular flow rates in early and late xylem. Research concerning the greening of parenchyma and the xylem organisation in the sinker of *Viscum album* is still missing.

THE MISTLETOE SHOOT

The hypocotyl of young mistletoe embryos that have achieved connection with the water conduits of the host during summer may come upright already in autumn. The shoot apex is drawn forth from the endocarp, grown empty and dry in the meantime, so that the rudiments of the opposite cotyledons become visible. In most cases, visible growth only begins when the next spring is coming. Two elongated leaves will arise on a more or less extended internode, decussate to the rudimentary cotyledons. Stem and leaf tissues grow firmer in texture as the vegetative period progresses, but no further organs develop at this time. A new twig finally develops in the following spring, its only pair of leaves decussate to those from the previous year (Figure 4A). With this simple growth gesture, *Viscum album* makes decussation as a typical characteristic held in common by most mistletoes (Kuijt, 1969). Because of the analogous appearance between mistletoe twigs and dicotyledonous seedlings, the mistletoe bush is occasionally referred to as an "aggregate of seedlings" (Grohmann, 1941).

(A)

(B)

Figure 4 A. *Viscum album* growing on apple tree at the beginning of the third vegetative period. B. Old and young twigs of *V. album* in May. A typical feature at this developmental stage is the negatively geotropic orientation of the pale green shoots of the current year. Photos: Raman.

Mistletoe leaves do not develop the typical bipolar differentiation into palisade and spongy parenchyma. Their tissue remains at a meristematic stage, with further growth in thickness and elongation in the following spring (Göbel, 1994). The normal case is that mistletoe leaves drop in the late summer of the second vegetative period, being still green and turgescent. Leaves that are three or four years old and correspondingly larger can, however, frequently be found on mistletoe growing on elms and fir trees. Goedings (1997) drew attention in this respect to the host dependence of some morphological and in conjunction with this also physiological properties of *V. album*.

From the second or third year of growth, new mistletoe twigs arise not only from the apical meristem but also from basal buds. Apical shoot meristems are no longer available for vegetative development from the fifth, sixth or seventh year of growth. From this time on they provide the basis for generative development and differentiate as highly compressed generative short shoots with flower buds. The result is a typical pseudo-dichotomy with two dominant forked shoots which grow from the axillae of the opposite leaves from the year before. Each of these can be accompanied by two lateral complementary shoots (Figure 4B) which grow from the axillae of the paired bud scales that initially protect every shoot (Figure 5A) but later drop off.

The transition from vegetative to generative organ development occurs during June when the primordia of the next year's twig generation are developed (Göbel, 1994; Ramm, 1995; Dorka 1996). When these arise in the leaf axils of developing young shoots, the intervention of the flowering impulse shows itself as soon as the primary pair of leaves has differentiated out, with the spherical structure of the inflorescence bud developing from the apical meristem (Figure 5A). At its base, the primordial meristem will rest for about nine months, and only begin to differentiate out flowering organs in the following spring.

Starting in the second half of May the currently developed twigs of *Viscum album* show characteristic growth movements (Göbel and Dorka, 1986; Dorka, 1996). These nutations, synchronised in time but largely independent in spatial terms, release the young shoots from their original, negatively geotropic orientation (Figure 4B). When growth movements come to an end in July, all the young mistletoe twigs have assumed their appropriate orientation in the spherical form of the bush. Mistletoe bushes may reach up to 1 meter in diameter during a life time of 20 or more years.

THE MISTLETOE FLOWER

In August, a short shoot becomes visible between the two mistletoe leaves, which in the typical case is bearing a terminal and two lateral generative buds (Figure 5B). Male flowers are enclosed in a single bract, whilst in a female inflorescence the two lateral flowers have one bract and the terminal flower another (Kuijt, 1969; Rispens, 1993). *Viscum album* is a dioecious plant, and bushes bearing female flowers are about four times more frequent in nature than those bearing male flowers. The genetic basis of dioecism, connected with differences in the distribution of sex-specific chromosomes, was established by Mechelke (1976).

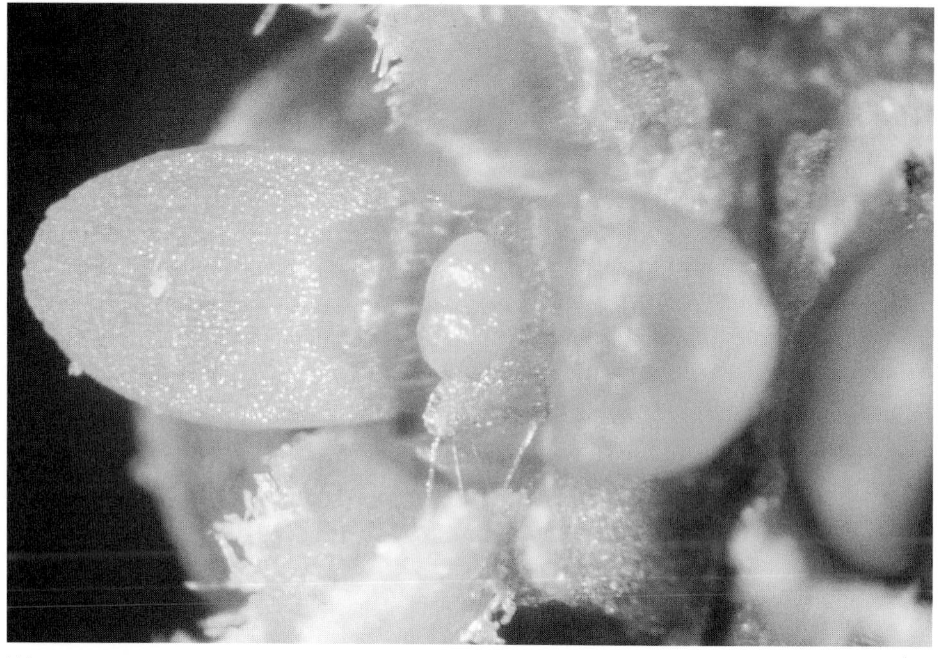

(A)

(B)

Mistletoe flowers are inconspicuous and simple in form. Male flowers are distinctly larger than female, and consist of four simple perigon leaves (Figure 6A). The stamens are fused into a kind of cushion, with the pollen grains produced in chambers embedded in these cushions. The female flower consists of a spherical green fruit primordium, with small yellowish perigon leaves, usually four in number, surrounding the papillous stigma (Figure 6B).

Male and female flowering organs differentiate out in spring and summer as the unfolding mistletoe twig finds its orientation in the spherical bush. Initially, a floral calyx grows from the meristem of the female flower, with the perigon leaves in its outer margin. Two carpels develop from the primordial meristem that has remained at the base, fusing with the surrounding axial tissue of the calyx and also with one another. A "central flask" will remain between, in which 7–9 embryo sac mother cells evolve, finally producing 1–5 embryo sacs. When the distal part of the flask grows into the papillous stigma by the end of September, all the organs of the female flower have differentiated out. The male flowers whose pollen grains have gone through the requisite maturation divisions will also be fully developed in autumn, now being in the binucleate stage with a vegetative and a generative nucleus (Pisek, 1923; Steindl, 1935; Zeller, 1976; Dorka, 1996). Both types of flowers rest until temperatures reach a level where flowering can begin.

Depending on geographical position and actual weather conditions, the main flowering period of *Viscum album* is in February or March. In warmer years and regions, however, the first flowers may open as early as January or even at the end of December. Pollination is by insects that are active also in winter. In field observations we have seen ants (*Formica*), different types of fly (Muscidae, Drosophilidae) and hoverflies (Syrphidae) on both male and female flowers. Honey bees (*Apis mellifica*) were only seen on male mistletoe flowers and are unlikely to count among mistletoe pollinators, which is also the view taken by Walldén (1961). The insects are clearly attracted by the scent arising from the nectaries at the base of the funnel-shaped male flowers, and by the nectar which is secreted all around the stigmata of female flowers.

DEVELOPMENT OF FRUIT AND EMBRYO

Successful pollination is followed by the development of mistletoe fruits. The pollen germinates on the stigma, with a pollen tube growing into the style and the tissues of the central body. As soon as germination starts, the generative nucleus divides, and the two new generative nuclei as well as the vegetative nucleus move into the pollen

◀ **Figure 5** A. Axillary bud with primordia of *V. album* twig. By mid-July the main elements of next year's shoot generation have already developed in the axils of the current year's mistletoe leaves. They are the two bud scales (top and bottom), and decussate to these the leaf primordia (left and right), with the spherical bud of the inflorescence between them (× 48). B. Typical shoot of *V. album* with female flower buds in November. Photos: Raman.

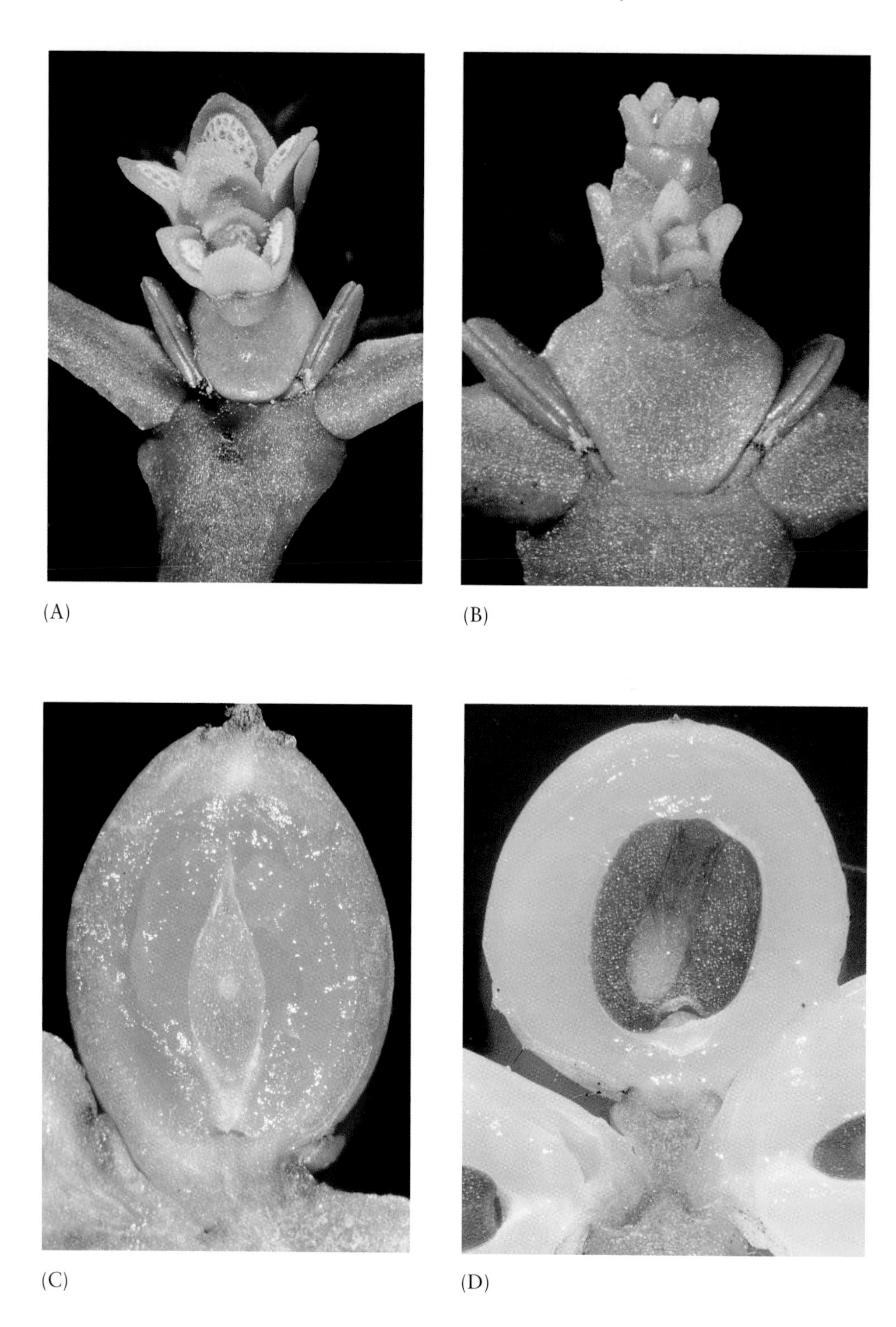
(A)
(B)
(C)
(D)

tube. When the tip of the pollen tube penetrates one of the embryo sacs, the vegetative pollen nucleus dissolves and further development is triggered by the two generative pollen nuclei. One fuses with the egg cell, with the resulting zygote resting for about two months. The other generative pollen nucleus unites with the endosperm nucleus; further cell divisions will start immediately and will lead to the formation of the endosperm. When the first division of the zygote comes in mid-May, it is already completely enclosed by the multiplying endosperm. The bicellular embryo stage goes through another rest period of about two weeks, and then development becomes more rapid (Pisek, 1923; Steindl, 1935).

The primordia of the cotyledons become visible as light-colored zones in the pale green endosperm in the beginning of July (figure 13), and the hypocotyl grows towards the periphery of the endosperm in August. By the end of September, a viable embryo has differentiated out in the mistletoe fruit, which is still green. *Viscum album shows polyembryonie, i.e. one seed may contain one or two, and, rarely, even three or four embryos.* Depending on the climate, and evidently also on other yet unknown factors relating to individual mistletoe bushes, the fruits begin to ripen in mid-October. The pericarp loses its green colour and appears white (figure 14). Sugars are produced in the colourless, translucent mesocarp, giving it a sweetish taste and making the outer parts in particular a favourite food for mistletoe-distributing birds.

Mesocarp development bases on the tissues of the flask-shaped central body, the cell walls of which turn mucilaginous and dissolve towards the end of June. Morphologically the exocarp arises from the walls of the calyx and the carpel tissues fused with it. With axial tissue also involved in this mode of fruit development, the correct botanical classification of mistletoe fruit is as a pseudo berry. And with the embryos developing from embryo sacs that lie without integuments naked in the central body tissue, mistletoe also does not have true seeds in the strictly botanical sense (Steindl, 1935).

FINAL REMARKS

Viscum album has some remarkable features that do not fit into the categories of classical botany. The haustorial system is comparable to the roots of higher plants only in so far as it serves to conduct water and minerals into the shoot. *V. album* does not have a true radicle, its sinker is green and has neither a protective root cap nor phloem structures. The cortical strands, on the other hand, do not only spread the haustorial system through the formation of secondary sinkers but also give rise

◀ **Figure 6** A. Male inflorescence of *V. album* (× 5). B. Female inflorescence of *V. album* (× 9). C. Fruit of *V. album* in early July. The mesocarp is slimy, and the primordia of the embryo become evident in the spindle-shaped endosperm (× 11). D. Ripe mistletoe fruit in November. The green colour of endosperm and embryo is contrasting with the whitish pericarp (× 8). Photos: Raman.

to shoot development. Because of this, the mistletoe's haustorial system is seen as an "organ *sui generis*", a term introduced by Tubeuf (1923) and repeatedly confirmed (Thoday, 1951; Sallé, 1983).

The shoot development of *Viscum album* is much reduced. Each twig of one year only consists of a single internode with one decussate leaf pair and a generative short shoot with three flowers. It takes three years to form the primordia of these few organs from an axillary meristem, to unfold stem and leaves, to blossom and finally to develop a vital embryo in the ripening fruit (Ramm, 1995). Urech and Ramm (1997) considered the greatly reduced shoot development seen in the mistletoe twig, and the absence of typical root-like growth in the sinker to be in relation with viscotoxin and mistletoe lectin formation and the distribution of these typical substances in the mistletoe bush.

Another characteristic feature is the green colour of the mistletoe's embryo and endosperm which persists during the winter time. Their photophysiological potential might be crucial for the survival of the embryo. Tubeuf (1923) could show that the viability of mistletoe embryos is greatly reduced if the bushes are kept in the dark. He called *Viscum album* a plant that is "green through and through". Indeed, the green colour both at the centre of the fruit and in the sinker indicates that *Viscum album* has a special relationship to light.

REFERENCES

Becker, H. (1986) Botany of European Mistletoe (*Viscum album* L.). *Oncology,* **43**, suppl. 1, 2–7.

Bellmann, P.G. (1963) Bericht über die bisherigen Mistelzüchtungsversuche. *Research report, Institute Hiscia,* Arlesheim (Switzerland).

Calvin, C.L., and Wilson, C.A. (1995) Relationship of the mistletoe *Phoradendron macrophyllum* (Viscaceae) to the wood of its host. *IAWA Journal*, **16**, 33–45.

Dorka, R. (1996) Zur Chronobiologie der Mistel. In R. Scheer, H. Becker, and P.A. Berg, (eds.), *Grundlagen der Misteltherapie*, Hippokrates, Stuttgart, pp. 28–45.

Fineran, B.A. (1995) Green tissue within the haustorium of the dwarf mistletoe *Korthalsella* (Viscaceae). An ultrastructural comparison between chloroplasts of sucker and aerial stem tissues. *Protoplasma,* **189**, 216–228.

Frochot, H., Pitch, M., and Wehrlen, L. (1978) Différences de sensibilité au gui (*Viscum album* L.) de quelques clones de peuplier (*Populus* sp.) *Proc. 103e Congr. Int. des Soc. savantes*, Nancy-Metz, Vol. 1, Biblioteque nationale, Paris, pp. 157–165.

Goedings, P. (1997) Über die Bildung von *Viscum album. Elemente d. Naturw.*, **67**, 1–23.

Göbel, T. (1994) *Erdengeist und Landschaftsseele.* Verlag am Goetheanum, Dornach.

Göbel, T., and Dorka, R. (1986) Zur Raumgestalt und zur Zeitgestalt der Weissbeerigen Mistel (*Viscum album* L.). *Tycho de Brahe Jahrbuch für Goetheanismus*, Tycho Brahe Verlag, Niefern-Öschelbronn, pp. 167–192.

Grazi, G. (1986) The role of birds in the dispersion and control of mistletoe in Arlesheim (Switzerland). *Abstracts of the Int. Meeting on V. album*, Heidelberg, April 1986.

Grazi, G. (1987) Mistelkultivierung im Laboratorium Hiscia. In R. Leroi, (ed.), *Misteltherapie, eine Antwort auf die Herausforderung Krebs; die Pioniertat R. Steiners und I. Wegmans*, Verlag Freies Geistesleben, Stuttgart, pp. 148–159.

Grazi, G., and Urech, K. (1983) La susceptibilité des chênes, des ormes et des mélèzes au gui (*Viscum album* L.). *Revue Scientique du Bourbonnais*, 6–12.

Grazi G., and Urech K. (1996) Meisen und Misteln. *Gefiederte Welt*, **120**, 206–207.

Grohmann, G. (1941) Ueber die Mistel. *Manuskripte & Notizen von Weleda-Mitarbeitern*, Arlesheim.

Hariri, E.B., Sallé, G., and Andary, C. (1991) Involvement of flavonoids in the resistance of two poplar cultivars to mistletoe (*Viscum album* L.). *Protoplasma*, **162**, 20–26.

Hariri, E.B., Jeune, B., Baudino, S., Urech, K., and Sallé, G. (1992) Elaboration d'un coéfficient de résistance au gui chez le chêne. *Can. J. Bot.*, **70**, 1239–1246.

Kissling, P. (1980): Clef de détermination des chênes médioeuropéens (*Quercus* L.). *Ber. Schweiz. Bot. Ges.*, **90**, 29–44.

Kuijt, J. (1969) *The biology of parasitic flowering plants*. University of California Press, Los Angeles.

Löffler, B. (1923) Beiträge zur Entwicklungsgeschichte der weiblichen Blüte, der Beere und des ersten Saugorgans der Mistel (*Viscum album* L.). *Tharandter Forstliches Jahrbuch*, **74**, 49–62.

Mechelke, F. (1976) Sexcorrelated complex heterozygosity in *Viscum album* L. *Naturwissenschaften*, **63**, 390–391.

Melchior, H. (1921) Ueber den anatomischen Bau der Saugorgane von *Viscum album* L. *Beitr. zur allgem. Bot.*, **2**, 55–87.

Pisek, A. (1923) Chromosomenverhältnisse, Reduktionsteilung und Revision der Keimentwicklung der Mistel (*Viscum album* L.). *Jahrbücher f. Wiss. Bot.*, **62**, 2–20.

Ramm, H. (1994) The Soil in which Mistletoe-Bearing Oaks Grow. *Annual Report Verein für Krebsforschung* (Arlesheim, Switzerland), pp. 7–9.

Ramm, H. (1995) Die Mistel in der Zeit. *Der Merkurstab*, **2**, 113–123. (Transl.: Ramm H. (1996) Mistletoe in the time stream. *J. Anthroposophical Medicine*, **13**, 79–89.)

Rispens, J.A. (1993) Die Zweihäusigkeit der weissbeerigen Mistel (*Viscum album* L.). *Elemente d. Naturw.*, **59**, 22–39.

Sallé, G. (1978) Origin and early growth of the sinkers of *Viscum album* L. *Protoplasma*, **96**, 267–273.

Sallé, G. (1979) The endophytic system of *Viscum album*: its anatomy, ultrastructure and relations with the host tissues. *Proc. 2nd Symposium on Parasitic Weeds*, North Carolina State University, Raleigh (USA), pp. 115–128.

Sallé, G. (1983) Germination and establishment of *Viscum album* L. In D.M. Calder, and Bernhardt, P., (eds.), *The Biology of Mistletoes*. Academic Press, London, pp. 145–159.

Scheffer, F., and Schachtschabel, P. (1979) *Lehrbuch der Bodenkunde*, Enke Verlag, Stuttgart.

Schulze, E.-D., Turner, N.C., and Glatzel, G. (1984) Carbon, water and nutrient relations of two mistletoes and their hosts. *Plant, Cell and Environment*, 7, 293–299.

Steindl, F. (1935) Pollen- und Embryosackentwicklung bei *Viscum album* L. und *Viscum articulatum* Burm. *Ber. Schweiz. Bot. Gesellschaft*, **44**, 343–386 (and plates).

Thoday, D. (1951) The haustorial system of *Viscum album*. *J. of Experimental Botany*, **2**, 1–19.

Tubeuf, K.F. von (1923) *Monographie der Mistel*, Oldenbourg, München.

Urech, K., and Ramm, H. (1997) Die Polarität der Mistel. *Merkurstab*, **50**, 157–168.

Walldén, B. (1961) Die Mistel an ihrer Nordgrenze. *Sv. bot. Tidskr.*, **55**, 526–543.

Weber, H.C. (1993) Untersuchungen zur Entwicklungsweise der Laubholzmistel *Viscum album* L. (Viscaceae) und über Zuwachsraten während ihrer ersten Stadien. *Beitr. Biol. Pflanzen*, **67**, 319–331.

Zeller, O. (1976) Die Jahresrhythmik der Laubholzmistel *Viscum album* ssp. *album*. *Beitr. Erw. d. Heilkunst*, **6**, 3–20.

7. THE BIOTECHNOLOGY OF *VISCUM ALBUM* L.: TISSUE CULTURE, SOMATIC EMBRYOGENESIS AND PROTOPLAST ISOLATION

SPIRIDON KINTZIOS and MARIA BARBERAKI

Department of Plant Physiology, Faculty of Agricultural Biotechnology, Agricultural University of Athens, Greece

INTRODUCTION

Until today, mistletoe has not been a particular focus of cell and tissue culture experiments, i.e. the removal of cell groups or tissue explants from a mistletoe donor plant and their subsequent cultivation on a nutrient medium supplemented with growth regulators and other substances. *In vitro* techniques are a major component of plant biotechnology, since they permit to artificially control several of the parameters affecting the growth and metabolism of cultured tissues. Researchers working with *V. album* world wide could benefit from the establishment of tissue culture protocols for this species in the following ways:

1. By altering the culture parameters, it might be possible to control the quantity, composition and timing of production of mistletoe extracts. In this way, problems associated with the standardisation of mistletoe extracts (Wagner *et al.*, 1986; Lorch and Tröger, this book) might be overcome. By feeding cultures with precursor substances for the biosynthesis of certain metabolites, a higher productivity may be achieved from cultured cells (*in vitro*) than from whole plants.
2. Potentially, entirely novel substances may be synthesised through biotransformation or by taking advantage of somaclonal variation, i.e. a transient or heritable variability of metabolic procedures induced by the procedure of *in vitro* culture.
3. The establishment of a callus culture is the first step required in order to obtain genetically modified cells or plants, e.g. crop plants with a viscotoxin-based resistance to pathogens such as *Plasmodiophora brassicae* (Holtorf *et al.*, 1998).
4. Due to its semiparasitic nature, it is very difficult to propagate mistletoe, which is accomplished with the aid of birds or insects, carrying distantly mistletoe seeds (Becker, 1986; Grieve, 1994; Ramm *et al.*, this book). Biotechnology could offer an alternative method for the production of considerable mistletoe biomass in a relatively short time, either through cell and callus proliferation, or plant regeneration *via* organogenesis or somatic embryogenesis.

Becker and Schwarz (1971) were the first to mention the possible use of mistletoe callus cultures as a source of bioactive products. In 1990, Fukui *et al.* reported on the induction of callus from leaves of *V. album* var. *lutescens*. They were able to identify 2 galactose-binding lectins in the callus which were originally observed in mistletoe leaves. They also found that the contact of the mistletoe callus with a callus of its host beech (*Fagus crenata*) did not result in contact inhibition for neither of the two plant species. In our laboratory we have recently conducted a number of experiments in order to define more precisely the conditions under which mistletoe callus cultures can be obtained at a satisfactory rate. The effect of different plant parts, explant handling, growth regulators and culture medium composition has been investigated. We also achieved to isolate mistletoe protoplasts from different sources.

COMPONENTS OF MISTLETOE TISSUE CULTURE

Plant Material

Apart from *V. album* ssp. *coloratum* var. *lutescens,* growing on beech (Fukui *et al.*, 1990), *V. album* ssp. *abietis*, one of the three main mistletoe varieties, growing on fir trees on Mount Parnitha (Attiki, Greece) were studied by our group. The plants were classified as "winter accessions" (plants collected from December to March) or "summer accessions" (plants collected from April to November). Donor plants were kept at 4°C before the removal of leaf and stem explants (1 cm long).

Explant Disinfection

Explant surface disinfection is a delicate and indispensable procedure, because every subsequent step in culture must take place under sterile conditions. An optimal disinfection protocol is a compromise between the minimisation of surface microorganisms and the avoidance of tissue damage by the disinfectant. For mistletoe, these requirements can be fulfilled by treating explants in a solution of either 1% (w/v) sodium hypochlorite for 15 min, or 0.1–0.2% (w/v) mercuric chloride for 12 or 20–25 min. Mistletoe leaf explants are generally more sensitive to surface sterilisation than stem explants.

When tissue-culturing *V. album*, explant browning can be an serious problem: explants develop necrotic-like areas which occasionally lead to explant decline and death, thus negatively affecting callus induction. However, depending on the disinfection protocol and the composition of the induction medium, a callus tissue can be formed on 4–12% of the declined explants and proliferate further. These callus tissues retain a black-brownish colour even after repeated subcultures.

The addition of ascorbic acid (at a concentration of 10 mg/l) to the culture medium greatly reduced the frequency of the necrotic symptoms, allowing thus for an increased callus induction (13–45% of cultured explants, depending on medium composition and explant source). These calli obtain a pink-whitish colour (Fig. 1A) and demonstrate a higher proliferation rate (expressed as callus fresh and dry weight) than the black ones (Fig. 1B). The formation of creamy-pink callus tissues has been observed by Fukui *et al.* (1990) as well.

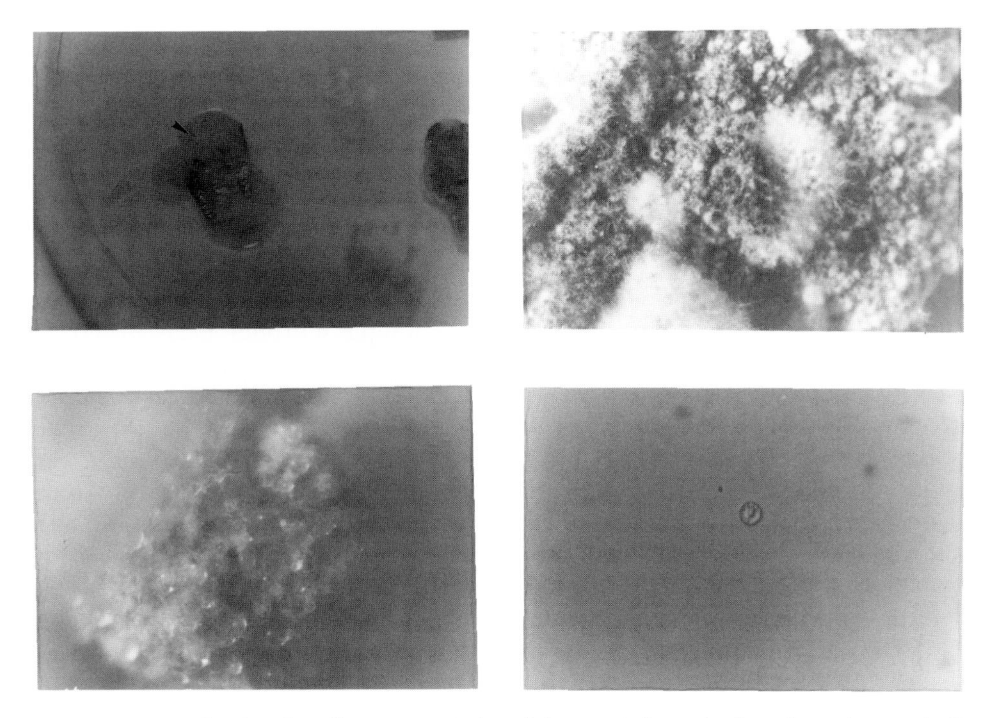

Figure 1 A. Pink-whitish callus tissues induced from mistletoe leaf explants (arrow indicates seed-like structures). B. Callus tissue (black-brownish type) covered with filamentous protrusions. C. Mistletoe callus with globular somatic embryos. D. A mistletoe protoplast isolated from leaf tissue.

Callus Induction

Callus induction takes place after two weeks (on a medium with ascorbic acid) or five weeks (on a medium without ascorbic acid). Callus induction and proliferation is affected by medium composition, explant source and the collection period of the donor plants. Fukui *et al.* (1991) used a half-strength MS (Murashige and Skoog, 1962) nutrient medium supplemented with the growth regulators α-napthylacetic acid (NAA), an auxin and kinetin, a cytokinin, for callus induction from *V. album* var. *lutescens*. In our study, all types of culture media consisted of full strength MS basal medium solidified with 0.8% agar and supplemented with 3% sucrose and either 2,4-dichlorophenoxyacetic acid (2,4-D) (an auxin), NAA, kinetin and benzyladenine (BA) (a cytokinin), alone or in combination, as plant growth regulators (PGRs), at various concentrations. Cultures were incubated under a photosynthetic photon flux density (PPFD) of 150 μmol m^{-2} s^{-1} (16/8, cool white fluorescent).

It is possible that use of culture media, such as MS, containing various salts at high concentrations is preferable for the tissue culture of mistletoe, the parasitism of which affords special adaptation to mineral nutrition (Becker, 1986; Grieve 1994).

Cultures could be established exclusively from "winter" accessions only. In a general sense, higher induction rates were observed from stem explants, reaching a maximum value of 45% on a medium supplemented with 2 mg/l 2,4-D. On the other hand, callus proliferation from leaf explants was definitely better, reaching a maximum average fresh weight of 1.78 g and a dry weight of 0.17 g five weeks after inoculation on a medium supplemented with 1 mg/l kinetin.

Morphological Features of Mistletoe Callus Cultures: Formation of Somatic Embryos

Apart from the aforementioned distinction between "black" and "white" callus forms, mistletoe tissue cultures exhibit some unique morphological structures. Yellowish filamentous protrusions, covering the callus tissues like a mould have been frequently observed (Fig. 1B). They sometimes originated from small, black seed-like structures (Fig. 1A). The function of these protrusions is unknown, but they could possibly be related to haustoria. Finally, somatic embryos at the globular stage were observed on calli cultured for over four weeks on an ascorbic acid-containing medium (Fig. 1C). Somatic embryogenesis is the process of embryo formation from somatic (sporophytic) tissues and comprises a number of distinct stages, i.e. the formation of globular embryos (proembryos), the further development to the heart-shaped stage, to the torpedo stage and finally to the cotyledonary stage (Torres, 1989). As the formation of secondary metabolites in intact plants depends on a well-defined tissue specialisation, it is possible that mistletoe somatic embryos, which exhibit a high level of tissue differentiation, could accumulate various substances of pharmaceutical importance at a satisfactory rate.

Protoplast Isolation

Protoplasts are plant cells having their cell wall artificially removed. In this way, they can be used in gene transfer experiments and for the creation of hybrid cells, i.e. cells resulting from the direct fusion of two protoplast cells that might have been derived from entirely different species. In our laboratory we have isolated protoplasts from both mistletoe leaf (Fig. 1D) and callus tissues, at a maximal yield of 100×10^4 protoplasts/ml, using an enzyme solution of pectinase (0.008–0.016% w/v) and cellulase (0.007–0.014% w/v). The total protoplast yield was higher from calli than leaves, but callus-derived protoplasts were far less viable, probably because older, brownish callus tissues were used for the isolation experiment.

CONCLUSION

Although documented research with tissue culture of *Viscum album* is limited to a small number of reports, the successful induction of callus and somatic embryos from this species allows for a discernible potential for the application of this methodology to facilitate the *in vitro* production of useful metabolites. Further progress in somatic embryogenesis and other advanced aspects of cell culture (e.g. protoplast culture and fusion, creation of autotetraploid lines) could lead to a significant involvement of biotechnology to the utilisation of mistletoe as a medicinal plant species.

REFERENCES

Becker, H. (1986) Botany of European mistletoe (*Viscum album* L.). *Oncology*, **43** (Suppl 1), 2–7.

Becker, H. and Schwarz, G. (1971) Callus cultures from *Viscum album*. A possible source of raw materials for gaining therapeutic interesting extracts. *Planta Medica*, **20**, 357–362.

Fukui, M., Azuma, J., and Okamura, K. (1990) Induction of callus from mistletoe and interaction with its host cells. *Bulletin of the Kyoto University Forests*, **62**, 261–269.

Grieve, M. (1994) Mistletoe (*Viscum album*). A modern herbal. In C.F. Leyel (ed.), *A Modern Herbal*. Tiger Books International, London, pp. 547–548.

Holtorf, S., Ludwig-Müller, J., Apel, K., and Bohlmann, H. (1998) High-level expression of a viscotoxin in *Arabidopsis thaliana* gives enhanced resistance against *Plasmodiophora brassicae*. *Plant Molecular Biology*, **36**, 673–680.

Murashige, T., and Skoog, F. (1962) A revised method for rapid growth and bioassays with tobacco tissue cultures. *Physiologia Plantarum*, **15**, 472–497.

Torres, K.C. (1989) *Tissue Culture Techniques for Horticultural Crops*. Chapman & Hall, New York.

Wagner, H., Jordan, E., and Feil, B. (1986) Studies on the standardization of mistletoe preparations. Oncology, **43** (Suppl 1), 16–22.

8. CHEMICAL CONSTITUENTS OF EUROPEAN MISTLETOE (*VISCUM ALBUM* L.)

Isolation and Characterisation of the Main Relevant Ingredients: Lectins, Viscotoxins, Oligo-/polysaccharides, Flavonoides, Alkaloids

UWE PFÜLLER

Institute of Phytochemistry, University Witten/Herdecke, Stockumer Strasse 10, 58453 Witten, Germany

INTRODUCTION

Extracts from the European mistletoe (*Viscum album* L.) exhibit both, cytotoxic properties and pronounced immunomodulating abilities (see Büssing, this book). Mistletoe extracts containing carbohydrate-binding proteins (lectins), thionins (viscotoxins), flavonoids and many other components, but also the mistletoe lectins (ML) by themselves, are targets of clinical studies in cancer treatment and other diseases.

The evergreen plant is growing on host trees such as apple tree, acer, robinia, poplar, wallow, and oak. The host tree and other natural factors like sex, local climate, harvesting time, parts and age of plant used *etc.* have a major influence on the content of proteins, polypeptides and carbohydrates (Hincha *et al.*, 1997).

CHEMICAL CONSTITUENTS OF EUROPEAN MISTLETOE

Lectins – Common Introduction

Lectins are sugar-binding proteins derived from plants, animals or bacteria which do not change chemically the recognised carbohydrate structures and do not exert antibody function (Sharon and Lis, 1989; Pusztai and Bardocz, 1995; Rhodes and Milton, 1998). The lectins show specificity for terminal and/or subterminal carbohydrate residues and have at least two sugar-combining sites. Therefore, lectins will specifically recognise and bind to carbohydrate residues on cell surface, to cytoplasmic and nuclear structures, and to components of extracellular matrix. The binding ability of lectins is usually described in terms of simple sugars inhibiting lectin binding to its targets. This is an oversimplification because the complex sugar sequences and structures recognised by a lectin are in most cases unknown. Lectin binding sites of a glycanor glycoconjugate consist usually of 1 to 4 saccharide moieties. Further, non-covalent interactions with hydrophobic or other residues of the target molecule may enhance the binding.

Lectins are important tools in histochemistry (Brooks *et al.*, 1997), structural analysis of oligosaccharide chains, detection of alteration of glycosylation, quantification of glycoconjugates, affinity purification of glycoconjugates, cytofluorimetric studies of cells, and cell separation (Rhodes and Milton, 1998). Some lectins are the mitogenic, while others are cytotoxic, or influence neuronal trafficking, pathogen–host interaction, and lectins are also carriers for drug delivery (Haltner *et al.*, 1997). Moreover, plant lectins, widely distributed in food and feed, play an important role in nutrition and health science (Brooks *et al.*, 1997; Haltner *et al.*, 1997; Pusztai and Bardocz, 1991).

Within the large group of plant lectins, a number of lectins from non-related species belong to a family of so-called ribosome inactivating proteins (RIP). The most prominent member of this group, ricin, was recognised over hundred years ago as a very toxic and haemagglutinating protein from the Castor bean. Besides ricin and the mistletoe lectins (ML), volkensin, abrin, modeccin and ebulin II are established members of the RIP lectin family (for review see Van Damme *et al.*, 1998a,b; Pusztai and Bardocz, 1991, 1995; Rhodes and Milton, 1998; Stirpe and Batelli, 1990; Barbieri *et al.*, 1993).

Commonly, the RIPs are divided in two groups: type 1 RIPs are single-chain proteins which do not bind to carbohydrates, and type 2 RIPs which consist of a lectin B subunit and a toxophoric A chain which is a RNA *N*-glycosidase that affects tRNA, and thus block protein synthesis (Stirpe and Batelli, 1990; Barbieri *et al.*, 1993) (Table 1). Type 1 RIPs and the A chains of type 2 RIPs show nearly the same enzymatic activity especially in cell-free systems (Barbieri *et al.*, 1993).

The type 2 RIPs have strongly related molecular subunit architectures. The toxophoric A chain is connected with the sugar-binding B chain *via* a disulfid bridge. X-ray and sequence studies of MLs and other members of the toxic lectin family demonstrate a high degree of structural identity but quite different biological activities. Whereas more than 30 type 1 RIPs have been described, only a few type 2 RIPs are characterised (Table 1). Further examples of type 2 RIPs are described by Van Damme *et al.* (1998a,b). In this regard, the isolation of lectins with the same molecular architecture like ricin from *Viscum album* was also a big surprise in the seventies (Luther, 1976; Franz *et al.*, 1977).

There is some evidence that type 2 RIPs evolved from type 1 and both from a common unknown ancestor (Van Damm *et al.*, 1998b). Further, RIPs are suggested to be products of the fusion of two different genes – the gene for the lectin B-chain and the gene for the *N*-glycosidase A-chain (Stirpe and Batelli, 1990; Barbieri *et al.*, 1993). It is not yet clear whether ML I, ML II and ML III, which have approximately similar molecular weights (about 60 kDa) and high degree of homology as compared to ricin, are products of one gene and distinctions between them are determined by post-translational changes, or they are products of several genes (Pusztai and Bardocz, 1995; Barbieri *et al.*, 1993). Recent investigations by Western Blot analysis revealed, that eluted antibodies specific for ML I A or B chain also recognised A and B chain of ML II and B-chain of ML III, suggesting homologies of epitopes of the three ML (Stein *et al.,* 1999d). Only the unglycosylated ML III

A-chain was not detectable with these antibodies. The European mistletoe belongs to the few plant families containing more than one class of lectins, the gal/galNAc-recognising ML I, ML II and ML III of the RIP 2 type, and one lectin with hevein basic structure, the glucNAc-oligomere binding lectin VisalbCBL.

Table 1 Characteristics of representative type 2 RIPs and schematic description of RIP structures.

type 2 RIP			*Type 1 RIP*	
(A)-S-S-(B)			(A)-$(SH)_{0-1}$	
26–31 kD	30–38 kD		25–32 kD	
Plant	*Family*	*Lectin*	*Tissue/Yield*[1]	*Sugar specificity*
Ricinus communis	Euphorbiaceae	Ricin (RCA II)	seeds/120	D-galactose
		Ricinus communis agglutinin (RCA I)	seeds/75	galNAc/ d-galactose
Abrus precatorius	Fabaceae	abrin	seeds/75	D-galactose
Viscum album	Viscaceae	ML I[2]	leaves, stems/3–170	D-galactose
		ML II	leaves, stems/0.1–35	D-galactose/ galNAc
		ML III[3]	leaves, stems/0–67	galNAc
Viscum album var. *coloratum*	Viscaceae	ML I-like	leaves, stems/up to 70	β-galactoside >galNAc
Phoradendron californicum	Viscaceae	PCL	leaves, stems/42	D-galactose
Adenia digitata	Passifloraceae	Modeccin	roots/20–180	D-galactose
Adenia volkensii	Passifloraceae	Volkensin	roots/0.2	D-galactose
Sambucus ebulus	Sambucaceae	Ebulin I	leaves/3.2	D-galactose

[1] fresh plant material; yields of ML I, ML II and ML III and PCL are referred to air dried material (mg/100 g).

[2] Other names are viscumin, *Viscum album* agglutinin (VAA I).

[3] Other names are *Viscum album* agglutinin II (VAA II) which may be identical with ML III or ML II/III.

Mistletoe Lectins (ML): ML I, ML II and ML III – Members of RIP 2 Family

ML I binds preferentially to β-galactosides, ML II to β-galactosides and N-acetyl-galactosides, whereas ML III recognises N-acetyl-galactosides (Table 1) (Franz, 1989, 1991; Pfüller, 1996). The nomenclature of ML is not uniform. Beside the designation given by our group, ML I, ML II, ML III (Franz *et al.*, 1981; Pfüller, 1996), others termed ML I "viscumin" (Olsnes *et al.*, 1982) or *Viscum album* agglutinin I (VAA I; Samtleben *et al.*, 1985); in this context, VAA II designates apparently a mixture of ML II/ML III (Samtleben *et al.*, 1985).

The ML are cytotoxic as they induce apoptosis, and stimulate immunocompetent cells to produce a set of cytokines (see Büssing, this book). The toxophoric A chains of the toxic lectins, and especially that of ML, are potent effector molecules for the construction of immunotoxins useful as site-directed immunosuppressive and cytotoxic agents (Tonevitsky *et al.*, 1991; Schütt *et al.*, 1989; Pfüller *et al.*, 1988). Further, especially ML I and ML III are new tools in histochemistry to label immunocompetent cells in lymph nodes and other tissues (Pfüller and Niedobitek, 1998; Schumacher *et al.*, 1994, 1995). In the following paragraphs, the isoform pattern of mistletoe lectins, their isolation, characterisation, structure-function relationship and therapeutically relevant properties will be reviewed referring to Franz (1989, 1991), Pfüller (1996) and Büssing (this book).

The isolectin pattern of RIP 2 Viscum album lectins

The haemagglutinating effect of mistletoe extracts were first described by Krüpe (1956) and by Bird (1954). In 1973, Luther *et al.* (Luther, 1976; Franz *et al.*, 1977) isolated a *Viscum* lectin by galactose treatment of erythrocytes incubated with a mistletoe extract. In 1981, Franz *et al.* (1981) were able to separate and characterise three different mistletoe lectins, termed ML I, ML II and ML III. The optimised isolation procedure (Eifler *et al.*, 1993) involves PBS extraction of fresh or air dried plant material, absorption of proteins by cation exchangers, elution of the proteins by salt gradient, affinity chromatography on lactosyl-sepharose 4B, elution of ML II/MLIII by PBS, elution of ML I by lactose and separation of ML II and ML III by Fast Protein Liquid Chromatography (FPLC) on Mono S. The main properties of ML are given in Table 2. Figure 1 represents a typical electrophoretic separation pattern of the three isolectin groups (Eifler *et al.*, 1993). Depending on seasonal factors and the host tree, mistletoe contains variable amounts of each isolectin groups with different sugar specificities ranging from galactose (ML I), galactose/galNAc (ML II) to galNAc (ML III), as demonstrated by Surface Plasmon Resonance Studies (Figures 2).

Mistletoes grown on pine trees contain mainly ML III and only minor amounts of ML I, while mistletoes growing on deciduous trees contain mainly ML I (Eifler *et al.*, 1993). Host tree and seasonal dependent lectin pattern of mistletoes are demonstrated in Figure 3. The isolectin composition of ML I seems to be related to host tree and frost impact (Pfüller, 1996; Eifler *et al.*, 1993) according to results on preparative isolation of isoforms with heavy (ML I-2) and light (ML I-1) A chain of ML I (Eifler, Pfüller and Pfüller, in preparation). In general, ML I isolated from

Table 2 Properties of RIP II type lectins of *Viscum album*.

Lectin	*MW*	*IP*	*Yield** (mg/100 g)*	*Titer***	*Sugar inhibition*	*Plant material (main origin)*
ML I****	115.000	6.7, 7.2, 7.6, 7.9	2–25	1:128	Gal >> galNAc	*Viscum album* from deciduous trees
ML I-1	65.000		0,1–5 ***	1:256	Gal >> galNAc	*Viscum album* from birch
ML I-2	67.000		0,1–1***	1:256	Gal >> galNAc	*Viscum album* from birch
ML II	63.900	6.5, 6.7	0,01–3	1:16	Gal/galNAc	deciduous trees
ML III	61.600	5.8, 6.1	2–20	1:32	GalNAc > gal	*Viscum album* from pine tree
rML	58.400	< 8.5	–	1:64/128	Gal >(>) galNAc	recombinant (*E. coli*)

Some batches of isolated ML I containing the ML I isolectins ML I-1 (isoform with light A chain) and ML I-2 (isoform with heavy A chain) are separated in pure isomers by FPLC. The data of the recombinant mistletoe lectin (rML) are reported by Möckel *et al.* (1998) and Zinke *et al.* (1998).

* Harvesting time October to February, Berlin and Dresden areas.

** Haemagglutination titer (1 mg ML/ml PBS, pooled human erythrocytes).

*** Depending on host tree and harvesting time.

**** ML I is a dimer in solution.

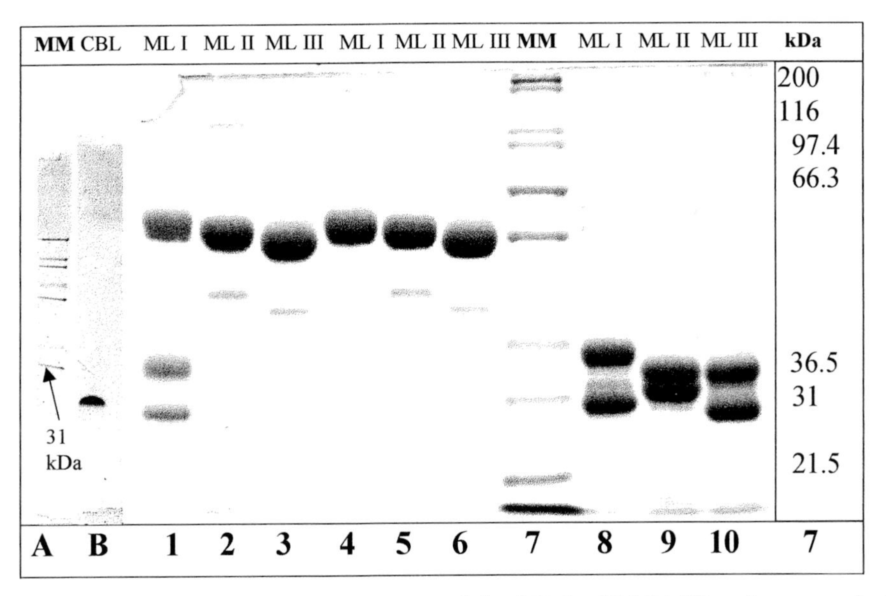

Figure 1 Electrophoretic separation pattern of the ML by SDS-PAGE under non-reducing (lanes 1–6, A, B) and reducing conditions (8–10), respectively. Samples representing lanes 4–6 were blocked prior to separation by iodine acetamide. Lane B indicates the chitin binding lectin (CBL) from Viscum album L., VisalbCBL. The ML B chains exhibit a higher molecular weight (about 32 to 36 kDa) as compared to the A chains (about 28.5 to 31 kDa). Molecular weight marker (MM) is represented by lanes A (for CBL) and 7.

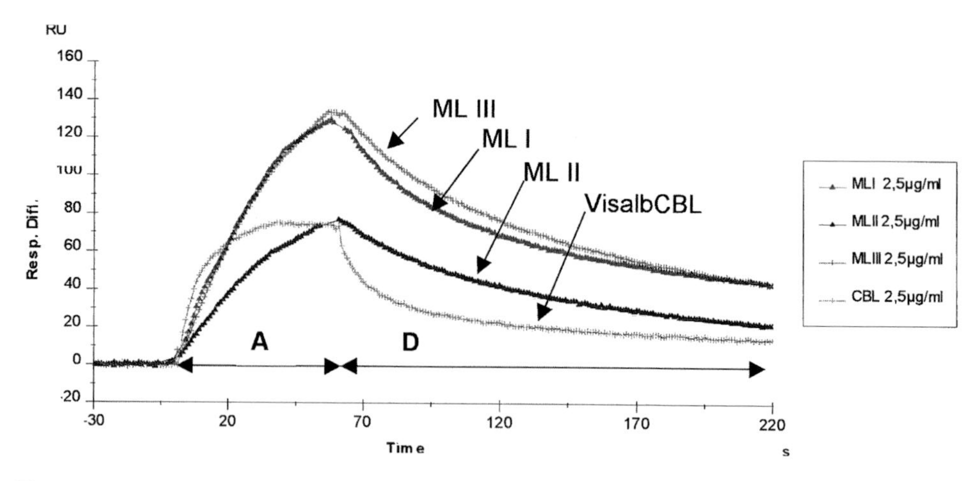

Figure 2 Recognition pattern of a complex glycoconjugate (asialofetuin) by ML I, ML II, ML III and VisalbCBL as demonstrated by surface plasmon resonance spectroscopy (BIA, biochemical interaction analysis). The lectins were applied to asialofetuin immobilised on sensorchip CM5 (Pharmacia) in equal concentrations. While the binding pattern of ML I and ML III are strongly related to each other and show only some differences in the dissociation step (D), ML II and VisalbCBL reveal different kinetics in the association (A) and dissociation (D) step of ligand (ASF) – analyte (lectins at 2.5 μg/ml HEPES-buffered saline) interaction. As compared to ML I or ML III, ML II has considerably lower binding constants for glycoconjugates, and core glucNAc residues are obviously less efficient targets for the VisalbCBL.

mistletoes grown on different deciduous trees contain mainly ML I-1. The preparative isolation and separation of single isoforms belonging to ML I, ML II and ML III is possible by chromatofocusing using FPLC-technique.

The differences in molecular mass, electrophoretic and chromatographic behaviour between mistletoe isolectin groups are mainly caused by the glycosylation degree and possibly, to some extent, by the amino acid sequence. All three lectins are glycosylated and carry plant typical mannose- and complex-type sugar side chains (Debray *et al.*, 1992, 1994; Zimmermann *et al.*, 1996) (Figure 4). A glycosylation of the ML III A chain of ML III isolated from ML III from different mistletoes (regarding to host tree, season and localisation) was never observed (Pfüller and Eifler, unpublished results).The glycoconjugate nature of ML allows also "passive" interactions of ML with mannose-, glucNac- and fucose- recognising lectins on tissue structures which are able to bind ML in different quantities reflecting the sugar composition of the glycan part.

Recently the amino acid sequences of ML I A chain and B chain were published, demonstrating high conservation of structure when compared to other type 2 RIP toxins (Eschenburg *et al.* 1998, Soler *et al.*, 1998; Krauspenhaar *et al.*, 1999). The X-ray structure of ML I was estimated to a degree of resolution of 3.0 Å (Figure 5), underlying high degree of structural identity with other known RIP 2 molecules. Noticeable, the dimeric character of ML demonstrated by gel permeation chroma-

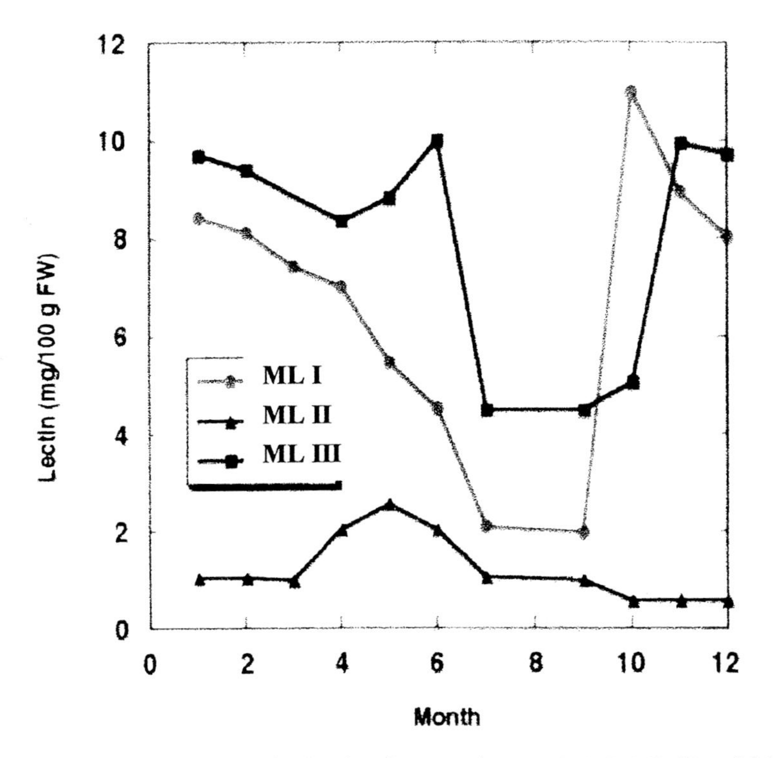

Figure 3 Seasonal variations in the levels of ML isolectins (ML I, ML II and ML III) as measured in the leaves from Viscum album L. grown on host tree Malus sylvestris. Leaves (1–2 years old) were harvested from January (1) to December (12). The amount of different ML (mg/100 g fresh plant weight, FW) was determined by FPLC and ELISA techniques. ML II, however, shows a peak in May, while the concentration of ML I and ML III strongly decreased during summer and showed the highest level during winter time.

tography (Ziska *et al.*, 1978) and by electron microscopy (Lutsch *et al.*, 1984) was confirmed by X-ray investigations (Sweeney *et al.*, 1993; Krauspenhaar *et al.*, 1999). Depending on concentration and composition of buffer solutions, only ML I but not ML II or ML III shows dimerisation (Franz *et al.*, 1981; Lutsch *et al.*, 1984).

Independently of high structural identity between ML I and ricin, no cross reactivity between both polyclonal and monoclonal anti-ricin and anti-ML antibodies was detected (Jäggy *et al.* 1995; Tonevitsky *et al.*, 1995, 1999). Anti-ML antibodies are important tools to describe structural features of ML and allow their detection in plant extracts. Hybridomas producing monoclonal antibodies (mab) against ML and their A and B subunits have been obtained by Jäggy *et al.* (1995) and Tonevitsky *et al.* (1995, 1999). Three groups of mAb displaying different affinities to ML and recombinant ML I A-chain were generated: (1) mAb against ML I-A and ML II-A, (2) mAb against ML II-A and ML III-A, and (3) mAb against A-subunits of ML I, ML II and ML III. Antigenic determinants of ML recognised by mAb MNA4, MNA9 and MTC12 contain no carbohydrate side chains (Tonevitsky *et al.*, 1999).

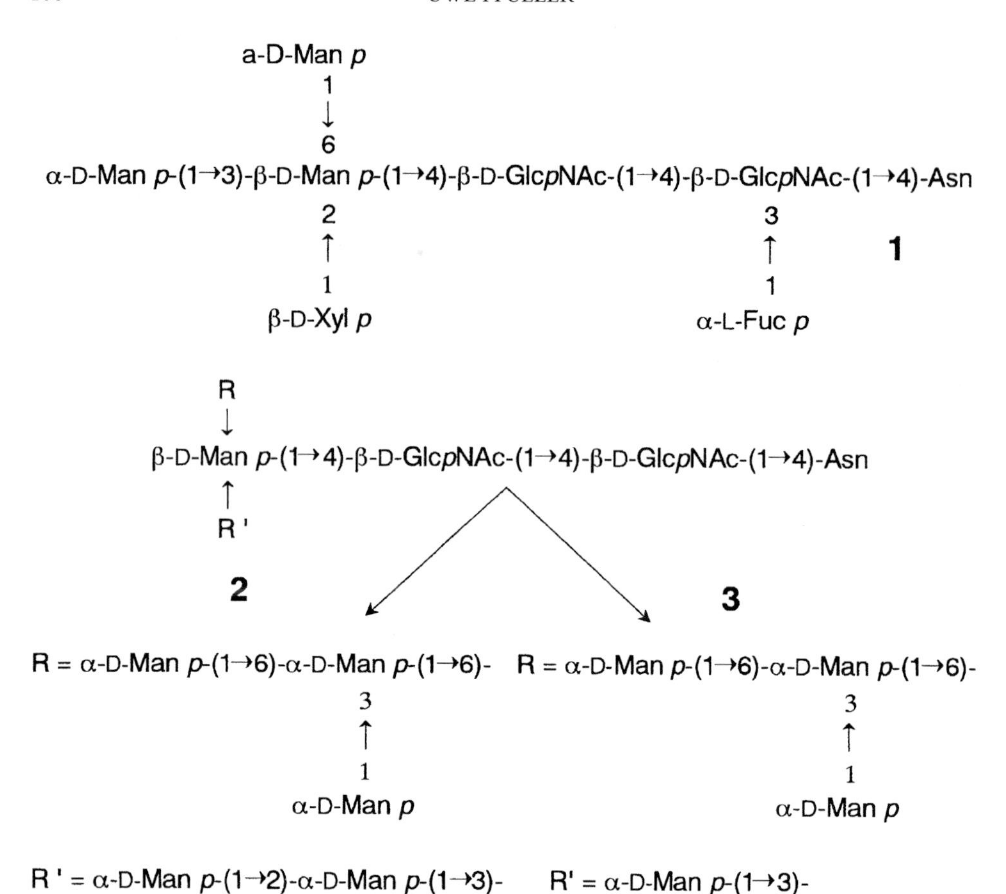

Figure 4 Carbohydrate side chains of ML I according to Debray *et al.* (1992). There is no indication what kind of glykan chains 1–3 are localised in the heavy or light A chains or in the B chain of ML I.

Sandwich ELISA test-systems allow to identify ML I, ML III, and, at favourable concentration ratios, of ML I and ML III to ML II, and also ML II at higher concentrations and relations to ML III > 5:1 (Tonevitsky *et al.*, 1995, 1999). Franz *et al.* (1983) did not observed cross reactivity of anti-ML (ML I toxoid) and anti-viscotoxin (A3) antibodies with ML I and viscotoxin A3 antigens and *vice versa*. However, Stein *et al.* (1998) demonstrated by Western blotting techniques that human anti-ML antibodies can be cross reactive to viscotoxins. Therefore further studies are necessary including complete pattern of ML and viscotoxins.

Biological activities of RIP 2 Viscum album lectins

The cytotoxic and immunomodulatory properties of mistletoe preparations and ML have been studied intensively in past years (see Büssing, this book). The ML were

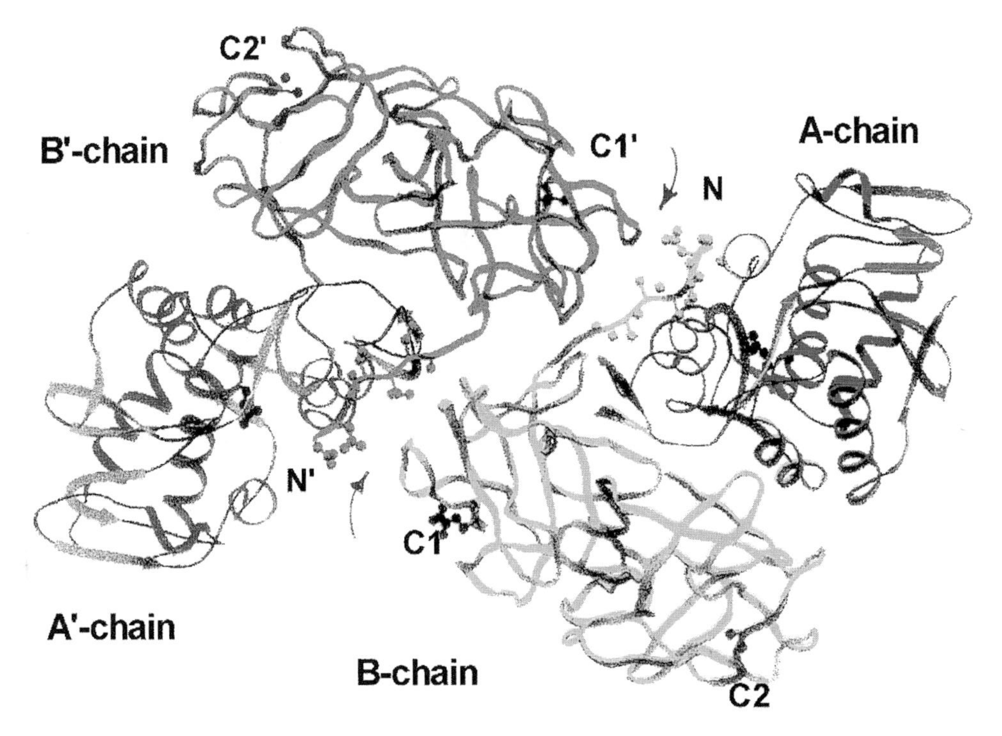

Figure 5 Quaternary structure of ML I as determined by X-ray cristallography demonstrates a dimeric composition. ($[A_L$-S-S-$B]_2$) of ML I with light A chain according to Sweeney *et al.* (1998). The B chains have two sugar combining sites C2, C1 and C2′, C1′, respectively. At the A chains, one enzymatic combining site N and N′, respectively, is localised. Dimerisation of two ML I monomers is performed by contacts mainly *via* the B chains including water molecules.

suggested by some groups to represent the biological activity of mistletoe extracts used to treat tumour patients. On the other hand, there is growing evidence that different components of mistletoe, such as lectin binders like glycoproteins and oligo/polysaccharides, viscotoxins and small molecular weight molecules, exert important biological effects in a direct or even indirect manner *via* stabilisation of lectins.

As shown in Table 3, which summarises the main biological activities of ML and their separated A and B subunits, the biological effects related to the mistletoe isolectin pattern of mistletoe extracts are a complex phenomenon characterised by cytotoxic and immunomodulatory effects (see Büssing, this book). ML are useful tools for the selective staining of microglia cells (Artigas *et al.*, 1992), lymph node compartments (Pfüller and Niedobitek, 1998) and Alzheimer's plaque glycoproteins (Schumacher *et al.*, 1994) While ricin and abrin are highly toxic in animals and humans when applied orally, there are no reports on similar effects of mistletoe preparations and extracts following oral uptake (Pusztai *et al.*, 1998a, 1998b;

Table 3 Biological activities of MLs and their subunits.

biological activity	*ML*	*dML I**	*A chain*	*B chain*
cell agglutination, haemagglutination	■■■■	----	----	■■■■
RIP activity**				
– cell free system	■■■■	■■■■	■■■■	----
– cell system	■■■■	■	■	----
[^{3}H]-thymidine uptake	■ ?	n.d.	■ ?	----
cytotoxicity	■■■■	■	----	----
Ca^{2+} influx	■■■	■	■	■■■
cytokine release	■■■■	■■■	(■)***	■■■■
NK cell activity	■■■	n.d.		■■■
macrophage activation	■■	----		■■
binding to				
– M-PIRE cells	■■■■	■	■	■■■
– microglia	■■■■	■	n.d.	■■

* partially inactivated ML I without sugar binding ability.
** inhibition of protein synthesis on ribosomal level.
*** oligomeric soluble self-aggregated A chain eventually is able to release cytokines.

Lavelle *et al.*, 1999; see Stein, this book). The acute toxicity of pure mistletoe lectins, especially ML I, is comparable to that of ricin after i.p. administration. The structurally related toxic lectins share pronounced cytotoxic properties but show a quite different behaviour in their abilities to modulate the immune system (for review see Barbieri *et al.*, 1993). The study of the molecular structure may help to understand their distinct effects and relevance in the discussion of the nutritive and antinutritive activity of plant lectins and their use or implications in biomedical sciences.

Mistletoe Lectins: The Hevein Domain Lectin VisalbCBL

Beside the type 2 RIPs from *Viscum album* L., Peumans *et al.* (1996) isolated a structurally non-related chitin-binding lectin from European mistletoe, designated VisalbCBL (Peumans *et al.*, 1996, 1998). This lectin is a homodimer of two identical subunits of 10.8 kDa. VisalbCBL shares a high degree of sequence similarity at its N-terminus with the hevein domain, and also the amino acid composition is similar to that of other chitin-binding hololectins.

Following the schematic representation of molecular structures (Figure 6), VisalbCBL consists of two subunit-forming hevein repeats forming a dimer with four chitin-binding sites (Peumans *et al.*, 1998). Basic properties of VisalbCBL are described in Table 4.

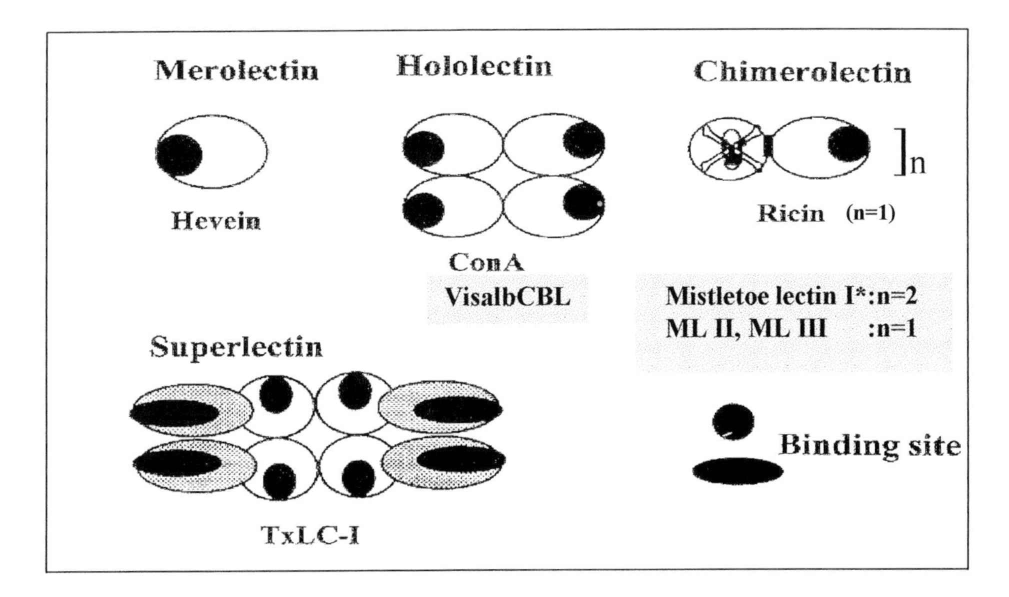

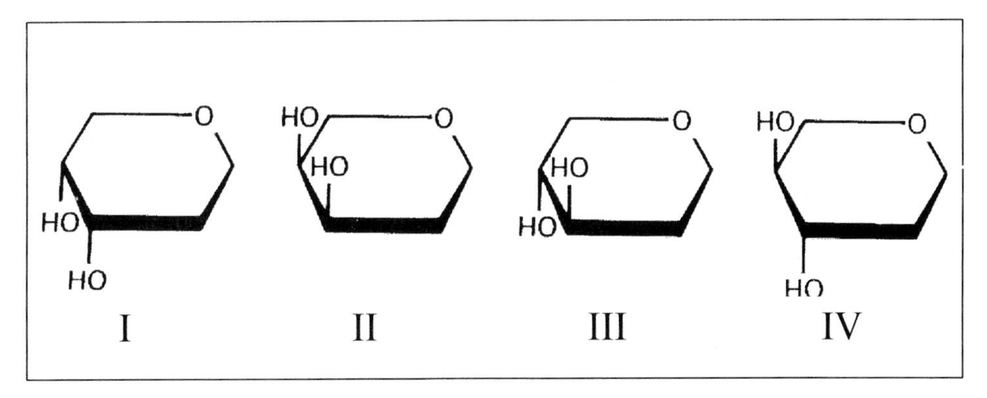

Figure 6 Structural classification of lectins (according to Peumans et al., 1998) and positions of the four mistletoe lectins, ML I, ML II, ML III and dimeric VisalbCBL (upper box). The lower box shows a possibility to classify lectins via their primary monosaccharide target. Class I contains fucose-recognising lectins, class II contains those recognising galactose/ galNAc (such as ML, ricin, jacalin, galectins), and class III those binding to glucose/mannose/ glucNAc (such as concanavalin A and VisalbCBL). Class IV is not yet occupied.
* ML I forms in most cases a dimer.

A hypothetical model of the molecular evolution of chitin-binding lectins including VisalbCBL is described by Van Damme *et al.* (1998b). The physiological role of VisalbCBL remains unclear. One may suggest that this lectin is part of the plants defence system against bacteria, fungi or insects as already described for other chitin-binding lectins (Van Damme *et al.*, 1998a,b). The direct and indirect

Table 4 Basic properties and data of the chitin-binding lectin VisalbCBL.

occurrence	green tissues, up to 0.001% in deciduous trees and < 0,00001% in pine tree
structure	$(P\text{-}P)_2$, molecular mass of protomer P: 11 kD
specificity	$(glcNAc)_n$, n: 2, 3, 4 (increasing binding)
titer	1:4 (1 mg/ml, pooled human erythrocytes)
glycosylation	no
cytotoxicity	> 100 μg/ml (Molt-4)
antibody cross reactivity	no cross reactivity with polyclonal and monoclonal antibodies against ML I, ML II and ML III

contribution of VisalbCBL which is detected in considerable amounts in most of commercially available mistletoe preparations (Pfüller *et al.*, unpublished results) to efficacy of these extracts is not yet investigated.

Vester Proteins, Non-lectin Proteins and Peptides

ML amounts to not more than 2% of total polypeptide and protein content of mistletoe. While polypetides of viscotoxin type (see below) are well characterised, there are only few information on structure and biological properties of other proteins present in the plant. An impression of number and quantity of that proteins was given by Schink (1990) following results of two-dimensional SDS-PAGE. Vester (1977) described the isolation of a protein complex (VP-16) with 14–125 kDa, exhibiting anti-tumour activity *in vivo* and strong cytotoxicity *in vitro* (for review see Luther and Becker, 1987). Franz (1985) classifies these proteins according to their molecular weights at least to mono- and dimeric ML and their subunits. However, we and others were unable to reproduce the results described by Vester (unpublished).

Starting from commercial mistletoe preparations, Kuttan and co-workers (Kuttan and Kuttan, 1992a,b) isolated polypeptide fractions with immunomodulating activities. Similarly, these polypetides remain to be characterised.

Sakurai and Okumura (Okumura and Sakurai, 1973; Sakurai and Okumura, 1979) isolated a cyclic amphiphilic pentapeptide from *Viscum album* var. *coloratum*, termed viscumamide. This amide or related substances has not yet been described for the European mistletoe. However, there is no report on its biological activity. The cyclopeptide (Figure 7) has pronounced membrane binding properties as demonstrated by interaction with artificial membranes using Surface Plasmon Resonance Spectroscopy and BIA-technique (Becker and Pfüller, unpublished) and isolated lymphocytes (Büssing *et al.*, unpublished results).

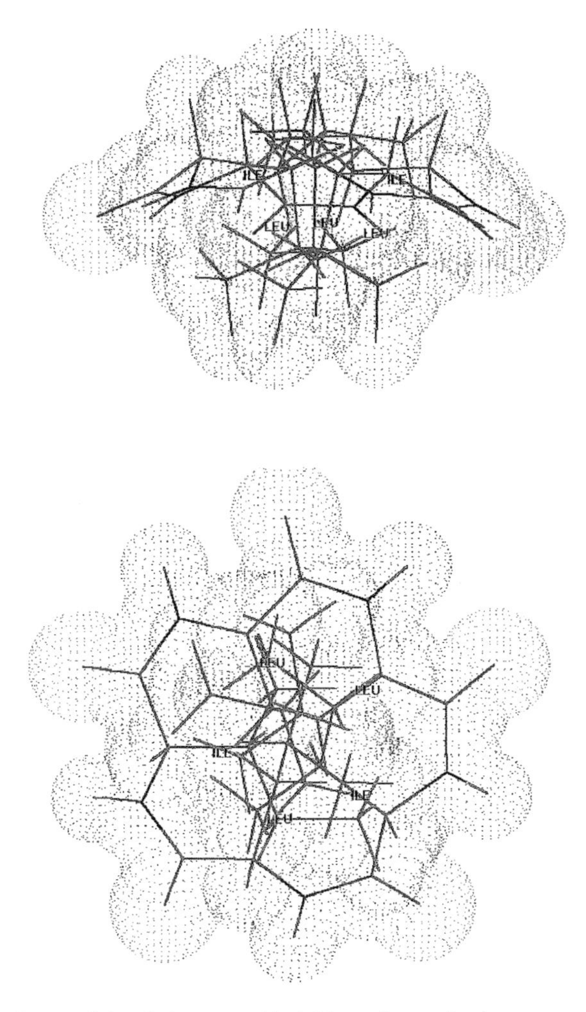

Figure 7 Molecular models of viscumamide (side and top view).

Thionins: Viscotoxins

Viscum album L. contains a group of amphipathic, strongly basic polypeptides, named viscotoxins. At least six isomeric viscotoxins were isolated by Samuelsson (1974) and Urech *et al.* (1995), mainly from *Viscum album* species. These peptide-toxins are strongly related to an already known group of amphiphilic basic peptides, thionins, which share a rather similar structure of 46 amino acids. They are common in plants like corn, wheat and others, and are suggested to be protective substances against infections by viruses, bacteria and fungi (Carrasco *et al.*, 1981).

Viscotoxins have a molecular weight of about 5 kD (isoelectric points at pH 9-11) and contain 3 to 4 disulphide bridges. The amino acid sequences of the most important of them, viscotoxin A2, A3, B, were published by Samuelsson (1974). Beside the sequence data, a molecular model of the viscotoxin A3 is given by Barre *et al.* (1996, 1997). The three-dimensional-models of viscotoxin A3 and of alpha 1-purothionin (from Durum wheat) revealed that both polypeptides are amphipathic, and thus may interact with membrane lipid bilayers. The viscotoxins as well as other thionins are very stable polypeptides even under denaturating conditions and tend to self-aggregate and to form protein-lipid-complexes, even soluble in simple hydrocarbons (Fernandez de Caleya *et al.*, 1974).

The physico-chemical properties of the membrane-permeabilising viscotoxins (Büssing *et al.*, 1999a,b) may explain some of their biological and pharmacological activities. However, it is unclear why amphipathic polypetides from mistletoe and other plants sharing great extend of structural identity show quite different biological behaviour. One may suggest that a change of only few amino acids in the biologically less active thionins may change the cytotoxic potential of the polypeptides, as suggested for viscotoxin B as compared to viscotoxin A_2 (Büssing *et al.*, 1999a). Therefore, the amphipathic properties, more or less equal in the studied polypeptides, must be completed by other structural features resulting in the comparable high solubility in water and outstanding biological activities of the viscotoxins.

Due to their tendency to form complexes with ML, the viscotoxins may modulate the interactions of ML with carbohydrates. One of the three disulphide bridges in the viscotoxins is well exposed and may thus be involved in SS/SH-exchange reactions with ML forming a covalent bond between both (Pfüller, unpublished observations).

Due to their outstanding properties and the occurrence in many commercially available mistletoe preparations, the viscotoxins are candidates for further investigations on the action of mistletoe extracts to the human organism (Stein *et al.* 1999a,b). On the other hand, the behaviour and modulating ability of thionins, widely distributed in food and feed, should be carefully discussed in regard of their potential impact for nutrition and therapy.

Alkaloids

In contrast to *Viscum album* var. *coloratum* and others, *Viscum album* L. obviously do not contain typical alkaloids. Reports on "alkaloid-like" compounds in *Viscum album*, including tyramine, phenylethylamine, choline and acetylcholine, have been summarised by Hegenauer (1966).

Khwaja *et al.* (1986) described alkaloids from *Viscum album* var. *coloratum* and *Viscum album* L. These "alkaloids" were defined only by the extraction procedure and their reaction with the not yet specific Dragendorff-alkaloid reagent. Due to their extreme lability, the substances remain uncharacterised until now.

Flavonoids

The existence of flavonoids in *Viscum album* has been described by Schindler (1955); a recent overview was given by Teuscher (1994). The following flavonoid glycosides were isolated by Becker *et al.* (Becker *et al.*, 1978; Becker and Exner, 1980) and Fukunaga *et al.* (1987):

2′-Hydroxy-4′,6′-dimethoxychalcone-4-O-glucoside
2′-Hydroxy-3,4′,6′-trimethoxychalcone-4-O-glucoside
2′-Hydroxy-4′,6′-dimethoxychalcone-4-O[apiosyl-(1→2)]glucoside
(2R)-5,7-Dimethoxyflavonone-4′-O-glycoside
(2S)-3′,5,7-Trimethoxyflavanone-4′-O-glycoside
(2S)-Homoeriodictyol-7-O-glucoside
Rhamnazin-3,4′-di-O-glucoside

The aglyca homoeriodyctiol, sakuranetin, rhamnazin, isorhamnetin and further six quercetin methylethers were detected in an extract of *Viscum album* subspecies *abietis* after hydrolysis by Becker and Lorch, respectively (Becker and Exner, 1980; Lorch, 1993). In the subspecies of *Viscum album* subspecies *laxum* 5-quercetin methylether, 5,7-Dimethoxy-4′-hydroxyflavon and sakuranetin were detected. Some other flavonoids were detected in *V. album* ssp. *coloratum* and other subspecies (see Becker and Park, this book).

Phenylpropanoids

Phenylpropanoids are related with flavonoids by the same biogenetic pathway. The most common phenylpropanoids, also known as cinnamic acid derivatives, are caffeic acid, ferulic acid and sinapic acid, which are found in *Viscum album* together with their degradation products, protocatechuic acid, syringic acid, vanillic acid, anisic acid and gentisic acid (Wagner *et al.*, 1984, 1986; Fukanaga *et al.*, 1987). From the leaves and stems of *Viscum album* L., syringenin-4′-O-glucoside (syringin), syringenin-4′-O-apiosyl-1→2-glucoside (syringoside), syringaresinol-4,4′-O-glucoside, the dimeric phenylpropanoids (lignans) 4″-di-O-glucoside (eleutheroside E) and syringaresinol-mono-O-glucoside were isolated. The isolated compounds, especially syringin and its aglykon syringenin are suitable for the HPLC- or HPTLC-fingerprint analysis of alcoholic or aqueous *Viscum album* extracts (see Lorch, this book).

Phytosterols

The ubiquitous phytosterols β-sitosterol, stigmasterol and their respective glycosides, are found in *Viscum album* L. (Ohta and Yagashita 1970).

Triterpenes

Viscum album is rich in triterpenes, β-amyrin, β-amyrin-acetate, betulinic acid (Fukunaga *et al.*, 1987), oleanolic acid, ursolic acid (Wagner *et al.*, 1984).

Mono-, Oligo- and Polysaccharides

Due to their manifold abilities to interact with various plant constituents influencing their solubility, stability and even biological behaviour by complex interaction processes, there is a growing interest to study plant carbohydrates as potential biologically active components. The immunomodulating effects of polysaccharides isolated from *Viscum album* were described by Büssing, this book, and Schietzel and Stein, this book.

Stems and leaves contain mainly methylester of 1→α4 galacturonic acid (Jordan and Wagner, 1986b). The berries contain rhamnogalacturonanes as basic structures to which individual branched (1→β6)-D-galactan chains are linked *via* O-4 of rhamnose residue, while arabinosyl residues or complex arabinan side chains are linked *via* O-3 group of galactose units. Beside a highly methylated homogalacturonan, a pectin with a molecular weight of 42 kDa (following hydrolysis of ester bonds), minor amounts of an arabinogalactan (110 kDa) were isolated from stems and leaves. In the berries, a high molecular weight rhamnogalacturonan (700 kDa) with arabinogalactan side chains was detected.

In 1999, Edlund and co-workers (Edlund, 1999, Edlund *et al.*, 1999) isolated and characterised polysaccharides from mistletoe berries. The isolated acidic (1,340 kDa) and neutral (30 kDa) oligo- and polysaccharides are composed of a rhamnose-galacturonic acid backbone with highly branched arabinose-galactose side chains (acidic oligosaccharides) and arabinogalactans together with minor amounts of xyloglucan (neutral polysaccharides). Especially the acidic carbohydrates strongly interacts with ML I, as demonstrated by haemagglutination assays, size exclusion chromatography and BIA (Biochemical Interaction Analysis). In contrast to ML I,. the high molecular weight acidic arabinogalactan stimulated the proliferation of $CD4^+$ T helper lymphocytes but not $CD8^+$ T lymphocytes or B lymphocytes (Stein *et al.*, 1999). Thus, also non-lectin components may contribute to the suggested antitumour effects of mistletoe extracts. Interactions between a high molecular weight acidic arabinogalactan and ML I were described by Jordan and Wagner (1986a,b). Further studies are necessary to elucidate biological activities of oligo- and polysaccharides from *Viscum album* and their manifold possibilities to interact with other active components (see Büssing, this book, and Stein and Schietzel, this book).

Polyalcohols

Richter (1992) isolated 1D-1-O-methyl-muco-inositol, a derivative of O-methylinositol widely distributed in higher plants, from *Viscum album*. Increasing amounts of this cyclitol were found during the cold season, suggesting a possible role as an cryoprotectant. The concentration of 1D-1-O-methyl-muco-inositol in the leaves of the plant grown on *Crataegus monogyna*, is up to 7% of dry weight during March as compared to 4% in July. Many inositol derivatives including 1D-1-O-methyl ethers inhibit the interaction of ML with galactosyl compounds (Pfüller, unpublished results). Other sugar alcohols detected were mannitol, quebrachitol, pinitol and viscumitol (Richter, 1992; Richter and Popp, 1992).

CONCLUDING REMARKS

Therapeutically applied mistletoe preparations are complex plant extracts containing a wide variety of substances ranging from high molecular weight proteins to low molecular weight compounds such as flavonoids and other. Among the biologically active components, the sugar–binding ML and amphiphilic membrane-active viscotoxins are at present prominent candidates in the ongoing discussion on the clinical relevance of biological effects of these drug extracts, such as cytotoxicity and immunomodulation. The large pool of further components and their role in pharmacological activity remain unclear yet. There is no doubt that some glycosylated proteins are responsible for stability of the lectins as *in vitro* detected for polysaccharides and non-sugar binding (glyco)proteins present in mistletoe extracts (Pfüller, unpublished results), influencing also their biological activities, and may also exert therapeutically relevant effects for their own.

ACKNOWLEDGMENTS

We greatly acknowledge to the Federal Ministry of Education, Science and Research (BMBF), KAI e.V, grant No. WIP 018328/M.

REFERENCES

Artigas, J., Bachler, B., Habedank, S., Taube, F., Franz, H., and Niedobitek, F. (1992) Comparative lectinhistochemical studies on paraffin- and glycol methacrylate-embedded CNS tissue specimen from AIDS autopsies. *Zentralblatt für Pathologie*, **138**, 272–277.

Barbieri, L., Battelli, M.G., and Stirpe, F. (1993) Ribosome-inactivating proteins from plants. *Biochimica et Biophysica Acta*, **1154**, 237–282.

Barre, A., Pfüller, U., and Rouge, P. (1996) Structure-Function relationships of amphipathic polypeptides: Viscotoxins of mistletoe, thionins and purothionins. In S. Bardocz, F.V. Nekrep and A. Pusztai, (eds), *COST 98. Effects of antinutrients on the nutritional value of legume diets*, Vol. 3, Luxembourg: Offices for Official Publications of the European Communites, pp. 129–136.

Barre, A., Pfüller, U., and Rouge, P. (1997) Molecular models for mistletoe viscotoxins and related plant thionins. Structural similarities with invertebrate defensins and toxins. In S. Bardocz, M. Muzquiz and A. Pusztai, (eds.), *COST 98. Effects of antinutrients on the nutritional value of legume diets*, Vol. 4, Luxembourg: Office for Official Publications of the European Communities, pp. 137–149.

Becker, H., Exner, G., and Schilling, G. (1978) Isolierung und Strukturaufklärung von 2′-Hydroxy-4′,6′-dimethoxychalkon-4-glukosid aus *Viscüm album* L. *Zeitschrift für Naturforschung*, **33c**, 771–773.

Becker, H. und Exner, J. (1980) Vergleichende Untersuchungen von Misteln verschiedener Wirtsbäume anhand der Flavonoide und Phenolcarbonsäuren. *Zeitschrift für Pflanzenphysiologie*, **97**, 417–428.

Bird, G.W.G. (1954) Observations on the interactions of the erythrocytes of various species with certain seed agglutinins. *British Journal of Experimental Pathology*, **35**, 252.

Bird, G.W.G. and Wingham J. (1973) The action of seed and other reagents on Erythrocytes. *Vox Sang.*, **24**, 48–57.

Brooks, S.A., Leathem, A.J.C., and Schumacher, U. (1997) *Lectin Histochemistry*, BIOS Scientific Publishers, Oxford.

Büssing, A., Stein, G.M., Wagner, M., Wagner, B., Schaller, G., Pfüller, U., *et al.* (1999a) Accidental cell death and generation of reactive oxygen intermediates in human lymphocytes induced by thionins from *Viscum album* L. *European Journal of Biochemistry*, **262**, 79–87.

Büssing, A., Vervecken, W., Wagner, M., Wagner, B., Pfüller, U., and Schietzel, M (1999b) Expression of mitochondrial Apo2.7 molecules and caspase-3 activation in human lymphocytes treated with the ribosome-inhibiting mistletoe lectins and the cell membrane permeabilizing viscotoxins. *Cytometry*, **37**, 133–139.

Carrasco, L., Vazques, D., Hernandez-Lucas, C., Carbonero, P., and Garcia-Olmedo, F. (1981) Thionis: Plant Peptides that Modify Membrane Permeability in Cultured Mammalian Cells. *European Journal of Biochemistry*, **116**, 185–189.

Debray, H., Wieruszeski, J.M., Strecker, G., and Franz, H. (1992) Structural analysis of the carbohydrate chains isolated from mistletoe (*Viscum album*) lectin I. *Carbohydrate Research*, **236**, 135–143.

Debray, H., Montreuil, J., and Franz, H. (1994) Fine sugar specificity of the mistletoe (*Viscum album*) lectin I. *Glycoconjugates Journal*, **6**, 550–557.

Edlund, U. (2000) Untersuchungen der Wechselwirkungen von Beerenpolysacchariden und Lektinen der weißbeerigen Mistel (*Viscum album* L.), Thesis, University Witten/Herdecke.

Edlund, U., Hensel, A., Fröse, D., Pfüller, U., and Scheffler, A. (2000) Polysaccharides from fresh *Viscum album* L. berry extract and their interaction with *Viscum album* agglutinin I. *Arzneimittel-Forschung/Drug Research* [in press].

Eifler, R., Pfüller, K., Göckeritz, W., and Pfüller, U. (1993) Improved procedures for isolation and standardization of mistletoe lectins and their subunits: lectin pattern of the European mistletoe. In J. Basu, M. Kundu, and P. Chakrabarti, (eds.), *Lectins: Biology, Biochemistry, Clinical Biochemistry*, Vol. 9. Wiley Eastern Limited, New Delhi, pp. 144–151.

Eschenburg, S., Krauspenhaar, R., Mikhailov, A., Stoeva, S., Betzel, C., and Voelter, W. (1998) Primary structure and molecular modeling of mistletoe lectin I from *Viscum album*. *Biochem Biophys Res Commun*, **247**, 367–372

Fernandez de Caleya, R., Gonzales-Pascual, B., Garcia-Olmedo, F., and Carbonero, P. (1972) Susceptibility of phytopathogenic bacteria to wheat purothionins *in vitro*. *Applied Microbiology*, **23**, 998–1000.

Franz, H., Haustein, B., Luther, P., Kuropka, U., and Kindt, A. (1977) Isolierung und Charakterisierung von Inhaltsstoffen der Mistel (*Viscum album* L.). *Acta Biol. Med. Germ.*, **36**, 113–117.

Franz, H., Ziska, P., and Kindt, A. (1981) Isolation and properties of three lectins from mistletoe (*Viscum album* L.). *Biochemical Journal*, **195**, 481–484.

Franz, H., Kindt, A. Eifler, R., Ziska, P., Benndorf, R., and Junghahn, I. (1983) Differences in toxicity and antigenicity between mistletoe lectin I and viscotoxin A3. *Biomedica and Biochimica Acta*, **42**, K21-K25.

Franz, H. (1985) Inhalsstoffe der Mistel (*Viscum album* L.) als potentielle Arzneimittel. *Pharmazie*, **40**, 97–104.

Franz, H. (1989) Viscaceae lectins. *Advances in lectin research*, **2**, 28–59.

Franz, H. (1991) Mistletoe lectins (2). *Advances in Lectin Research*, **4**, 33–50

Fukunaga, T., Kajikawa, I., Nishiya, K., Watanabe, Y., Takeya, K., and Itokawa, H. (1987) Studies on the constituents of the European mistletoe *Viscum album* L. *Chem Pharm Bull*, **35**, 3292–3297

Haltner, E., Borchard, G., and Lehr, C.-M. (1997) Use of Lectins for Drug Delivery – Absorption Enhancement by Lectin-mediated Endo- and Transcytosis. In J.M. Rhodes and J.D. Milton, (eds.), *Methods in Molecular Medicine – Lectin Methods and Protocols*. Humana Press Inc., Totowa, New Jersey, pp. 567–581

Hegnauer, R. (1966) *Chemotaxonomie der Pflanzen*, Vol. 4. Birkhäuser Verlag, Basel, pp. 430–438.

Hincha, D.K., Pfüller, U, and Schmitt, J.M. (1997) The concentration of cryoprotective lectins in mistletoe (*Viscum album* L.) leaves is correlated with leaf frost hardiness. *Planta*, **203**, 140–144.

Jäggy, C., Musielski, H., Urech, K., and Schaller, G. (1995) Quantitative determination of lectins in mistletoe preparations. *Arzneimittel-Forschung/Drug Research*, **45**, 905–909.

Jordan, E. and Wagner, H. (1986) Structure and properties of polysaccharides from *Viscum album* (L.). *Oncology*, **43** (Suppl. 1), 8–15.

Khwaja, T.A., Dias, C.B., and Pentecost, S. (1986) Recent studies on the anticancer activities of mistletoe (*Viscum album*) and its alkaloids. *Oncology*, **43** (Suppl. 1), 42–50.

Krauspenhaar, R., Eschenburg, S., Perbandt, M., Kornilov, V., Konareva, N., Mikailova, L., *et al.* (1999) Crystal structure of mistletoe lectin I from *Viscum album. Biochem Biophys Res. Commun.*, 257, 418–424.

Krüpe, M. (1956) Blutgruppenspezifische pflanzliche Eiweißkörper (Phythämagglutinine). Enke Verlag, Stuttgart.

Kuttan, G. and Kuttan, R. (1992a) Immunological mechanism of action of the tumor reducing peptide from mistletoe extract (NSC 635089) cellular proliferation. *Cancer Letters*, **66**, 123–130.

Kuttan, G. and Kuttan, R. (1992b) Immunomodulatory activity of a peptide isolated from *Viscum album* extract (NSC 635 089). *Immunological Investigations*, **21**, 285–296.

Lavelle, E.C., Grant, G., Pusztai, A., Pfüller, U., and O'Hagan, D.T. (2000) Mucosal immunogenicity of plant lectins in mice. *Immunology*, **99**, 30–37.

Lorch, E. (1993) Neue Untersuchungen über Flavonoide in *Viscum album* L. *abietis, album* and *austriacum. Zeitschrift für Naturforschung,* **48c**, 105–107.

Luther, P. (1976) The agglutination of human erythrocytes and mouse ascites tumor cells by extracts from mistletoe (*Viscum album* L.)]. *Acta Biol. Med. Ger.*, **35**, 123–126.

Luther, P. und Becker, H. (1987) Die Mistel – Botanik, Lektine, medizinische Anwendung. Springer-Verlag Berlin, Heidelberg.

Lutsch, G., Noll, F., Ziska, P., Kindt, A., and Franz, H.: (1984) Electron microscopic investigation on the structure of lectin I from *Viscum album* L. *FEBS Letters,* **170**, 335–338.

Ohta, N. and Yagashita, K. (1970) Isolation and structure of new flavonoids, flavoardorinin A, flavoyadorinin B and homoflavoyadorinin B in the leaves of *Viscum album* var. *coloratum* epiphysic on *Pyrus communis. Agric. Biol. Chem.*, **34**, 900–907.

Okumura, Y. and Sakurai A. (1973) Chemical studies on the mistletoe. II. The structure of viscumamid, a new cyclic peptide isolated from *Viscum album* Linn. var. *coloratum Ohwi. Bull. Chem. Soc. Jpn.*, **46**, 2190–2193.

Olsnes, S., Stirpe, F., Sansvig, K., and Pihl, A. (1982) Isolation and characterization of Viscumin, a toxic lectin from *Viscum album* L. (mistletoe). *Journal of Biological Chemistry*, **257**, 13263–13270.

Peumans, W.J., Verhaert, P., Pfüller, U., and Van Damme, E.J. (1996) Isolation and partial characterization of a small chitin-binding lectin from mistletoe (*Viscum album*). *FEBS Letters*, **396**, 261–265.

Peumans, W.J., Verhaert, P., Pfüller, U., and Van Damme, E.J.M. (1998) The chitin-binding mistletoe (*Viscum album*) agglutinin. In S. Bardocz, U. Pfüller and A. Pusztai, (eds.), *COST 98. Effects of antinutrients on the nutritional value of legume diets,* Vol. 5, Luxembourg: Office for Official Publications of the European Communites, pp. 63–68.

Pfüller, G. and Niedobitek, F. (1998) Binding of Mistletoe Lectins, Subunits and Commercial Mistletoe Preparations to Human Lymph Node Tissue. In S. Bardocz, U. Pfüller and A. Pusztai, (eds.), *COST 98. Effects of antinutrients on the nutritional value of legume diets,*

Vol. 5, Luxembourg: Office for Official Publications of the European Communites, pp. 125–138.

Pfüller, U., Franz, H., Pfüller, K., Junghahn, I., and Bielka, H. (1988) Selective inactivation of mistletoe lectin I and ricin using ethylammonium nitrate – A molten salt liquid at room temperature. In J. Kocourek and D.L.J. Freed, (eds.), *Lectins: Biology, Biochemistry, Clinical Biochemistry*, Vol. 6, Sigma Chemical Company, St. Louis, Missouri, pp. 299–304.

Pfüller, U. (1996) Immunmodulation durch Mistelinhaltsstoffe. In R. Scheer, H. Becker, P.A. Berg, (eds.), *Grundlagen der Misteltherapie. Aktueller Stand der Forschung und klinische Anwendung*. Hippokrates Verlag, Stuttgart, pp. 170–182.

Pusztai, A. and Bardocz, S. (1991) *Plant Lectins,* Cambridge University Press, Cambridge.

Pusztai, A. and Bardocz, S. (1995) *Lectins – Biomedical Perspectives*, Taylor and Francis, London.

Pusztai, A., Grant, G., Gelecsér, E., Ewen, S., Pfüller, U., Eifler, R., *et al.* (1998a) Effects of an orally administered mistletoe (type-2 RIP) lectin on growth, body composition, small intestinal structure, and insulin levels in young rats. *Nutritional Biochemistry*, **9**, 31–36.

Pusztai, A., Grant, G., Gelecsér, E., Ewen, S.W.B., and Bardocz, S. (1998b) Nutritional and metabolic effects of mistletoe lectin ML I (type 2 RIP) in the rat. In S. Bardocz, U. Pfüller and A. Pusztai, (eds), *COST 98. Effects of antinutrients on the nutritional value of legume diets.* Vol. 5, Luxembourg: Office for Official Publications of the European Communites, pp. 164–167.

Rhodes, J.M. and Milton, J.D. (1998) *Lectin Methods and Protocols.* Humana Press, Totowa, New Jersey

Richter, A. (1992) Viscumitol, a dimethylether of *muco*-inositol from *Viscum album. Phytochemistry,* **31**, 3925–3927.

Richter, A., and Popp, M. (1992) The physiological importance of accumulation of cyclitols in *Viscum album* L. *New Phytol.*, **121**, 431–438.

Sakurai, A. and Okumura, Y. (1979) Synthesis of viscumamide and its analogs. *Bull. Chem. Soc. Jpn.* **52**, 540–543.

Samtleben, R., Kiefer, M., and Luther, P. (1985) Characterization of the different lectins from *Viscum album* (mistletoe) and their structural relationships with the agglutinins from *Abrus precatorius* and *Ricinus communis.* In T.C.J. Bog-Hansen and J. Breborowicz, (eds.), *Lectins. Biology, Biochemistry, Clinical Biochemistry*, Vol. 4. Walter de Gruyter & Co., Berlin, New York, pp. 617–626.

Samulesson, G. (1974) Mistletoe toxins. *System. Zool.*, **22**, 566–569.

Schindler, H. (1955). *Inhaltsstoffe und Prüfungsmethoden homöopathisch verwendeter Arzneipflanzen.* Editio Cantor, Aulendorf.

Schink, M. (1990) Vergleichende Untersuchungen über das Proteinmuster der europäischen Mistel (*Viscum album* L.) mit Hilfe der Zwei-Dimensionalen Polyacrylamidgelelektrophorese. Thesis, University Hohenheim.

Schütt, C., Pfüller, U., Siegl, E., Walzel, H., and Franz, H. (1989) Selective killing of human monocytes by immunotoxins containing partially denatured mistletoe lectin I. *International Journal of Immunopharmacology*, **11**, 977–980.

Schumacher, U., Adam, E., Kretzschmar, H., and Pfüller, U. (1994) Binding of mistletoe lectins I, II and III to microglia and Alzheimer plaque glycoproteins in human brains. *Acta Histochemica*, **96**, 399–403.

Schumacher, U., Stamouli, A., Adam, E., Peddie, M., and Pfüller, U. (1995) Biochemical, histochemical and cell biological investigations on the actions of mistletoe lectins I, II and III with human breast cancer cell lines. *Glycoconjugates Journal*, **12**, 250–257.

Sharon, N. and Lis, H. (1989) *Lectins*, Chapmann and Hall, London.

Soler, M.H., Stoeva, S., and Voelter, W. (1998) Complete amino acid sequence of the B chain of mistletoe lectin I. *Biochem Biophys Res Commun*, **246**, 596–601.

Stein, G.M., von Laue, H.B., Henn, W., and Berg, P.A. (1998) Human anti-mistletoe lectin antibodies. In S. Bardocz, U. Pfüller, A. Pusztai (eds.), *COST 98. Effects of antinutritients on the nutritional value of legume diets.* Vol. 5, European Communities, Luxembourg, pp. 168–175.

Stein, G.M., Schaller, G., Pfüller, U., Schietzel, M., and Büssing, A. (1999a) Thionins from *Viscum album* L.: influence of viscotoxins on the activation of granulocytes. *Anticancer Research*, **19**, 1037–1042.

Stein, G.M., Schaller, G., Pfüller, U., Wagner, M., Wagner, B., Schietzel, M., *et al.* (1999b) Characterisation of granulocyte stimulation by thionins from European mistletoe and from wheat. *Biochem Biophys Acta*, **1426**, 80–90.

Stein, G.M., Edlund, U., Pfüller, U., Büssing, A., and Schietzel, M. (1999c) Influence of polysaccharides from *Viscum album* L. on human lymphocytes, monocytes and granulocytes *in vitro*. *Anticancer Res*, **19**, 3907–3914.

Stirpe, F. and Batelli, M.G. (1990) Toxic proteins inhibiting protein synthesis. In W.T. Shier and D. Mebs, (eds.), *Handbook of Plant Toxinology*. Marcel Dekker Inc., New York, pp. 279–308.

Sweeney, E.C., Palmer, R.A., and Pfüller, U. (1993) Crystallization of the ribosome inactivating protein ML I from Viscum Album (Mistletoe) complexed with ß-D-Galactose. *Journal of Medical Biology*, **234**, 1279–1281.

Sweeney, E.C., Tonevitsky, A.G., Palmer, R.A., Niwa, H., Pfüller, U., Eck, J., *et al.* (1998) Mistletoe lectin I forms a double trefoil structure, *FEBS Letters*, **431**, 367–370.

Teuscher, E. (1994). *Viscum album*. In R. Hänsel, K. Keller, H. Rimpler, and G. Schneider, (eds.), *Hagers Handbuch der Pharmazeutischen Praxis,* Volume 6, 5th ed., Springer Verlag, Berlin, pp. 1160–1183.

Tonevitsky, A.G., Toptygin, A.Yu., Pfüller, U., Bushueva, T.L., Ershova, G.V., Gelbin, M., *et al.* (1991) Immunotoxin with mistletoe lectin I A-chain and ricin A-chain directed against CD5 antigen of human T-lymphocytes; comparison of efficiency and specificity. *International Journal of Immunopharmacology*, **13**, 1037–1041.

Tonevitsky, A.G., Rakhmanova, V.A., Agapov, I.I., Shamshiev, A.T., Usacheva, E.A., Prokoph'ev, S.A., *et al.* (1995) The interactions of anti-ML I monoclonal antibodies with isoforms of the lectin from *Viscum album*. *Immunology Letters*, **44**, 31–34.

Tonevitsky, A.G., Agapov, I., Temiakov, D., Moisenovich, M., Maluchenko, N., Solonova, O., *et al.* (1999) Study of heterogeneity of lectins in mistletoe preparations by monoclonal antibodies to their A-subunits. *Arzneimittel-Forschung/Drug Research*, **49(II)**, 970–975.

Urech, K., Schaller, G., Ziska, P., and Giannattasio, M. (1995) Comparative study on the cytotoxic effect of viscotoxin and mistletoe lectin on tumour cells in culture. *Phytotherapy Research*, **9**, 49–55.

Van Damme, E.J.M, Peumans, W.J., Barre, A., and Rouge, P. (1998a) Plant lectins: a composite of several distinct families of structurally and evolutionary related proteins with diverse biological roles. *Critical Reviews in Plant Science*, **17**, 575–692.

Van Damme, E.J.M., Peumans, W.J., Pusztai, A., and Bardocz, S. (1998b) *Handbook of Plant Lectins*, J. Wiley and Sons, Chichester, New York, Weinheim Brisbane, Singapore, Toronto, pp. 417–421.

Vester, F. (1977) Über die kanzerostatischen und immunogenen Eigenschaften von Mistelproteinen. *Krebsgeschehen*, **5**, 106–114.

Wagner, H., Feil, B., und Bladt, S. (1984) *Viscum album* – die Mistel. Analyse und Standardisierung von Arzneidrogen durch Hochleistungsflüssigkeitschromatographie (HPLC) und andere chromatographische Verfahren (III). *Deutsche Apotheker Zeitung*, **124**, 1429–1432.

Wagner, H., Feil, B., Seligmann, O., Petricic, J., and Kalogjera, Z. (1986) Phenylpropanes and lignans of *Viscum album* cardioactive drugs V. *Planta Medica*, **2**, 102–104.

Zimmermann, R., Wahlkamp, M., Göckeritz, W., und Pfüller, U. (1996) Glykosylierungsmuster der Mistellektine. In R. Scheer, H. Becker, P.A. Berg, (eds.), *Grundlagen der Misteltherapie. Akueller Stand der Forschung und klinische Anwendung*. Hippokrates Verlag, Stuttgart, pp. 85–94.

9. BIOLOGICAL AND PHARMACOLOGICAL PROPERTIES OF *VISCUM ALBUM* L.

From Tissue Flask to Man

ARNDT BÜSSING

Krebsforschung Herdecke, Department of Applied Immunology, University Witten/Herdecke, Communal Hospital, Herdecke, Germany

Each indrawn breath is the exhaled hope of others.

INTRODUCTION

Application of *Viscum album* L., the white berry mistletoe, is one of the most widely used unconventional remedies in Europe to treat hypertension and arteriosclerosis, arthrosis and/or arthritis, and, even more important, cancer. Studies in cancer patients have shown significant increases of immunological parameters. Some 50 clinical studies, including historical, retrospective, prospective and randomized trials, reported extended survival times, improved quality of life, or tumour regression with mistletoe therapy. Although the quality of the trials is in most cases poor, several studies suggested a positive effect (for review see Kiene, 1989, 1996; Kleijnen and Knipschild, 1994). However, the majority of these studies do not fit the current standards of research. Gabius and co-writers (1994) stated an obvious discrepancy between the popularity of mistletoe extracts and their classification as a non-conventional treatment modality with unproven efficacy in oncology.

This review will present an overview on the biological and pharmacological properties of whole plant extracts from *Viscum album*, the most active compounds, and the clinical effects of mistletoe treatment.

HYPERTENSION

Experimental Investigation

During the 18th century, *Viscum album* was applied for "weakness of the heart" and oedema, and later on to treat hypertension and arteriosclerosis. In Japanese folk medicine, mistletoe (*Taxillus kaempferi*) has been prescribed as a hypotensive drug (Nanba, 1980). Also the American mistletoe (*Phoradendron flavescens*) was suggested to possess circulatory depressant actions (Hanzlik and French, 1924).

In dogs, Crawford (1905, 1911, 1914) reported a temporary fall of blood pressure, even if both *vagus* nerves were cut, followed by a marked and sustained raise together with increases in pulse rate and diuresis by the intravenous injection of an extract from American mistletoe (*Phoradendron flavescens*); the initial fall of blood pressure was prevented by the acetylcholine inhibitor atropine. Thus, Crawford and Watanabe (1914) suggested the presence of choline or acetylcholine. Both, *Phoradendron flavescens* and *Phoradendron villosum* produced temporary hypotension, and an initial increase in respiration and pulse rate, peripheral vasoconstriction, reduction of the volume of peripheral organs, and decrease of the cardiac volume and urine output (Hanzlik and Frech, 1924). After that, a quick recovery of the blood pressure was observed, while the pulse rate remained accelerated; however, respiration was reduced in rate and amplitude, or even stopped. As also the intestines and uteri of non-pregnant animals responded with increased tonus and contractility, Hanzlik and French (1924) suggested a direct stimulation of muscle cells by the drug.

The hypotensive effects of European mistletoe were investigated by Gaultier and Chevalier in man and animals (Gaultier, 1907, 1910). Injection of an aqueous extraction from *Viscum album* leaves in frogs produced bradycardia, negative dromotropic effects resulting in irregularities of heart rhythm, and arrest of the heart in systole (Ebster and Jarisch, 1929). Also in cats and rabbits, intravenous application of *Viscum album* resulted in hypotension, bradycardia, arrhythmia and cardiac arrest in systole (Ebster and Jarisch, 1929; Jarisch und Henze, 1937; Henze and Ludwig, 1937; Jarisch, 1938; Enders, 1940; Zipf, 1950), and constriction of the intestine and uterus (Enders, 1940). A continuous decrease of the arterial blood pressure was associated with a constant raise of the venous blood pressure until cardiac arrest (Enders, 1940). 3 to 4 mg of the *Viscum album* extract (VA-E) killed cats and rabbits (Enders, 1940); however, rats were killed by an intravenously applied aqueous VA-E from leaves at 60 to 800 mg/kg body weight (BW) (Ebster and Jarisch, 1929). These results indicate a distinct affection of heart activity by compounds from *Viscum album*. However, affection of the heart by *Viscum album* was suggested not to depend on the sympathic or parasympathic system but to be due to a direct affection of the heart muscle cells (Enders, 1940).

Apart from the hypotensive effect of aqueous VA-E from the dried leaves of mistletoe grown on different host trees in dogs, cardioxicity was prevented by oral application, and by heating (Pora *et al.*, 1957), indicating a proteinic-nature of the relevant compounds. Indeed, oral application of *Viscum album* concentration 3× higher than those applied intravenously did not result in cardiac arrest (Enders, 1940), indicating the degradation of the toxic compound in the stomach and/or intestine.

It became evident, that the effects of mistletoe on blood pressure and heart activity may be due to distinct compounds present in *Viscum album*. The hypotension-inducing substance present in *Viscum album* was suggested to be acetylcholine (Müller, 1932). Indeed, Winterfeld and Kronenthaler (1942) verified the presence of acetylcholine and choline. Later on, Winterfeld and Rink (1948) were able to attribute the cardiotoxic effects of VA-E to a compound termed "viscotoxin". The lethal concentration of this toxin was found to be 0.5 mg/kg BW in rabbits (Winterfeld and Bijl, 1948), and 0.2 to 0.4 mg/kg BW in rats (Zipf, 1950). The animals died with dyspnoe, cramps, and systolic heart arrest. The cardiotoxic component was recognised by Zipf (1950) to be identical with the necrosis-inducing compound described by Koch (1938a). Samuelsson (1959) reported that the transient hypotensive effects of VA-E were caused by choline and γ-aminobutyric acid, while the main effect was a toxic manifestation which could be attributed to the viscotoxins. Later on, also flavonoids such as quercitrin, rhamnetin-3-O-rhamnoside, rhamnocitrin-3-O-rhamnoside, homo-flavoyadorinin-B and acetylcholine, isolated from Japanese mistletoes (*Taxillus yadoriki* DANSER, *Hyphear tanakae* HOSOKAWA, and *Viscum album* LINNE var. *coloratum* OHWI), were suggested to be responsible for the temporary hypotensive response (Fukunaga *et al.*, 1989).

Viscotoxin when applied intravenously in cats at 35 μg/kg BW, and also a thionin from *Phoradendron serotinim*, termed Phoratoxin, at 400 μg/kg BW, resulted in vasoconstriction of blood vessels, reflex bradycardia, negative inotropic effect on the heart and thus, hypotension (Rosell and Samulesson, 1966). However, viscotoxin-induced bradycardia was completely prevented by bilateral vagotomy, indicating its reflex origin, while the decrease of the contractile force of the heart was only transiently influenced by vagotomy, indicating that viscotoxin exerts its negative inotropic effect directly upon the heart muscle (Rosell and Samuelsson, 1966). Indeed, a thionin from *Phoradendron tomentosum macrophyllum* (Phoratoxin B) was found to irreversible depolarise the membranes of frog skeletal and heart muscles as well as papillary heart muscle (Sauviat *et al.*, 1985).

Clinical Results

To treat hypertension and arteriosclerosis, it was recommended to extract *Viscum album* in cold water overnight (1 tea spoon of the drug per cup of water) and to drink this drug three times a day (Ripperger, 1937), while others used 30–40 g of the drug applied in 1 l wine (Pic and Bonnamour, 1923). Several reports indicate beneficial effects of *Viscum album* treatment in patients with hypertension (Goetsch, 1930; Strauss, 1931; Orlowski, 1932; Mattausch, 1938), however, most of them are retrospective treatment observations without clear descriptions of the results or treated patients.

In a study reported by Pora *et al.* (1957), 100 patients with hypertension were treated for 14 days with 1 g of a powder from leaves of a mistletoe grown on *Malus domestica*. In response to the treatment, the blood pressure markedly decreased (however, only data from 7 and 9 patients were presented), and the patients general well-being increased. In none of the patients, cardiac or respiratory side-effects were observed.

In an open study including 120 patients with slight or medium hypertension, oral application of (unspecified) VA-E resulted in a significant decrease of the systolic blood pressure (12 to 15 mm Hg) within 6 weeks of treatment, while the heart frequency did not change (Bräunig *et al.*, 1993). Drops (600 mg of a drug not further defined), juice (2,700 mg) and tablets (2,850 mg) were the same effective.

Due to the lack of clearly described studies, the evaluation of data on the blood pressure reducing properties of VA-E is unsatisfactory. It was stated that mistletoe treatment of hypertension did not result in rapid release of symptoms and is effective only by long-term treatment (Mattausch, 1938). Since in most cases a significant and stable reduction of high blood pressure is required, the clinical relevance of the mild anti-hypertensive effects of orally applied VA-E remains unclear.

DIABETES

Treatment of Diabetes mellitus was traditionally relied on dietary measures which included the use of plant therapies. Extracts from mistletoes were used in Nigerian folk medicin (Obatomi *et al.*, 1994) and West Indies (Peters, 1957). In fact, aqueous extracts from the leaves of *Loranthus bengwensis* L. grown on lemon (*Citrus limon* L. Brum f., Rutaceae) and guava trees (*Psidium guajava* L., Myrtaceae) significantly decreased serum glucose levels in non-diabetic and streptozotocin-induced diabetic rats, while mistletoe parasitic on jatropha (*Jatropha curcas* L, Euphorbiaceae) did not (Obatomi *et al.*, 1994). Similarly, an extract from *Viscum album* supplied as 6.25% by weight of the diet for 9 d in streptocotocin diabetic mice resulted in relief of polydipsia, hyperphagia and body weight loss, while the plasma glucose or insulin concentrations remained unchanged (Swanston-Flatt *et al.*, 1989). Further, an aqueous extract from dried *Viscum album* leaves and stems evoked a stimulation of insulin secretion from clonal pancreatic B cells (Gray and Flatt, 1999). These treatments warrant further evaluation.

ARTHROSIS

The VA-E *Plenosol* is an aqueous extract from mistletoes grown on poplar, and is used to treat patients with arthrosis and/or arthritis of several joints (Elsner, 1940; Legel, 1942; Drüen, 1943; Hoffmann-Axthelm and Zellner, 1954; Vasold and Händel, 1954; Sommer, 1957; Müller, 1962; Groh, 1965; Schimmel, 1971; Reis, 1986; Irmler, 1989; Kienholz, 1990; Kuban, 1991; Zell *et al.*, 1993). It is strictly recommended to apply the extract intracutaneously, as subcutaneous applications were suggested to be less effective, and to increase the concentration individually in response to the local reaction (Elsner, 1940; Vasold and Händel, 1954; Sommer, 1957).

Within this extract, standardised on "necrosis-inducing units" as measured in mice, an antibody-inducing "phytotoxin" and a necrosis-inducing compound was suggested (Koch, 1938a). Although these compounds were not investigated in further detail, it is tempting to speculate that the observed effects might be ascribed to at least mistletoe lectin and viscotoxins.

Application of the drug results in a release of severe pain, increase of motility and improved endurance of the affected joints; the side effects which may occur are inflammatory local reactions, slight increase of body temperature, swelling of local lymph nodes, night-sweating, headache, and exhaustion, (Elsner, 1940; Vasold and Händel, 1954; Müller, 1962; Groh, 1965; Schimmel, 1971; Müller and Müller, 1978a,b; Reis, 1986; Irmler, 1989; Kienholz, 1990; Zell *et al.*, 1993), without affections of the blood pressure (Elsner, 1940). After injection, transient lymphocytosis and eosinophilia were observed (Drüen, 1943). Within 4 weeks of treatment, an increase of relative B cell numbers and CD4/CD8 ratio was recognised; however, other T cell subsets, complement components and immunoglobulins did not significantly change (Zell *et al.*, 1993). The underlying mechanisms resulting in anti-inflammation and reduction of pain are still unclear. The local reaction after the weal setting is suggested to result in a signalling *via* vegetative nerves to spinal ganglions, and subsequent increase of blood circulation in the segmental periphery (Elsner, 1940; Groh, 1965). Legel (1942) suggested a histamine release in the area of injection which may inactivate choline esterase and thus, may result in an accumulation of acetylcholine.

An improvement of burden in patients with arthrosis deformans was observed in several joints even after injection in the near area of a single joint (Roderfeld, 1950), indicating that the underlying mechanisms may also involve distinct systemic effects. An involvement of β-endorphin and encephalins or an inhibition of pro-inflammatory cytokines might be assumed, but was never proved by published data.

Most published studies on the effects of *Plenosol* treatment in arthrosis and/or arthritis are case reports or retrospective analyses with poor quality (in most cases no controls). In large groups of patients (ranging from 90 to 319 patients), several authors report an improvement of symptoms by the application of *Plenosol* in 74% to 91% (Elsner, 1940; Vasold and Händel, 1954; Müller, 1962; Grisar, 1969; Sickel,

Table 1 Results of arthrosis treatment with *Plenosol* (according to Kuban, 1991).

Gonarthrosis	*Prior to treatment*		*After treatment*			
(n = 21)	*Disturbance of motility*		*Disturbance of motility*		*Labour pain*	
	not present	3	not present	18	Improvement	8
	mild	15	mild	1	mild	13
	severe	3	no improvement	2	severe	0
Periarthrosis humeroscapularis	*Prior to treatment*		*After treatment*			
(n = 12)	*Disturbance of motility*		*Disturbance of motility*		*Labour pain*	
	not present	0	not present	3	Improvement	0
	mild	3	mild	9	mild	11
	severe	9	no improvement	0	severe	1

1971; Reis, 1986; Kienholz, 1990). Most of these patients were reported to be treated with several other therapy modalities without any significant benefit. In patients with Gonarthrosis (osteoarthritis of the knee) and Periarthrosis humeroscapularis, application of the drug resulted in a general improvement of burden in all patients (Table 1) (Kuban, 1991). However, patients with neuralgia and "rheumatism of the muscles" were observed not to benefit from this therapy (Elsner, 1940).

In 483 patients with arthrosis of the knee joint, Groh (1965) compared the therapeutical results of different therapy modalities such as physical therapy, injection of cortison into the joint, i.c. injection of *Plenosol*, and combinations thereof. Within 300 patients evaluated, the therapeutical results, such as relief from stress, decongestion, immobilisation, and carefully dosed movement without fully stress, were improved by the injection of *Plenosol* or the combination of *Plenosol* and cortison, as compared to cortison injection or physical therapy alone.

In a randomised, double-blind, placebo-controlled study enrolling 61 patients with non-active gonarthrosis, a pasture containing mistletoe and comfrey (*Symphytum officinalis*) was applied to the patients. Within 8 weeks, both placebo pasture lacking active compounds and verum pasture (Syviman^{R} N) resulted in a significant improvement of pain (Schmidtke-Schrezenmeier *et al.*, 1992). However, especially evening pain decreased in the verum group as compared to the control group. Thus, it is not mistletoe alone which is effective but also depending on its intracutaneous application at the affected joints. It remains to be clarified whether this intracutaneous treatment of arthrosis and/or arthritis with *Viscum album* is an unspecific segment therapy or also involves distinct systemic effects.

CANCER

Mistletoe was believed to influence life and death. Recent scientific research has confirmed the folklore with evidence that mistletoe extracts (1) induce apoptotic killing of cultured tumour cells and lymphocytes, (2) stimulate the immune system, and (3) protect DNA against chemotherapy and radiation (vor review see Büssing, 1999).

Cytotoxicity

The most prominent effect of VA-E is their cytotoxic property. The first investigations of cytostatic and/or cytotoxic effects were conducted with plants and found that synergistic activity of the alkaloid colchicin and a dialysate of *Viscum album* (i.e., no viscalbin alkaloid present) resulted in greater growth-inhibition to roots and root-hairs of wheat germ than a comparable amount of dialysed VA-E or colchicum (Havas, 1937). Simultaneous administration of colchicum and a pressed sap of *Viscum album* that was rich in viscalbin, produced no such effects, however, the total weight of the shoots increased (Havas, 1937). After application of sap from *Viscum album*, Crown-Gall tumours normally induced by *Agrobacterim tumefaciens* were also rejected (Havas, 1939).

Animals

Intracutaneus administration of high drug concentrations (Koch, 1938a) to mice and rabbits was less toxic compared to an intravenous application. However, intra- or peritumoural injection in murine Ehrlich ascites carcinoma resulted in tumour necrosis without significant affections of the healthy tissue, and in several cases, complete remissions were observed (Koch, 1938b). Using a dried VA-E and an isolated compound termed "Toxomelin", Chernov (1954, 1955) observed decreased growth of tumour cell lines (i.e., M1 sarcoma, Sarcoma 180, Ehrlich ascites sarcoma). In rabbits, Toxomelin reduced the growth rate of Brown-Pearse tumours and slightly decreased incidence of metastases (Chernov, 1955). In a series of experiments with mice, 6 per group, subcutaneous injection of a mixture of tumour ascites combined with increasing concentrations of VA-E (*Iscador* and *Plenosol*) that were incubated for 30 minutes at 4°C, prevented the development of Ehrlich ascites carcinoma in a dose-dependent manner (Seeger, 1965a-c). Histological analyses revealed lymphocytosis, demarcation, encapsulation of necrotic tumours, and necrosis of the tumour in several cases. Although not confirmed, these effects were hypothesized to be the result of increased activity of the reticulo-endothelial system induced by a proposed phytotoxin present in the drug extracts (Seeger, 1965a-c). This antigenic phytotoxin resembled acetylcholin, but might be mistletoe lectin (ML).

Several experiments using tumour-bearing animals showed impressive reduction of tumour growth and/or increased survival with the application of mistletoe therapy (Selawry *et al.*, 1959, 1961; Franz, 1986; Nienhaus and Leroi, 1970; Nienhaus *et al.*, 1970; Beuth *et al.*, 1991; Drees *et al.*, 1996; Mengs *et al.*, 1998). Intraperitoneal injection of a saline VA-E (2 h at 40°C) and the fermented VA-E (*Iscador*) effectively inhibited development of several tumour types (Carcinoma 755, Leukemia 1210, Ehrlich carcinoma, Sarcoma 180) inoculated subcutaneously in mice (Selawry *et al.*, 1959). Tests for tumour inhibition of several fractions from *Viscum album* found that the most active fraction was present in the native protein fraction (Selawry *et al.*, 1961). Prophylactic subcutaneous injections of a fermented VA-E (*Iscador* Q i.p.) or selected basic protein fractions from *Viscum album* prior to transplantation of Sarcoma 180 cells into mice significantly reduced or prevented tumour growth (Nienhaus and Leroi, 1970; Nienhaus *et al.*, 1970). Encapsulation of the tumour in firm connective tissue was accompanied by an increased weight of thymus, indicating an activation of immunocompetent cells. However, other studies on the subcutaneous treatment with the fermented drug extract *Iscador* did not reveal significant inhibition of subcutaneous transplants of B16 melanoma or leukemia L1210 in BDF_1 mice (Khwaja *et al.*, 1986) and leukemia L5222 in BDIX rats (Berger and Schmähl, 1983). These conflicting results were thought to be related to differences in tumour strains, mode of drug administration, and/or age of animals (Khwaja *et al.*, 1986); however, one may also suggest that the results may be related to differences in the biological activity of the compounds present in a given drug extract.

The purified ML I, administered subcutaneously twice and four times per week, significantly decreased the number of lung (L-1 sarcoma) and liver colonies (RAW 117-H 10 lymphosarcoma) in BALB/c mice after tumour cell inoculation (Beuth *et al.*, 1991). ML I used either subcutaneously (0.014, 0.14 and 1.4 μg/kg BW) or orally (70 μg/kg BW) reduced the number of lung metastases in mice (C57 Bl-6) injected with Lewis lung tumor cells, and raised the level of serum TNF-α (Kubasova *et al.*, 1998). However, ML I treatment of mice receiving radiation prior to or after the injection of tumor cells strongly increased the number of lung metastases (Figure 1). In another study (Kunze *et al.*, 1997, 1998), ML I at 1 ng/kg BW applied twice week subcutaneously for 15 month was ineffective to inhibit N-methyl-N-nitrosurea-induced urinary bladder carcinogenesis in rats.

Intravenous application of VA-E in allogenic BALB/c mice (n = 5 per group) receiving intravenously immunogenic B16 melanoma cells reveiled a significant reduction of melanoma cells in the pulmonary lavage at necropsy 3 weeks later (Mengs *et al.*, 1998), while, however, the decreased number of melanoma cells in the lung tissue did not significantly differ from the number of melanoma cells in placebo-treated mice (Weber *et al.*, 1998). Anyway, the number of lung macrophages in bronchoalveolar fluid and thymocytes in the thymus gland increased in response to VA-E (Mengs *et al.*, 1998; Weber *et al.*, 1998). Thus, it is tempting to speculate that intravenous VA-E-treatment may prevent the development of injected melanoma cells in the lung by an activation of macrophages.

Recent reports indicate that even an oral uptake of high ML I concentrations may have an inhibitory effect on tumours. In NMRI mice injected subcutaneously Krebs II lymphosarcoma cells and prefed on diets containing 0.42, 0.83 and 1.67 mg ML I per g diet (6 g/mouse average daily food intake), the tumour growth decreased in a dose-dependent manner within 10 d (Pryme *et al.*, 1998b). ML I stimulated a hyperplastic growth of the small intestine as observed also in mice fed on PHA (Pryme *et al.*, 1998a). The ML did bind strongly to the surface of jejunal villi with a moderate degree of endocytosis, while the lumenal surface of the crypts was strongly stained and endocytosis was apparent; apoptotic bodies were detected usually in the proximal half of the villi (Ewen *et al.*, 1998). Furthermore, rats fed on diet containing 67 or 200 mg ML I per kg BW for 10 d induced hypertrophy of the pancreas (with depression of circulating insulin level), lungs, and small intestine, and increased systemic catabolism (Pusztai *et al.*, 1998). These animals exhibited higher TNF-α and IL-1β level as compared to the controls (Pusztai *et al.*, 1998), indicating that immunomodulation by the ML may not be restricted to parenteral application alone. The lymphosarcoma-bearing mice fed on lactalbumin and ML I at 10 mg showed a complete histological dissappearance of the transplanted tumour cells in 3 of 5 animals (Ewen *et al.*, 1998). Thus, in contrast to its parenteral toxicity, ML when applied orally did not display overt toxicity symptoms but was capable of affecting tumour growth, probably by depriving the tumour of nutritional and growth factors by maximal crypt hyperplasia.

Taken together, VA-E and defined components were reported to inhibit tumour development in the animals after adequate application of the drug. However, in most cases it is unclear, whether the remission of highly immunogenic tumours after injection of tumor cells reflects a clinical relevant anti-tumour activity or not. Effectiveness may also depend on the mode application and composition of the drug.

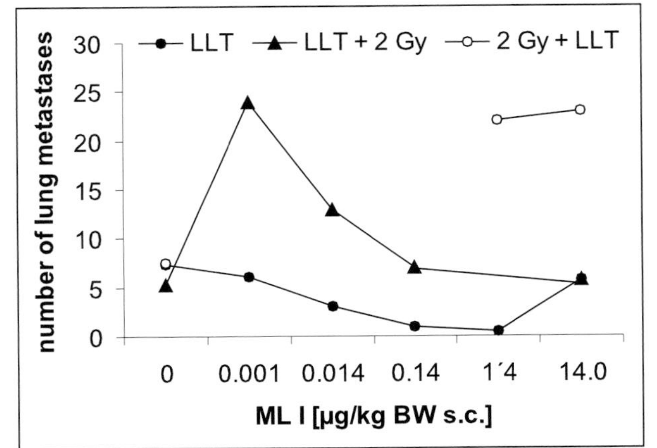

A

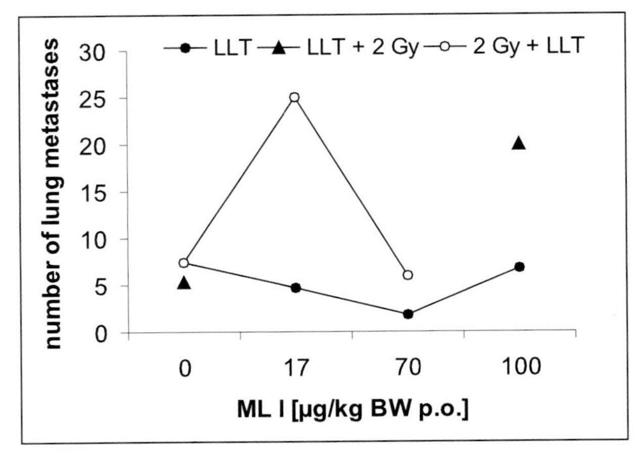

B

Figure 1 Effects of ML I applied subcutaneously (A) and orally (B) 4× with 5-day intervals in female C57 Bl-6 mice receiving Lewis lung tumor cells (LLT, 5×10^5) intramuscularly and radiation (2 Gy), as indicated. The number of lung metastases (mean value of 3–7 mice) was measured in animals sacrified after 19 d. Modified according to Kubasova *et al.* (1998).

Cell cultures

Various mistletoe preparations have been tested for their impact on the growth of cultured murine and human tumour cell lines, fibroblasts, and lymphocytes (Selawry *et al.*, 1961; Vester *et al.*, 1968; Khwaja *et al.*, 1980b, 1986; Hülsen and Mechelke, 1982; Hülsen *et al.*, 1986; Ribéreau-Gayon *et al.*, 1986a,b; Kuttan *et al.*, 1988; Doser *et al.*, 1989; Jung *et al.*, 1990; Dietrich *et al.*, 1992; Jurin *et al.*, 1993; Janssen *et al.*, 1993; Beuth *et al.*, 1994b; Zarkovic *et al.*, 1995, 1998; Büssing *et al.*, 1995b, 1996d; Büssing and Schietzel, 1999; Urech *et al.*, 1995; Schaller *et al.*, 1996; Drees *et al.*, 1996; Hostanska *et al.*, 1998; Staak *et al.*, 1998; Lenartz *et al.*, 1998). Overall, the studies found that the degree of cell growth inhibition varied according to the cell type, method of preparation, host tree, and subspecies of mistletoe plants. In fact, the ML content varies within the season and within different host trees (Franz, 1989).

In mice and cell cultures, fermented VA-E were less cytotoxic than the fresh plant extracts (Bloksma *et al.*, 1982; Khwaja *et al.*, 1986; Jung *et al.*, 1990, Büssing *et al.*, 1996d), but the drug extracts containing the squeezed sap were more cytotoxic than aqueous extracts (Büssing *et al.*, 1996d; Büssing and Schietzel, 1999). These differences may depend on the sensitivity of the cell line, since fermented drugs contain low amounts of ML and are less cytotoxic to the leukemic Molt 4 cells than fresh plant extracts but slightly more potent in inhibiting the growth of hepatoma tissue culture (HTC) cells (Ribéreau-Gayon *et al.*, 1986). Using purified ML, the Molt 4 cells were more sensitive to the MLs than the HTC cells (Ribéreau-Gayon *et al.*, 1986) or Yoshida sarcoma cells (Urech *et al.*, 1995).

According to anthroposophic philosophy, mistletoe from the apple tree is recommended to treat women (i.e., with breast cancer). Although this sounds peculiar, uptake of [^{3}H]-thymidine in the DNA of mitogen-activated lymphocytes from healthy females was more effectively suppressed by the addition of a VA-E from apple trees than that of healthy men (Büssing *et al.*, 1995b). Furthermore, sex-specific differences in the responses to VA-E were reported by Hülsen and co-workers (Hülsen *et al.*, 1992; Hülsen and Born, 1993). Because of the complexity of actions, several researchers have attempted to define and purify the main biologically active components present in *Viscum album*.

Mechanisms of Cell Death: Apoptosis and Necrosis

Apoptosis or programmed cell death is the mechanism by which superfluous or damaged cells are removed in most organ systems. This applies to embryogenesis, organogenesis, maintenance of normal tissue structure, responses to mild chemical or physical damage, neutrophils following ingestion of bacteria, and AIDS pathogenesis. Apoptotic cells are driven into death by an active process that involves activation of several calcium-dependent endonucleases which bind to the internucleosomal spacer, resulting in the fragmentation of the DNA. These cells are morphologically characterised by condensation and fragmentation of cell nuclei, cytoplasmic densification, nuclear membrane blebbing, and breakdown of the nucleus into discrete fragments; however, integrity of cytoplasmic membranes and organelles is preserved. On the other hand, necrosis is a "non-specific" mode of cell death induced by cell membrane affections (complement attack, severe hypoxia, hyperthermia, lytic viral infection, and several toxic chemicals) which cause ions and water efflux, resulting in disruption of cytoplasmic and nuclear membranes, swelling of mitochondria, and floculation of chromatin (for review see Schwartz and Osborn, 1993).

A wide variety of stimuli can trigger cell death, such as T cell receptor signalling, binding of the APO-1/Fas ligand, glucocorticoides, TNF-α or transforming growth factor-β to their receptors on appropriate cells, DNA strand breakage *etc.* (reviewed by Schwartz and Osborn, 1993; Green and Scott, 1995; Green and Martin, 1995). The "decision" of a cell to die is under the control of a number of different pathways considered as cellular "sensors" of apoptosis-inducing signals (such as p53 and others), which in turn trigger the central mechanisms leading to cell death (for review see Green and Martin, 1995). Most of the defects in apoptosis; which may contribute to the transformed state of a cell, are suggested to focus on the triggering or sensing of apoptosis-inducing signals (Green and Martin, 1995).

There is growing evidence that the susceptibility of malignant neoplasms to undergo apoptosis in response to different therapeutic modalities may be used in predicting clinical response. Arends *et al.* (1994) observed that high apoptotic rates in immortalised rat fibroblasts injected subcutaneously into immunosuppressed mice resulted in slowly growing fibrosarcomas with high ratios of apoptosis to mitosis and little necrosis, while lines with low apoptotic rates *in vitro* generated rapidly expanding tumours with high mitotic rates, extensive necrosis, and little apoptosis relative to mitosis. Thus, any treatment or condition that favour apoptosis may have desirable effects.

Whole plant extracts

VA-E differ in regard of their cytotoxic activity (Büssing *et al.*, 1996d; Büssing and Schietzel, 1999). This effect is probably not host tree-specific but dependent on the manufacturing process and thus, on biologically active compounds which in turns may depend on the host tree and time of harvest. VA-E were recognized to induce apoptosis of cultured cell lines within 24 to 72 h (Janssen *et al.*, 1993; Büssing *et al.*, 1996c,d, 1997, 1998a,b; Büssing and Schietzel, 1999). In human lymphocytes, the apoptosis-inducing property of the *Iscador* extracts and *ABNOBAviscum* extracts strongly correlated with their ML content, while, however, the ML-rich *Helixor* extracts did not (Büssing and Schietzel, 1999). Although Koch (1938a) observed the VA-E from winter harvests to possess a higher cytotoxic potential (probably due to a higher ML content) than those from summer harvests, however, no clear-cut correlation between defined components and the apoptosis-inducing potential of the aqueous extracts from the winter and summer harvest was found (Büssing and Schietzel, 1999). In fact, the winter extract from pine tree contained significantly higher amounts of ML but was the same effective the summer extract containing a lower amount of ML. It is tempting to speculate the involvement of other components which may modulate the activity of ML.

Defined components

During the 1960s, *Vester* and co-workers isolated and purified carcinostatic protein fractions from *Viscum album* (Vester and Nienhaus, 1965), which were recognised, in part, later on as ML and other proteins (Franz, 1986). The ML differ in molecular weight and carbohydrate specificity (Franz *et al.*, 1981; Franz, 1986). The ML were found to be potent inducer of apoptosis (Janssen *et al.*, 1993; Büssing *et al.*, 1996c, 1997, 1998a,b, 1999b,f; Hostanska *et al.*, 1996–1997; Möckel *et al.*, 1997; Bantel *et al.*, 1999), while the viscotoxins were found to affect cell membranes and thus induce accidental (necrotic) cell death with membrane permeabilisation, degradation of cytoplasm and chromatin, swelling of mitochondria with loss of their cristae, and generation of reactive oxygen intermediates (ROI) within 1 to 2 h, followed by secondary apoptosis-associated events (Büssing *et al.*, 1998c, 1999b,c,f).

Whatever the underlying mechanisms are, there may be at least two distinct pathways of ML-mediated cell death: (1) direct induction of apoptosis in response to an inhibition of protein synthesis by the enzymic ML A chain (reviewed by Büssing, 1996e), and (2) indirect induction of apoptosis in Fas^+ tumour cells by ML-activated $FasL^+$ T cells (Büssing *et al.*, 1999e).

Immunomodulation by Mistletoe

Based on the studies of Koch (1938b) and Seeger (1965a-c), the biological effects of *Viscum album* are defined not only by direct cytotoxicity, but also indirect immune activation. The first reports that suggest VA-E may modulate the immune response described splenomegaly and increased clearance of colloidal carbon by phagocytes after intravenous or intraperitoneal injection of the fermented VA-E *Iscador* (Zschiesche, 1966; Bloksma *et al.*, 1982) and increased thymus weight in mice (Nienhaus and Leroi, 1970; Nienhaus *et al.*, 1970). Other studies found augmentation of delayed type hypersensitivity (DTH) of mice immunized with red blood cells from sheep (SRBC); however, the response was present only when SRBC were mixed with *Viscum album* and the subcutaneous or intracutaneous injections were *Iscador* (Bloksma *et al.*, 1979, 1982; Khwaja *et al.*, 1980). Another indication that VA-E activated immune function became evident by the augmentation of splenic plaque-forming cells (PFC) three days after i.p. immunization with SRBC mixed with this drug extract (Bloksma *et al.*, 1979). Thus, intraperitoneal immunization of mice with SRBC induced a humoral response (PFC) and sensitized for DTH. Further experiments found that neither fresh plant juice from mistletoe grown on apple trees in the summer nor the polysaccharides from *Viscum album* alone induced DTH to SRBC; however, the combined extract and polysaccharides stimulated the antibody-formation (PFC) (Bloksma *et al.*, 1982). The fermented drug extract *Iscador* M, when applied subcutaneously, induced an early nonspecific inflammation with footpad swelling in mice that disappeared at 72 h, but an unfermented VA-E caused minor swelling 6 h after injection and gradually increased to a dose dependent optimum at 72 h. Application of polysaccharides from *Viscum album* did not result in paw swelling (Bloksma *et al.*, 1982).

Since the fresh plant extract used for the *Iscador* drug is fermented by *Lactobacillus plantarum*, it was unclear if the observed immune responses were due to contamination by microorganisms. However, *Iscador* and non-viable lactobacilli did not stimulate DTH to SRBC but stimulated the humoral response, however, viable lactobacilli stimulated the cellular response but failed to influence the PFC response (Bloksma *et al.*, 1979). Further, *Iscador* rendered bacterium-free by centrifugation retained its adjuvanticity for DTH, while sterile-filtration of the drug extract reduced inflammatory and adjuvant properties (Bloksma *et al.*, 1979). Thus, it is unclear whether these soluble components are of plant or microbial origin. Subsequent reports from Bloksma and collegues (1982) suggest that the activities of the fermented drug extract *Iscador* on immune-related reactions also were attributed to the fresh plant extract with polysaccharides and lactobacilli. However, nowadays *Iscador* extracts are completely free of bacterial contamination.

Recently published findings of Antony and co-workers (1999) indicate that metastatic B16F10 tumour-bearing mice treated with both, spleen cells activated *in vitro* with VA-E (*Iscador* M) or from mice treated intravenously with the extract showed a significant inhibition of tumour nodule formation and increased survival. These findings underline the notion that several effects of *Viscum album* treatment may be based on an activation of immune cells rather than direct cytotoxicty.

Cellular immune system

In breast cancer patients, Hajto (1986) observed an induction of fever, lymphocytopenia, and an increased number of juvenile granulocytes and zymosan-induced oxidative burst of granulocytes 6 h after intravenous application of the VA-E *Iscador* M. Within 24 h, uptake of [^{3}H]-thymidine in the DNA of mitogen-stimulated lymphocytes, natural killer (NK) and antibody-dependent cell-mediated cytotoxicity (ADCC) in the peripheral blood, and numbers of large granular lymphocytes (LGL) increased but returned to baseline levels by 48 to 72 h. After further *in-vivo* experiments with rabbits, the effects were ascribed to ML I since i.v. administration of ML I or its lectin B chain enhanced NK cytotoxicity and the number of LGL from 24 to 72 h (Hajto *et al.*, 1989). A single i.v. injection of ML I or mistletoe extract *Iscador* resulted in an increased number of LGL, granulocytes, and oxidative burst stimulated with opsonized zymosan within 24 h in 3 rabbits (ML I at 0.8 ng/kg BW). The effect was also observed in 14 breast cancer patients treated with *Iscador* (0.33 mg/kg BW corresponding to 1.65 ng/kg ML I). Similar, but less impressive results were observed in 17 breast cancer patients who received a s.c. injection of an unfermented VA-E (*Iscador* Q FrF at 0.17 mg/kg corresponding to 12 ng/kg ML I). From these studies, an "optimal" ML I concentration of 1 ng/kg BW ML I was postulated. In a later study by Hajto and co-workers, however, this "optimal" immunomodulatory reponse was not confirmed in healthy individuals who received the ML I s.c. (Hajto *et al.*, 1996). In a set of experiments in BALB/c mice treated s.c. with purified ML I or whole plant extract *Eurixor*, Beuth and co-workers observed that the number and activity of peritoneal macrophages increased as did the number of peripheral monocytes, interleukin-2 receptor-positive lymphocytes, thymus weight, and thymocyte numbers (Lyt-2/CD8+ cells, L3T4/CD4+ cells, and Lyt-2/CD8 L3T4/CD4-double positive cells) (Beuth *et al.*, 1991; 1993a, 1994a, 1995, 1996; Stoffel *et al.*, 1997).

Using both aqueous (*Helixor*) and fermented drug extracts (*Iscador*), Hülsen and co-workers observed increased NK cells activity against leukemia K 562 cells when pre-incubated for 20 h with the drug (Hülsen *et al.*, 1989; Hülsen *et al.*, 1992; Hülsen and Born, 1993). With subcutaneous administration of the whole plant extract with a defined ML I content (*Eurixor*) to 40 breast cancer patients, an increase of peripheral T helper cells, NK and T cells with expression of interleukin-2-receptor α chains (CD25), and concomitantly acute phase proteins such as complement component C3, C-reactive protein, and haptoglobin was observed within 4 weeks in 10 patients (Beuth *et al.*, 1992, 1993b). In 36 breast cancer patients treated with *Eurixor*, no significant changes were observed in the lymphocyte subsets; however, plasma levels of β-endorphin increased at six and twelve weeks after immunization in a subgroup of 25 responders (Heiny and Beuth, 1994).

By the investigation of long-term treated tumour patients (n = 23), the amount of lymphocytes and the number of natural killer (NK) cells increased within 7 month of subcutaneous application of VA-E (*Helixor*), while the number of other lymphocyte subsets (i.e. CD19+ B cells, CD4+ T helper cells, CD8+ CD28− suppressor cells, CD8+ CD28+ cytotoxic cells) and the proportion of CD25+ cells within T cells showed a statistically remarkable trend (Büssing *et al.*, 1999g). For CD19+ B cells, CD4+ T helper cells, CD8+ CD28+ cytotoxic cells and CD16+/CD56+ NK cells, we observed statistically remarkable peaks within the 2nd and the 3rd month of therapy, indicating a sufficient anti-mistletoe antigen response within this time range. Here, no effects of an immunosuppression, as defined by a decrease of defined lymphocyte subsets was observed by an adequate escalation of applied VA-E.

A strong proliferation response of lymphocytes from tumour patients and healthy individuals treated with VA-E was approved several groups (Schultze *et al.*, 1991; Fischer *et al.*, 1997a,b; Stein *et al.*, 1998; Stein and Berg, 1998b). The peaks of [^{3}H] thymidine uptake in the cells were interindividually different, varying from 3 to 10 weeks in tumour patients after the onset of therapy (Stein *et al.*, 1998); later on, this reactivity decreased in 6 out of 8 tested patients, indicating a desensitization of the immune response towards the *Viscum album* antigens. Few individuals did not showed a proliferation response towards the applied VA-E. However, lymphocytes from tumour patients never exposed to *Viscum album* antigens showed no specific proliferation response (Schultze *et al.*, 1991).

Another interesting finding came from the group of Kabelitz who observed a stimulation of a defined T cell subset by a heat-treated VA-E as measured by [^{3}H]-thymidine uptake. In response to this extract, T cells with $\gamma\delta$ T cell receptor were expanded, i.e. cells with the variable T cell receptor elements Vγ9 and Vδ2 (Fischer *et al.*, 1996). A strong stimulation of this small T cell subset which represents about 1 to 10% of peripheral T cells was also observed by heart-treated *Mycobacterium tuberculosis* (Fischer *et al.*, 1996), indicating that *Viscum album*-derived ligands share features with mycobacteria-derived ligands for this T cell subset. However, the function of $\gamma\delta$ T cells remains unclear.

The *in vivo* effects of VA-E are highly dependent on the mode of application. Specifically, subcutaneous injection of VA-E from quercus host tree (*Iscador*) in healthy individuals resulted in a moderate redness, induration, swelling, local warmth, and painfiul sensations within 1 to 10 h (Gorter *et al.*, 1998). Histological analyses of injection sites 10 d after the beginning of the treatment revealed normal surface epithelium and epidermis, while the corium and subcutaneous fat tissue showed a superficial and deep, dense, perivascular infiltrate with 60% T lymphocytes (50% $CD4^+$ T cells and 50% $CD8^+$ cells) and 40% macrophages (Gorter *et al.*, 1998). However, since no controls were investigated, no valid conclusion can bedran from these findings.

Intravenous application of high concentrations of aqueous VA-E did not affect hematopoiesis (Böcher *et al.*, 1996) but increased the number of monocytes and juvenile granulocytes within 24 h after VA-E administration (Büssing *et al.*, 1996f). However, the absolute number of neutrophils, NK, T helper, and T-suppressor/cytotoxic cells did not change significantly (Büssing *et al.*, 1996f). The lack of ML cytotoxicity in a clinical situation might be due to the induction of anti-ML-antibodies during therapy (Stettin *et al.*, 1990; Stein *et al.*, 1997b, 1998) and inhibition of MLs by serum glycoproteins/-lipids (Ribéreau-Gayon *et al.*, 1995, 1997). Thus, a clinically relevant ML-mediated cytotoxicity might be induced only in the case of intratumoural or intrapleural injection.

Pleurodesis and elimination of tumour cells in response to the intrapleural administration of VA-E in patients with malignant pleural effusions (Böck and Salzer, 1980; Böck, 1983a,b; Salzer, 1986; Salzer and Popp, 1990; Stumpf and Schietzel, 1994) was associated with an increase of eosinophils, T helper cells, and NK cells (Salzer, 1986; Salzer and Popp, 1990), while in the investigations of our group no significant changes of lymphocyte subsets (i.e., $CD4^+$ T helper cells, $CD8^+$ T suppressor/cytotoxic cells or $CD16^+$/$CD56^+$ NK cells) were observed; however the decline of tumour cells was associated with a transient increase of macrophages and eosinophils only in the responder group (Stumpf and Büssing, 1997b).

Intratumoural application of VA-E did not produce significant changes of peripheral immunocompetent cells, but slightly increased granulocytes, modestly decreased monocytes (Scheffler *et al.*, 1996), and slightly increased $CD8^+$ $CD28^-$ putative suppressor cells (Stumpf *et al.*, 1997a). Nevertheless, the effects of VA-E on tumour-infiltrating immune cells may be more relevant than on peripheral immune cells, but no investigations have been reported at this time.

Induction of cytokines

Immunostimulation of immunocompetent cells by VA-E was further evidenced by a release of TNF-α nd IL-6 in the supernatants of VA-E-stimulated peripheral mononuclear cells, indicating an activation of monocytes/macrophages, and a release of T cell-associated cytokines INF-γ and IL-4 with individual variations (Stein *et al.*, 1996b, 1998a; Stein and Berg, 1998b). Similar responses were observed in cultured cells from healthy individuals treated subcutaneously with VA-E (Stein and Berg, 1998b). At present, however, no valid conclusions can be drawn whether *Viscum album* treatment drives the immune response towards a "cytotoxic" Th1-

associated reaction (IL-2, IL-12, IFN-γ) or towards a "humoral" Th2-associated direction (IL-4, IL-5, IL-10). Although especially the VA-E containing micelles from the cell membranes of the plant (*Abnoba*) predominantly induced IFN-γ/IL-2 and of TNF-α/IL-6, while IL-4/IL-5 was induced only in few cases, a large proportion of investigated patients (about 38%) showed a release of both Th-1- and Th2-associated cytokines (Stein *et al.*, 1998a). The simultaneous secretion of Th1- and Th2-associated cytokines was also observed in healtyh individuals treated with an aqueous VA-E, while a dominance of Th2-associated cytokines was observed (Stein and Berg, 1998b). It remains to be clarified (1) whether a Th1 response is favourable in tumour patients, (2) whether a VA-E-directed Th2 response may favour suppression of cellular anti-tumour response or not, and (3) whether the anti-*Viscum album* antigen response and anti-tumour response are independent effects. Anyway, these results indicate that a modulation of the Th1/Th2 balance is highly dependent on the individual tested and on the stimulating components present in the applied drug.

Humoral immune system

During therapy with VA-E, antibodies are produced against antigens from *Viscum album*, predominantly ML (Stettin *et al.*, 1990; Stein *et al.*, 1997b, 1998a,b; Stein and Berg, 1998b). In most cases, these anti-ML antibodies are of IgG class, IgA was found in some cases, and IgE only in rare cases, while IgM antibodies were detected only infrequently (Stettin *et al.*, 1990; Stein *et al.*, 1998a,b; Stein and Berg, 1998b).

Human anti-ML antibodies were found to detect different epitopes of ML and other components from VA-E (Stein *et al.*, 1998b), i.e. an epitope of about 45 kD present in at least two VA-E from pine mistletoe (*Iscador* P and *Helixor* P). This epitope remains to be characterized.

Subcutaneous application of VA-E obviously results in a presentation of antigens from *Viscum album*, i.e. ML, to T and B cells in the lymph nodes. In fact, maximal anti-ML antibody responses were observed within 3 to 10 weeks and 4 to 12 weeks, respectively, after the VA-E therapy started (Stein *et al.*, 1998a; Stein and Berg, 1998b); within this time range, the number of B cells and T helper cells showed a peak (Büssing *et al.*, 1999g). Figure 2.

The major task of $CD8^+$ cytotoxic T cells is the lysis of MHC class I^+ virally infected cells and tumour cells, while $CD4^+$ cytotoxic T cells may have an immunomodulatory role as they eliminate activated MHC class II^+ immune cells by Fas/FasL interaction to prevent overreactions in ongoing immune responses and to remove potentially hazardous immune cells (Hahne *et al.*, 1995). In this regard it is of importance to notice that apoptotic $CD4^+$ and $CD8^+$ cells showed increased level of intracellular IL-4 (Stein *et al.*, submitted for publication). In agreement with this observation, Fas-mediated apoptosis of lymphoid cells was reported to result in a rapid production of IL-10 in these cells (Gao *et al.*, 1998). In a physiological context, this may indicate that apoptosis and tolerance are linked through the production of anti-inflammatory Th2-associated cytokines which may inhibit Th-1-associated immune reactions, and thus may prevent deleterious immune responses, such as strong local immune reactions induced by s.c. application of VA-E.

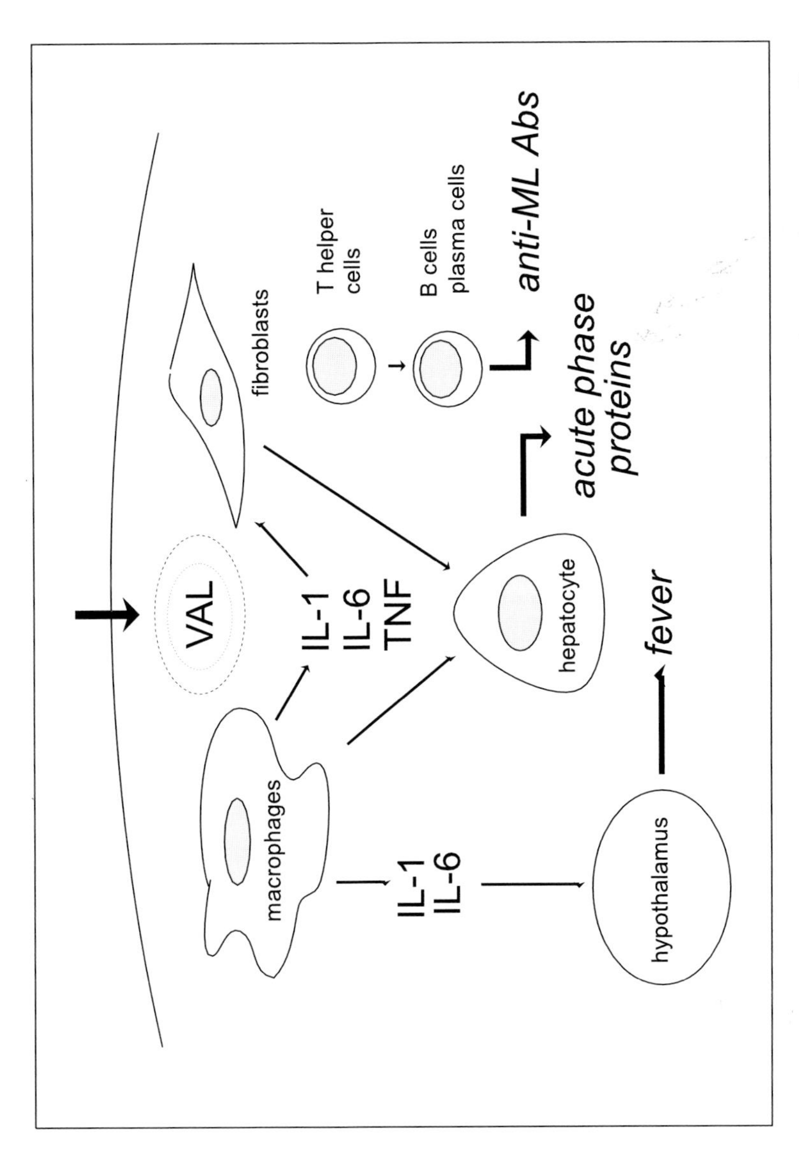

Figure 2 Schematic presentation of immune reactions after subcutaneous application of extracts from *Viscum album* L. (VAL).

DNA Stabilization

A study enrolling Turkish medicinal herbs found that fruits of *Viscum album* (bought from a herbalist market) significantly increased the number of DNA strand breaks as measured by COMET assay (alkaline single cell gel electrophoresis), while the extract, however, did not produce a positive response in the microsomal activation assay (Ames test) (Basaran *et al.*, 1996). These conflicting results were not approved by other groups with commercially available aqueous VA-E extracts from fresh plant material (Büssing *et al.*, 1994, 1995a,c, 1996a; Mengs *et al.*, 1997; Mengs, 1998), and with crude, heated or ethanolic extracts from Korean mistletoe (Ham *et al.*, 1998). Moreover, VA-E contain DNA stabilization properties for peripheral blood mononuclear cells. This effect has been demonstrated with the significant reduction of spontaneous and cyclophosphamide (CP)-induced sister chromatid exchange (SCE)-inducing DNA lesions of human peripheral blood mononuclear cells with the whole plant extract *Helixor* A (Büssing *et al.*, 1994, 1995c, 1996a). In rapidly proliferating amniotic fluid cells, the fermented VA-E *Iscador* P significantly reduced the SCE level only at very high concentrations (Büssing *et al.*, 1995b). Concomitantly, the drug extract *Helixor* A also protected CP-mediated depression of activation-associated surface molecules, specifically interleukin-2 receptor α chains (CD25) and transferrin receptors (CD71) on T cells (Büssing *et al.*, 1995c). However, purified components, such as the ML and viscotoxins, did not prevent the activation marker depression on T cells mediated by CP (Büssing *et al.*, 1995c), indicating that the whole plant extract is effective, not the purified compounds. In leukemic Jurkat T cells, however, the simultaneous addition of CP and VA-E resulted in a more severe decline of cell numbers than CP alone (Büssing *et al.*, 1996a). In mice infected with mamma carcinoma cells and treated with CP and VA-E (*Isorel*), the number of lung metastases was severely reduced compared to animals treated with CP or VA-E alone (Büssing *et al.*, 1996b). Thus, in the murine model and in cultured leukemic cells, VA-E produced no protective effects.

Using a fermented VA-E (*Iscador* M), Kovacs and co-workers (1991) observed an improved incorporation of [^{3}H]-thymidin in the DNA of UV-damaged lymphocytes from breast cancer patients after subcutaneous application. Based on sedimentation of DNA strand breaks on an alkaline sucrose gradient, VA-E treatment further increased DNA repair of radiation and CP-induced damage of lymphocytes in breast cancer patients (Kovacs *et al.*, 1995, 1996). Murine models illustrated the protective effects of VA-E for CP-treated and γ-radiated mice with increased survival and reduced leukocytopenia (Kuttan and Kuttan, 1993b; Narimanov *et al.*, 1992). Recently published studies have shown that administration of a fermented VA-E (*Iscador* M) significantly reduced sarcomas induced in the C57BL/6 mice by 20-methylcholanthrene administration (Kuttan *et al.*, 1996, 1997). Only one animal among 15 developed sarcoma within an observation period of 160 days in contrast to all control-mice that developed sarcoma within 80 days.

Irregardless of the mechanism of action, the results indicate that VA-E reduces DNA damages induced by carcinogens and probably inhibits tumour development. Whether these effects are of benefit for the patients especially during chemotherapy, is under investigation in an ongoing controlled and randomized clinical study.

Apart from this, cellular abnormalities such as uni-, bi- and multipolar anaphases, resulting in tetra- and aneuploidy, asynchronic bimitoses, and micronuclei were detected in the root tips of onions (*Allium cepa)* after treatment with the drug extract produced under the conditions described by Winterfield and co-workers (Acatrinei, 1966, 1967, 1969). These effects were thought to be related to a destruction of the spindle apparatus by the mistletoe extract. The findings are in agreement with those of our group (Büssing *et al.*, 1998b), as we observed increased frequencies of telomeric associations and C-anaphases in lymphocytes treated with aqueous drug extracts from mistletoe from fir trees and by ML III at 10 ng/ml. Clearly, a correlation exists with chromosomal affections and apoptotic changes induced by VA-E (Table 2). Using a whole plant extract from VA-E (*Lektinol*) Mengs *et al.* (1997) did not observed gene mutations and chromosomal aberrations, as measured in bacteria and mammalian cells.

Using non-toxic concentrations of ML I, ML II and ML III (0.001, 0.014, 0.14 ng/ml), the number of micronuclei in the human lung carcinoma cell line Calu-1 did not change (Köteles *et al.*, 1998), while ML I at $\geq$ 74 ng/ml induced these cytogenetic damages. After irradiation of Calu-1 cells, the number of micronuclei decreased within 1 h in response to ML I and ML II at 0.001 and 0.014 ng/ml, while ML III, however, showed a slight increase of these cytogenetic damages. Further, intraperitoneal treatment of irradiated rabbits (1 Gy) with ML I at 1 ng/kg BW irradition revealed a strong decrease of micronuclei within 24 h (Köteles *et al.*, 1998). As the induction of micronuclei was associated with a decrease of viability in lymphocytes, lymphoblasts and the Calu-1 cells, one may suggest that radiation-damaged cells are more sensitive to the ML-mediated apoptosis. Again, a correlation between cytogenetic damage and cell death was evident. Whether this deletion of abberation-carrying cells may have clinical relevance in the treatment of irradiated cancer patients has to be clarified in clinical studies.

MECHANISM OF ACTION: BIOLOGICALLY ACTIVE COMPONENTS

Viscum album contains a variety of biologically active components (see both, Becker and Pfüller, this book). While the cytotoxic properties of VA-E were linked to the ML and to the viscotoxins, the immunomodulating properties were attributed not only to the ML but also to alkaloids, polysaccharides and oligosaccharides (i.e. rhamnogalacturonan), viscotoxins, a 5 kDa peptide, the vesicles from *Viscum album*, and an undefined non-lectin antigen present in a fermented extract from pine trees. The compounds that exert the protective effects of the plant extracts remain to be characterized. The role of flavonoids and phenolic acids are conflicting, as some of them may act as anti-carcinogens or inhibit the growth of tumour cells, whereas others act as co-carcinogens, are mutagenic or able to induce DNA damage.

Viscotoxins

The viscotoxins consist of a group of basic polypeptides with a molecular weight of about 5 kDa and were isolated first by Winterfeld and co-workers (1948,a,b, 1956)

Table 2 Apoptotic changes and chromosomal affections in lymphocytes treated with VA-E.

	medium control	*Helixor A (100 μg/ml)*	*VaAR (100 μg/ml)*	*ML III (10 ng/ml)*
% Apo2.7+ cells	12%	22%	59%	35%
% telomeric associations	2.06	18.05	23.65	14.28
% C-anaphases	5.15	5.55	13.98	14.28
carbohydrates (μg/ml)	/	300	600	–
proteins (μg/ml)	/	15	30	–
ML (μg/ml)	/	1.4	3.6	0.001
viscotoxins (μg/ml)	/	< 0.5	< 0.5	–

Human lymphocytes were incubated with ML III-rich VA-E produced from mistletoe grown on fir trees (*Helixor* A and VaAR) at final concentrations of 100 μg/ml, and ML III at 10 ng/ml. VaAR is an experimental, non-fermented aqueous plant extract produced by the manufacturer *Helixor* using a new preparation method. Expression of mitochondrial membrane protein Apo2.7 was analyzed by flow cytometry as described (Büssing *et al.*, 1998a). Similar results were obtained using Annexin-V and the DNA intercalating dye propidium iodide (data not shown). For analysis of C-anaphases and telomeric associations, a minimum of 72 to a maximum of 100 Giemsa-stained metaphases were examined by light microscopy as described previously (Multani *et al.*, 1996).

and later by Samuelsson and co-workers (1959, 1961, 1970). Fractionations of the crude viscotoxins led to the isolation of several homogenous, cytotoxic components termed B, A_2, A_3 and 1-PS (Samuelsson and Pettersson, 1970), and A_1 and U-PS (Schaller *et al.*, 1996) or I, II, III, IVb (Konopa *et al.*, 1980; Woynarowski and Konopa, 1980). The 46 to 50 amino acid polypeptide chains are rich in lysin and arginin, contain six cystein residues at identical positions, and are formed by cleavage of larger precursors (Samuelsson, 1974; Schrader and Apel, 1991). Due to their positive charge, they may form complexes to nucleic acids (Woynarowski and Konopa, 1980).

The cytoxicity of the viscotoxins was assessed mainly by parenteral administration in animals. In cats, intravenous application of viscotoxins at 0.1 mg/kg was lethal for all animals whereas 0.5 mg/kg was lethal for 50% of mice when applied intraperitoneally (Samulesson, 1974). At lower concentrations, application of viscotoxins resulted in hypotension, bradycardia, and negative inotropia of the heart. Thus, an obvious similarity of viscotoxins and cobra cardiotoxins was observed (Rosell and Samuelsson, 1966; Samuelsson, 1974). Few studies have examined the cytotoxicity of viscotoxins to cultured cell lines. Konopa *et al.* (1980) reported that the isolated polypeptides exhibited cytotoxic activity against the human tumour cell lines KB and HeLa in a concentration range (ED_{50}) of 0.2 to 1.7 μg/ml. The Yoshida sarcoma cells were more sensitive to the viscotoxins and less sensitive to the ML as compared to the leukaemia cell line Molt-4 (Urech *et al.*, 1995).

Although the primary structure of the viscotoxins share a high degree of similarity, the viscotoxins significantly differ in cytotoxicity (Table 3). Against Yoshida sarcoma cells, the viscotoxins A_3 and 1-PS are most effective, followed by A_1 and A_2, while in lymphocytes A_2 and 1-PS were the most effective; the least potent are viscotoxins B and U-PS (Urech *et al.*, 1995, 1996; Büssing *et al.*, 1999c). According to Konopa *et al.* (1980), the viscotoxins IVa/IVb (A_3) and III (A_2) were most effective against KB cells whereas viscotoxins I and II (B) were less effective. Qualitative and quantitative differences in the VT pattern of the three European *Viscum album* subspecies have been described. *Viscum album* ssp. *austriacum* lacks viscotoxins A_1 and

Table 3 Cytotoxicity of viscotoxin fractions to cultured cell lines and lymphocytes.

ED_{50} (μg/ml)		*ED_{50} (μg/ml)*		*%Annexin-V^+ Eth^+ cells (ROI generation)*	
viscotoxins	*KB cells*[a]	*viscotoxins*	*Yoshida cells*[b]	*viscotoxins*	*lymphocytes*[c]
I	1.7	A_1	0.87	A_1	9
II (B)	1.5	A_2	1.08	A_2	12.6
III (A_2)	0.6	A_3	0.31	A_3	8.5
IVa	0.2	B	4.58	B	1.4
IVb (A_3)	0.2	1-PS	0.44	1-PS	13.1
crude	1.0	U-PS	4.04	crude	9.3

[a] Konopa *et al.* 1980; [b] Schaller *et al.*, 1996; [c] Büssing *et al.* 1999c.

A_2, *Viscum album* ssp. *album* lacks 1-PS and U-PS, but *Viscum album* ssp. *abietis* contains all six viscotoxins (Schaller *et al.*, 1996).

The viscotoxins exert their cytotoxicity by a rapid permeabilisation of the cell membrane. Within 1–2 h, treatment of lymphocytes with viscotoxins resulted in a loss of cell membrane asymetry with translocation of phosphatidyl serine from the inner to the outer leaflet of the cell membrane, uptake of the DNA-intercalating dye propidium iodide, generation of reactive oxygen intermediates (ROI), swelling of mitochondria with loss of their cristae, and degradation of cytoplasma and chromatin (Büssing *et al.*, 1998c, 1999b,c,f). These rapid changes indicate accidental (necrotic) cell death, with an induction of apoptosis in the "surviving" cells later on (Büssing *et al.*, 1999c,f).

The main important question is how the viscotoxins induce cell death. The capability of thionins to form pores in biological membranes is supported by their 3-dimensional structures, as the amphiphilic thionins may interact with amphiphilic phospholipids (Osario e Castro *et al.*, 1989, 1990; Angerhofer *et al.*, 1990; Teeter *et al.*, 1990; Vernon, 1996). There are several effects observed when cells were treated with *Pyrularia* thionin: the early responses of an increased order parameter of the phospholipids, an increase in membrane permeability, depolarisation of the cell membrane, and probably opening of Ca^{2+} channels (Vernon, 1996); long term responses include the formation of membrane blebs, and finally activation of endogenous phospholipase A_2 and adenylate cyclase (Evans *et al.*, 1989; Vernon, 1996). It is suggested that the viscotoxins affect membranes, resulting in rapid cell membrane permeabilisation and subsequently an excess generation of ROI that are no longer detoxified by cellular components with redoxidation potential, such as glutathione and membrane cardiolipin. Indeed, neither reduced glutathione nor N-acetyl-L-cysteine (5 and 10 mM) were able to prevent VT-mediated cytotoxicity in cultured human lymphocytes (Büssing *et al.*, 1999c,f). However, other polycationic and/or amphipathic molecules, such as poly-L-arginine, poly-L-lysine, poly-L-glutamic acid, protamine sulfate, purothionin, and mastoparan I and II, did not sufficiently kill the lymphocytes (unpublished own results), indicating that the polycationic structure alone is not sufficient to affect the cells. The cytotoxicity of viscotoxins was prevented only by cleavage of disulphide bounds; this may affect interaction of viscotoxins with membrane lipids.

Apart from cell death, another relevant property of the viscotoxins is the enhancement of the *Escherichia coli*-induced phagocytosis and oxidative burst of human granulocytes (Stein *et al.*, 1999a,b). This effect may be due to the polycationic structure of viscotoxins, as they were observed (in contrast to their cytotoxic properties) also with polycationic substances such as protamine sulfate, purothionin, histone, poly-L-arginine, poly-L-lysine, but not with the poly-anionic poly-L-glutamic acid (Stein *et al.*, 1999a).

Mistletoe Lectins

ML are ribosome-inactivating proteins, composed of two polypetide chains (A and B) linked by disulfide bond. ML differ in specificity and molecular weight from between

50 kDa to 63 kDa. ML I binds to D-galactose, ML III to N-acetyl-D-galactosamine, and ML II to N-acetyl-D-galactosamine and D-galactose (Franz *et al.*, 1981; Pfüller, this book). A number of plant proteins such as the toxic lectins from *Ricinus communis*, *Abrus precatoris* and *Viscum album*, which share a high grade of structural homology, have been identified that catalytically damage eukaryotic ribosomes, and the cells are consequently unable to perform the elongation step of protein synthesis (reviewed by Stirpe *et al.*, 1992; Büssing, 1996). These "ribosome-inactivating proteins" (RIPs) possess carbohydrate-binding B chains linked by hydrophobic bonds and disulfide bridges to the catalytic A chain.

The lectin domains of RIPs can bind to any appropriate carbohydrate domain on cell surface receptors, enabling the protein to enter the cell by receptor-mediated endocytosis (Endo *et al.*, 1988; Stirpe *et al.*, 1982, 1992). Subsequently, the catalytic A chain of the ML inhibits protein synthesis (Stirpe *et al.*, 1980; Olsnes *et al.*, 1982; Endo *et al.*, 1988; Stirpe *et al.*, 1982, 1992), and the cells undergo apoptosis (reviewed by Büssing, 1996). In agreement with these observations, also ricin and abrin were recognised to induce apoptosis (Griffiths *et al.*, 1987; Hughes *et al.*, 1996; Oda *et al.*, 1997). Results of Endo and co-workers (Endo *et al.*, 1987, 1988; Endo and Tsurugi, 1988) indicated that the ML I A chain is similar to the A chain of ricin: a *N*-glycosidase that releases adenine from position 4324 of 28 S RNA of 60 S ribosomal subunits. This irreversible modification of the ribosomes impairs its ability to interact with elongation factor 2 (EF-2) during the translocation reaction and inhibits the elongation process of the polypeptide chains (Benson *et al.*, 1975; Montanaro *et al.*, 1975). In addition, it has been suggested that ricin inhibits protein synthesis through disturbance of EF-1-dependent aminoacyl-tRNA binding to ribosomes (Furutani *et al.*, 1992). Beside the impairment of macromolecule synthesis, RIPs were suggested to possess polynucleotide:adenosine glycosidase activity and thus, resulting in depurination of DNA and RNA (Barbieri *et al.*, 1997). This may explaine the observed inhibition of RNA and DNA synthesis in response to VA-E and ML (Hülsen *et al.*, 1986; Metzner *et al.*, 1987; Dietrich *et al.*, 1992; Göckeritz *et al.*, 1994; Urech *et al.*, 1995). Protein synthesis was found to be inhibited by ML and other RIPs before that of DNA and RNA (Sargiacomo and Hughes, 1982), indicating that the effect on protein synthesis is of major importance for the fate of the cells. Similar findings were reported for whole plant VA-E (Hülsen *et al.*, 1986). A 50% inhibition of [^{3}H]-thymidine uptake was obtained with 0.06 ng/mL ML I, 0.015 ng/mL ML II and 0.010 ng/mL ML III (Dietrich *et al.*, 1992). However, other researchers observed comparable cytotoxicity for ML I and ML III to mitogen-stimulated lymphocytes since their IC_{50} was about 0.4 ng/mL (Göckeritz *et al.*, 1994). In mice trreated i.p. with the ML, the LD_{50} was was 28 μg/kg BW for ML I, 1.5 μg/kg BW for ML II, and 55 μg/kg BW (Franz, 1986).

Cytotoxicity of ML was inhibited by plasma proteins (Franz *et al.*, 1977; Ribéreau-Gayon *et al.*, 1995, 1997), the specific carbohydrates (Ziska *et al.*, 1978; Luther *et al.*, 1980; Ribéreau-Gayon *et al.*, 1997), and $CaCl_2$ (Büssing *et al.*, 1999g). Although the underlying mechanisms of these Ca^{2+} effects are unclear, one may suggest a stimulation of apoptsis-preventing "calcium-sensing receptors".

Cell death mechanisms

The first effect observed in lymphocytes treated with the ML or its lectin B chain is a rapid receptor-mediated "signal" that increases the content of intracellular Ca^{2+} (Göckeritz *et al.*, 1994; Büssing *et al.*, 1996; Wenzel-Seifert *et al.*, 1997), which is involved in certain activation pathways leading to endonuclease activation and subsequently DNA fragmentation. After the interaction of lectin B chain with an appropriate surface receptor, and endocytosis of the protein, the A and B chains dissociate in the cytosol (Figure 3). While the A chain enzymatically inhibits the ribosomes, the cytosolic actions of the B chains are unclear. Recent experiments with intensively purified A and B chains from ML I confirmed that neither the B chain nor the A chain alone was able to induce apoptosis (Vervecken *et al.*, 2000). However, the enzymic A chain was reported to induced a slight blastogenic transformation of lymphocytes (Metzner *et al.*, 1987). In contrast, Hostanska *et al.* (1997) reported an induction of apoptosis by the enzymic A chain of ML I but no effect with the lectin B chain. One cannot exclude the possibility that cross-contaminations of isolated ML chains with the hololectin or the opposite chains may be a reason for the conflicting results.

Subsequently, the ML induce expression of mitochondrial membrane molecules Apo2.7, generation of ROI, cytochrome C release, induction of caspase-3 and caspase-9, Bcl-2 protein degradation, binding of Annexin-V to phosphatidyl serine exposed on the outer leaflet of the cell membran, and loss/fragmentation of DNA (Janssen *et al.*, 1993; Büssing *et al.*, 1996c, 1997, 1998a,b, 1999b,f; Hostanska *et al.*, 1996–1997; Möckel *et al.*, 1997; Bantel *et al.*, 1999). These apoptosis-associated changes were also observed in lymphocytes treated with protein and/or RNA synthesis inhibitors cycloheximide and actinomycin D, and protein transport inhibitor brefeldin A (Büssing *et al.*, 1999b).

The caspases belong to a family of cystein proteases contributing to apoptosis through direct disassemmbly of cell structures (reviewed by Green and Reed, 1998). In human lymphocytes, we observed both, casapase-3^+ cells with and without simultaneous expression of mitochondrial Apo2.7 molecules, indicating that the caspase cascade may be activated in response to mitochondrial triggers (Büssing *et al.*, 1999f). However, leukemic Molt-4 cells were more sensitive to ML I as compared to differentiated lymphocytes; here, almost all cells were Caspase-3^+ Apo2.7^+. An activation of "death receptors" with Fas-associated death domains (FADD) such as APO-1/Fas, TNF-RI (tumour necrosis factor receptor type 1) and TRAIL (TNF receptor related apoptosis inducing ligand), which play a major role in apoptotic signalling, is unlikely as ML-induced apoptosis was observed in both, Fas-sensitive and -resistant Jurkat T cells, and also in BJAB cells stably transfected with a dominant-negative FADD mutant (Bantel *et al.*, 1999). Moreover, blocking of Fas-molecules on human lymphocytes did not prevent ML-induced cell death (Büssing *et al.*, in preparation).

However, several aspects of ML-induced cell death are unclear, such as the impact of protein synthesis inhibition, DNA damages or altered chromosomal stability, and consequently, the involvement of nuclear p53, Bcl-2 proteins, and IL-4. Since cycloheximide, actinomycin D, brefeldin A, ricin and ML I induced apoptosis (Kochi and

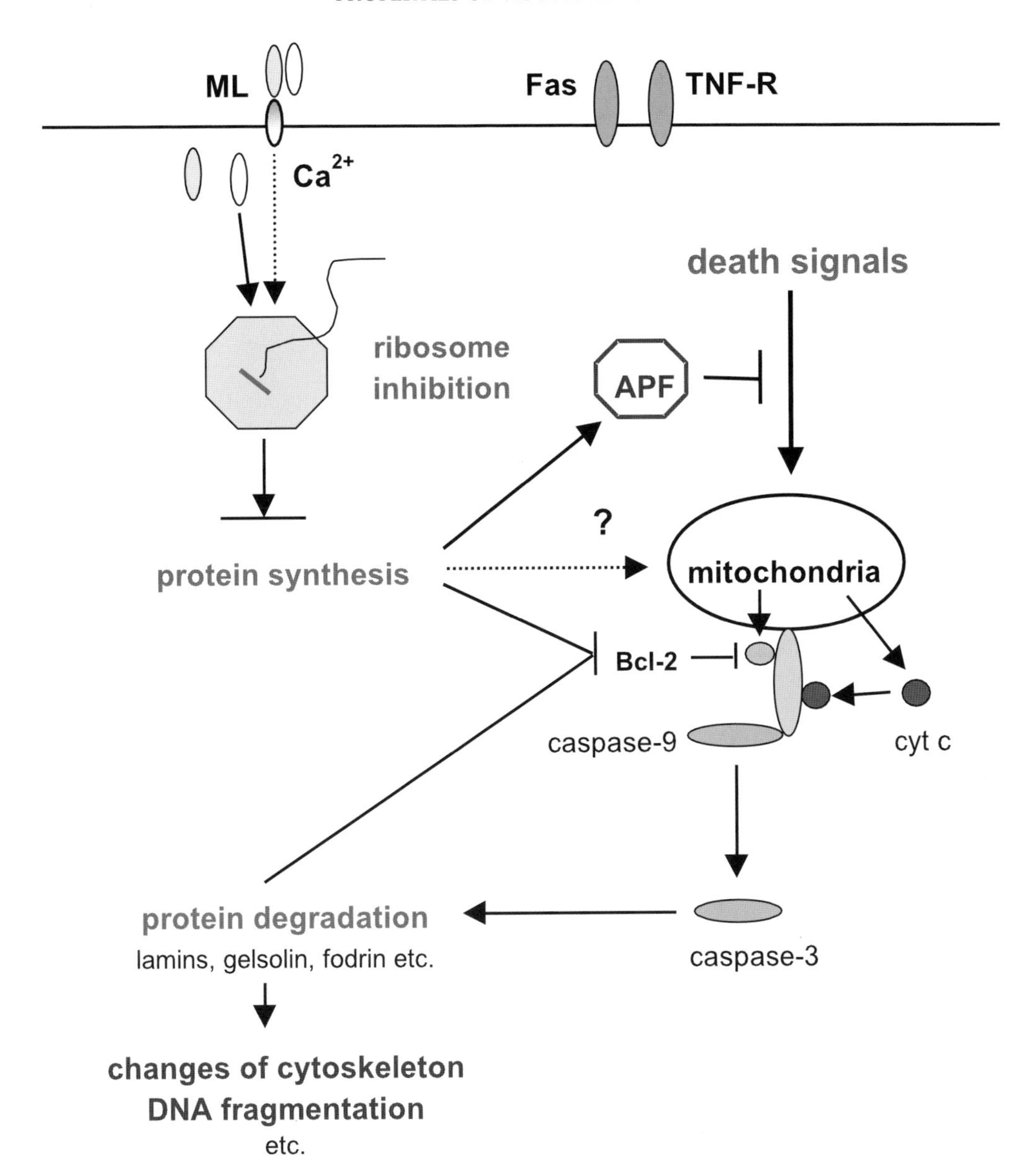

Figure 3 Schematic presentation of apoptosis-associated changes induced by the interaction of ML with appropriate surface receptors, internalisation of the protein, and dissociation of A and B chains in the cytosol. Inhibition of protein synthesis may result in decreased level of suggested so-called "apoptosis-preventing factors" (APF) which may inhibit the delivery of "death signals" to the mitochondria. Mitochondrial release of cytochrome c (Cyto c) activates caspases by binding of Apaf-1 (apoptosis-activating factor 1), inducing it to associate with procaspase-9, and thereby initiating the proteolytic caspase cascade that culminates in protein degradation and DNA fragmentation. Bcl-2 is suggested to regulate apoptosis by binding of Apaf-1 and by regulating the induction of mitochondrial permeability transition (reviewed by Kroemer *et al.*, 1997; Green and Reed, 1998).

Collier, 1993; Martin, 1993; Büssing *et al.*, 1999b), one may suggest that each obstacle resulting in an inhibition of protein synthesis or transport will result in an apoptotic cell death, and thus, survival of cells may depend on the constant production of "survival promotors" and/or "death suppressing factors". ML-affected protein synthesis may inhibit homolougs of the Bcl-2 family or protein kinases which may have an anti-apoptotic effect by the phoshorylation of apoptosis-regulating proteins, as suggested by Bantel *et al.* (1999). In fact, Apo2.7^{+} caspase-3^{+} cells were induced also by the protein kinase C inhibitor staurosporine. Also, ML III decreased the intracellular level of Bcl-2 proteins and p53 proteins (Büssing *et al.*, 1998b). Moreover, in response to ML I, Apo2.7 molecules and caspase-3 were detected only in cells with low level of Bcl-2 proteins (Bcl-2^{lo}), while these death markers were not observed in Bcl-2^{hi} cells (Büssing *et al.*, in preparation), indicating degradation of Bcl-2 proteins by the caspases during apoptosis (Büssing *et al.*, 1999f).

Induction of chromosomal instability

In lymphocytes treated with ML III, an increase of telomeric associations (TAs) and C-anaphases was observed (Büssing *et al.*, 1998b). Cancer cells, which lack normal p53 and retinoblastoma protein functions, show chromosomal instability leading to TAs, ring chromosomes and dicentric chromosomes (Healy, 1995; Sharma *et al.*, 1996). TAs have been considered as a cellular manifestation of telomeric loss. The telomeres on both ends of a chromosome serve as a cap and protect the chromosomes from fragmentation, help in the attachment to the nuclear membrane, and in the pairing of homologues during meiotic division (Pathak *et al.*, 1994a). Loss of even a single telomer could render a chromosome instable. Thus, one may assume that the cells with higher p53 level will be killed by the ML *via* apoptosis, while cells with low level of p53 proteins survive but may harbour chromosomal affections, which are incompatible with normal life. The exact contribution of these obstacles to the onset of apoptosis are yet unclear.

Selectivity of killing

The lectin chains of the various RIPs differ in their cellular interactions. This is suggested by the different lesions each toxin causes in animals (Stirpe *et al.*, 1992), with ricin at high concentrations damaging primarily Kupffer and other macrophagic cells, whereas modeccin and volkensin affect both parenchymal and non-parenchymal liver cells. Since the type 1 RIPs are lacking a lectin subunit, these single chain RIPs are less toxic than type 2 RIPs; however, they are highly toxic to some cells, for instance macrophages and trophoblasts, possibly due to their high pinocytic activity (Stirpe *et al.*, 1992). Obviously, the sensitivity of cells to the lectin-mediated cytotoxicity differs.

A selective killing of CD8^{+} lymphocytes with a "memory" phenotype (CD62L^{lo}) was observed by low concentrations of the galNAc-binding ML III (Büssing *et al.*, 1998a,b). The reason why ML III at 10 ng/ml selectively killed this subset as com-

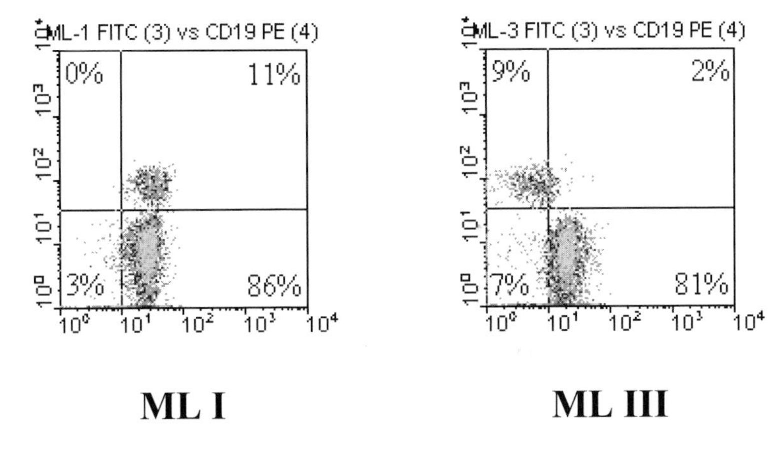

Figure 4 Staining of human CD19⁺ B cells (*y*-axis) with FITC-labelled ML I (*x*-axis) and FITC-labelled ML III (*x*-axis). Results are representative for 8 independent experiments.

pared to their CD8⁺ CD62L^{hi} counterparts, CD4⁺ T helper cells and CD19⁺ B cells is unclear. One explanation could be that the binding sites for ML III may differ in CD8⁺ CD62L^{hi} "naive" cells and CD8⁺ CD62L^{lo} "memory" cells. In fact, especially the CD8⁺ cells bind a higher number of ML I than CD4⁺ T cells and CD19⁺ B cells, while the binding capacity of these subsets for ML III were similar (Büssing *et al.*, 1999h). However, CD8⁺ cells were as sensitive to ML I-mediated cell death as compared to CD4⁺ T cells, but more sensitive to ML III (Büssing *et al.*, 1998). Also, analysis of ML-binding to the CD62L subsets within the CD4⁺ T cells and CD8⁺ cells did not explain the previously observed differences in the selectivity of ML III killing. Yet, there is at present no convincing rationale for these conflicting results. One may suggest that differences in the ML-cell surface affinity and/or intracellular uptake of the toxic proteins, and their subsequent degradation could be determinating factors. This has to be addressed in further investigations. Surprisingly, B cells did not bind adequately ML III but were able to bind ML I (Figure 4). One may suggest a lack of adequate "receptors" with galNAc domains on these cells. Anyway, this may explain the lower sensitivity of B cells to ML III-mediated cytotoxicity. In fact, in leukemic B cells a strong Apo2.7 expression was induced only by ML I but not by ML III (Büssing *et al.*, 1999h). In the light of this finding, the possible impact of ML I- and ML III-rich VA-E in the treatment of B cell neoplasia has to be investigated carefully (Büssing *et al.*, 1999d).

Induction of Fas ligand

Apart from the onset of apoptotic cell death by the ML, Fas ligand (FasL, CD95L) and TNF-R1 (CD120a) expression increased after incubation with ML I and ML III in the surviving lymphocytes, i.e. CD4⁺ T cells, CD8⁺ cells, and CD19⁺ B cells, while the Fas expression decreased (Büssing *et al.*, 1999e). The FasL belongs to the TNF family and induces apoptosis through its cell surface receptor, Fas (Apo-1, CD95). FasL plays a pivotal role in lymphocyte cytotoxicity but is also involved in the downregulation of immune reactions (reviewed by Nagata and Golstein, 1995). The molecule is expressed in immune privileged body sides such as the retina and testis, and most abundantly in T cells (Suda *et al.*, 1995). Once FasL is expressed on activated T cells, they may kill Fas^+ target cells. Malfunction of the Fas system may cause lymphoproliferative disorders and accelerates autoimmune diseases; its exacerbation may cause tissue destruction (Nagata and Golstein, 1995). The observed effect of a FasL-induction by ML may reflect "activation" of surviving cells which did not result in a proliferation response as measured by the expression of interleukin-2 receptors (CD25) and transferrin receptors (CD71), or nuclear Ki-67 antigens (Büssing *et al.*, 1999e).

However, not only tumour-specific cytotoxic T cells may kill Fas-sensitive tumour cells, also $FasL^+$ tumour cells may counteract this attack by the elimination of Fas^+ cytotoxic T cells *via* FasL/Fas-mediated apoptosis. On the other hand, the expression of the Fas molecule slightly decreased on the surface of ML-treated surviving lymphocytes, and thus, these cells might be less sensitive against a counterattack of $FasL^+$ tumour cells. Further, ML I and ML III did not induce FasL expression in cultured Molt-4 cells, T-CLL and B-CLL cells (Büssing *et al.*, 1999e); however, other tumour cell lines have to be tested.

Stimulation of immunocompetent cells

Apart from an induction of pro-inflammatory cytokines such as IL-1, IL-6, TNF-α (Hajto *et al.*, 1990; Männel *et al.*, 1991; Ribéreau-Gayon *et al.*, 1996; Joller *et al.*, 1996), and gene expression of IL-10, IFN-γ, and GM-CSF (Hostanska *et al.*, 1996) by monocytes/macrophages and fibroblast/keratinocytes stimulated with ML I, recent findings indicate a ML I-induced clonogenic growth of $CD34^+$ haematopoietic progenitor cells when co-cultured with several growth factors (Vehmeyer *et al.*, 1998). These properties may explain the observed release of juvenile granulocytes and monocytes from the bone marrow in response to an intravenous application of VA-E (Hajto, 1986; Büssing *et al.*, 1996f). But are these juvenile cells functionally competent?

The phagocytic activity of human leukocytes reportedly increased in response to the B chain of ML I, however, ML I and A chain had no effect even though they inhibited spontaneous migration of macrophages (Metzner *et al.*, 1985). ML I at 50 and 100 μg/ml induced superoxide anion in human neutrophils and enhanced menadione-dependent release of H_2O_2 in rat thymocytes treated with > 1 μg/ml (Timoshenko and Gabius, 1993, 1995). These observations may have no clinical relevance because the concentrations were highly toxic to the cells. Using a physiological stimulus such as *Escherichia coli*, as opposed to viscotoxins, ML I and ML III did not enhance the oxidative burst of human neutrophils but slightly impaired cell function (Stein *et al.*, 1999a,b; Stein and Schietzel, this book).

Flavonoids and Phenylpropanoids

Flavonoids such as quercetin and its derivatives are diphenyl propanoids widely distributed in dietary plants, and also present in *Viscum album* (see Becker, this book). They are suggested to play a dual role in mutagenesis and carcinogenesis. In fact, quercetin was recognized to induce apoptosis in various malignant and non-malignant cell lines (Wei *et al.*, 1994; Kuo, 1996; Plaumann *et al.*, 1996; Fujita *et al.*, 1997; Csokay *et al.*, 1997; Weber *et al.*, 1997), but also to exert mutagenic properties, as measured by mutagenicity assays such as Ames test, Comet assay, micro-nucleus test, and DNA fingerprint analysis (Hardigree and Epler, 1978; MacGregor and Jurd, 1978; Sahu *et al.*, 1981; Hatcher and Bryan, 1985; Rueff *et al.*, 1986; Malinowski *et al.*, 1990; Suzuki *et al.*, 1991; Anderson *et al.*, 1997). Indeed, quercetin induced both, SCE-inducing DNA lesions in cultured peripheral blood mononuclear cells and apoptosis, while the phenylpropanoid syringin induced SCE only at higher but less cytotoxic concentrations (Figure 5). Thus, both phenomenons are different end-points of cellular affection by these compounds. However, quercetin may also act as an protectant against hydrogen peroxide-induced DNA strand breaks (Duthie *et al.*, 1997).

Although quercetin is highly mutagenic *in vitro*, it is not necessarily carcinogenic when administered *in vivo*. While quercetin was found to induce micronuclei *in vitro* in V79 cells and human lymphocytes, it failed to induce them *in vivo*, as measured in bone marrow polychromatic erythrocytes in mice (Caria *et al.*, 1995). In fact, chronical application of high doses (50 mg/kg BW) to rodents for 12 month was not carcinogenic, an effect which was shown to be dependent on the rapid metabolic inactivation of the flavonoid by catechol-o-methytransferase (Zhu *et al.*, 1994). Also, no tumour or hyperplasia were observed in F344 rats fed with 5% quercetin in the diet for 4 weeks followed by N-butyl-N-(4-hydroxybutyl) nitrosamine for 29 weeks, and no lesions were detected in rats given 5% quercetin diet only (Hirose *et al.*, 1983), indicating that the flavonoid is not carcinogenic in rat urinary bladder. On the other hand, in F344/N rats fed with quercetin (about 40 to 1,900 mg/kg BW/d) for 2 years, toxic and neoplastic lesions were seen in the kidney, including increased severity of chronic nephrophathy, hyperplasia, and benign tumours of the renal tubular epithelium (Dunnick and Hailey, 1992). In the nitrosomethylurea (NMU) model of rat pancreatic carcinogenesis, quercetin was recognized to exert a promoting and progressing effect: dietary quercetin in rats treated with NMU resulted in a significantly higher number of rats with dysplastic foci (pancreatic nodules and focal acinar cell hyperplasias) (Barotto *et al.*, 1998). Furthermore, carcinoma *in situ* and one microcarcinoma were found in these animals. Mitosis was significantly increased and apoptosis was diminished in focal acinar cell hyperplasias of the quercetin group (Barotto *et al.*, 1998). Zhu and Liehr (1994) observed an increase of estradiol-induced tumourigenesis in Syrian hamsters by the coadministration of estradiol plus 3% quercetin, i.e. increased number of tumour nodules and incidence of abdominal metastases. Also Ishikawa *et al.* (1985) reported an enhancing effect of quercetin on 3-methylcholanthrene carcinogenesis in C57B1/6 mice, an effect suggested to be associated with an increased mutation rate in the host.

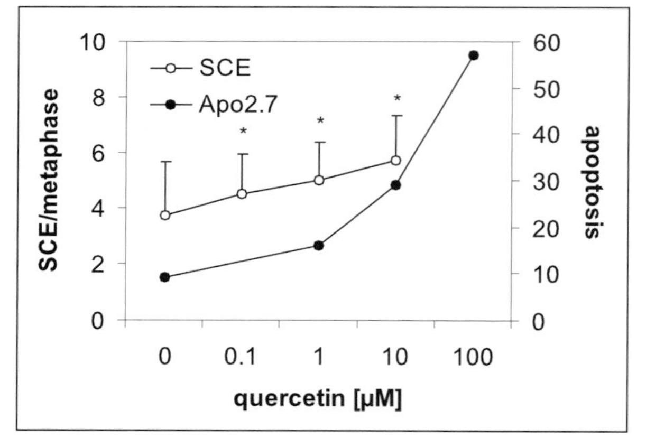

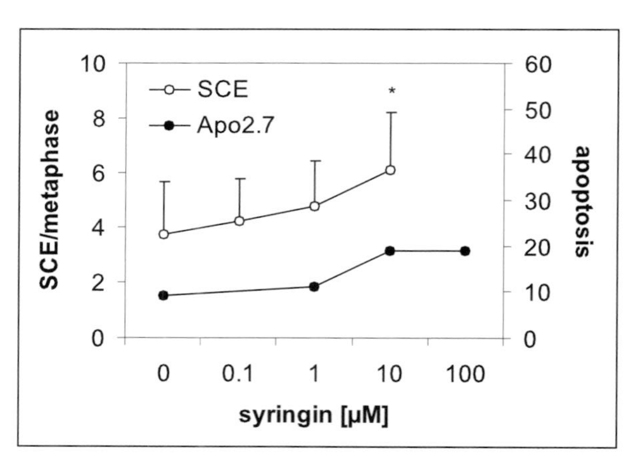

Figure 5 DNA lesions, as measured by the mean number (± SD) of sister chromatid exchanges (SCE) per 25 metaphase (SCE; ○), and induction of apoptosis, as measured by the expression of mitochondrial Apo2.7 molecules ($Apo2.7^{+}$; ●) in a representative experiment. To measure SCE, PHA-activated peripheral blood mononuclear cells were incubated with quercetin and syringin at 0, 0.1, 1, 10 and 100 μM for 72 h, while the investigation of apoptosis was performed on isolated lymphocytes after a 24 h incubation period as described (Büssing *et al.*, 1996a). * $p < 0.05$ (Wilcoxon's signed rank test)

Co-carcinogenicity of the flavonoids, however, is still controversial. The putative genotoxic metabolites of quercetin vary for different genetic end-points considered, and the fate of flavonoids might partly account for the conflicting data about their genotoxicity *in vivo* and carcinogenic activity (Rueff *et al.*, 1986). Oliveira *et al.* (1997) addressed the question of exposure to low levels present in the diet and approved an adaptive response by low doses of quercetin to challenging doses of quercetin, hydrogen peroxide and mitomycin C, using induction of chromosomal aberrations in V79 cells as the end point. Moreover, oral uptake of quercetin at concentrations that were about 10^3 times greater than the estimated average human intake of total flavonoids did not show mutagenicity/carcinogenicity in mice, as measured by micronucleus test and Ames Salmonella tester strain TA 98 (Aeschenbacher *et al.*, 1982). Calomme *et al.* (1996) stated that quercetin should not merely be regarded as a genotoxic risk factor in the human diet, since its mutagenicity may be inhibited by accompanying compounds including other flavonoids, and since quercetin itself exhibited an anitmutagenic effect against 2-aminofluorene in the Ames test (Calomme *et al.*, 1996), and produced anti-genotoxic effects in combination with food mutagens such as 3-amino-1-metyl-5H-pyrido (4,3-b)indole (Trp) and 2-amino-3-methylimidazo (4,5-f)quinoline (IQ) effects in human lymphocytes and sperm (Anderson *et al.*, 1997).

The mutagenic or anti-mutagenic effects of quercetin are further dependent on the mutagen tested and its activation (Ogawa *et al.*, 1985). In fact, quercetin enhanced the mutagenicity of tricyclic aromatic amines (aminofluorene, aminoanthracene, aminophenanthrene) and their acetamides, whereas the mutagenicity of aniline derivatives, biphenyl derivates, and bi- and tetra-cyclic amino derivatives are depressed (Ogawa *et al.*, 1985, 1987a,b,c). Here, quercetin may promote N-hydroxylation and deacetylation in the microsomes, and inhibits deacetylation in the cytosol. The mutagenicity-modulation of heterocyclic amines (Trp-P-1, Trp-P-2, Glu-P-1, Glu-P-2) by quercetin was liable to be affected by the content of S9 in a mammalian metabolic activation system (Ogawa *et al.*, 1987b). Mutagenicity and mutagenicity-enhancing effects of flavonoids, as measured in the Ames test, seem to depend on the number of hydroxyl groups substituted at the 3′, 4′ and 5′ position of the B ring, and the presence of a free hydroxy or methoxy group in the 7 position of the A ring, while the presence of a hydroxyl group at the 2′ position in the B ring of the flavonoid molecule markedly decreases its mutagenic activity (Ogawa *et al.*, 1987a; Czeczot *et al.*, 1990).

An inhibition of quercetin's mutagenic activity was observed by the addition of metal salts ($MnCl_2$, $CuCl_2$, $FeSO_4$, and $FeCl_3$), probably by facilitating catalytic oxidation of the flavonoid, while ascorbate, superoxide dismutase, NADH and NADPH enhanced the mutagenic activity of quercetin (Hatcher and Bryan, 1985). Addition of liver homogenate (S9 mix) may enhance the mutagenic activity of quercetin by scavenging superoxide radicals, thus inhibiting its autoxidation, and possibly by reducing quinone oxidation products of quercetin (Hatcher and Bryan, 1985). However, other reports point out that in mammalian cells quercetin is unable to induce SCE and point mutations, and that a putative clastogenic effect of the flavonoid is abolished by the addition of liver homogenate (van der Hoeven *et al.*, 1984).

Modulation of the detoxifying systems by flavonoids and their metabolites may be one of the key factors to explain the conflicting findings. In fact, polyphenolic flavonoids such as quercetin, myriacetin and kaempferol were observed to decrease the content of nuclear antioxidant defense glutathione (GSH) and glutathione S-transferase (Sahu and Gray, 1996), and thus can lead to oxidative DNA damage, which may be responsible for their mutagenic effects. However, depletion of reduced GSH by quercetin occurred prior to death of lymphocytes, Caco-2, HepG2, and HeLa cells (Duthie *et al.*, 1997), indicating that oxidative stress by itself may induce apoptosis, or that oxidative DNA damages induces arrest of the cell cycle *via* accumulation of the tumour-suppressor protein p53. Indeed, cell death induced by quercetin was observed in G_1 and S phase of the cell cycle of leukemic cell lines (Yoshida *et al.*, 1992; Wei *et al.*, 1994; Larocca *et al.*, 1996), an effect associated with a suppression of growth-related genes histone H4, cyclin A and B, and p34cdcc2 (Yoshida *et al.*, 1992). In the non-tumour cell line C3H10T1/2CL8 induction of apoptosis and induction of the p53 protein occurred out of the G_2/M phase of the cell cycle (Plaumann *et al.*, 1996). The G_2/M arrest seems to be p53-dependent as it did not occur in p53 knockout fibroblasts. Also in MDA-MB468 human breast cancer cells, the block of cell cycle at the G_2/M phase was associated with a prevention of the accumulation of newly synthesized p53 protein (Avila *et al.*, 1994).

The exact underlying mechanisms leading to apoptosis are unclear. The initial interaction of quercetin with DNA may have a stabilizing effect on its secondary structure, but prolonged treatment leads to an extensive disruption of the double helix (Alvi *et al.*, 1986). Quercetin may also block signal transduction pathways by inhibiting protein tyrosine kinases and serine/threonine protein kinases (Hagiwara *et al.*, 1988; Ferriola *et al.*, 1989; Kang and Liang, 1997), 1-phosphatidylinositol 4-kinase and 1-phosphatidylinositol 4-phosphate 5-kinase resulting in a reduction of inositol-1,4,5-trisphosphate (Nishioka *et al.*, 1989; Kang and Liang, 1997) which should decrease the release of Ca^{2+} from intracellular sources. An early down-regulation of the c-myc and Ki-ras oncogenes (Csokay *et al.*, 1997; Weber *et al.*, 1997) is suggested to be part of the antiproliferative action of quercetin in K562 human leukemia (Csokay *et al.*, 1997). Larocca *et al.* (1996) postulated that quercetin exerts its growth inhibitory action by interaction with type II estrogen binding sites and subsequent induction of transforming growth factor (TGF)-β1 expression and secretion.

Findings of Yokoo and Kitamura (1997) elucidated a novel action of quercetin as an apoptosis inhibitor. Pretreatment with quercetin protected mesangial cells from hydrogen peroxide (H_2O_2)-induced apoptosis. A similar effect was observed in other cell types including LLC-PK1 epithelial cells and NRK49F fibroblasts. This cyto-protective effect was found to be mediated *via* suppression of the tyrosine kinase-c-Jun/activator protein-1 (AP-1) pathway triggered by oxidant stress. However, quercetin did not inhibit caspase-3 activation and Apo2.7 expression in ML I-treated human lymphocytes (Büssing, unpublished results).

Although the bioavailability of quercetin and its derivatives from *Viscum album* is unclear, one may not ignore the fact, that flavonoids from VA-E applied repeatedly to tumour patients may accumulate in the blood. However, Lorch was unable to detect quercetin in mistletoe grown on appletree, pine tree or fir tree: VA-E from fir tree contain predominantly homoeriodyctiol-glycoside, and in mistletoe from pine 5,7-dimethoxy-4-hydroxyflavanon (personal communication). Both flavonoids were found in about 3 mg/g dried plant material. However, whole plant extracts from *Viscum album* were recognised not to induce mutagenic and/or genotoxic effects (see Stein, this book).

Phenolic Acids

Phenolic acids such as caffeic acid (CA) and ferulic acid (FA) are widely distributed in plant material in both free and combined forms and, as such, are components of the human diet. However, less is know on their effects when applied parenterally. The physiological relevance of CA and FA is suggested to be represented by their antioxidant action *in vivo* (Iwahashi *et al.*, 1990; Graf, 1992; Nardini *et al.*, 1997, 1998). This effect seems to be due to the ability of CA to reduce GSH depletion and to inhibit lipid peroxidation during t-butyl-hydroperoxide-induced oxidative stress (Nardini *et al.*, 1998). It can be concluded that CA exerts an antioxidant action inside the cell. Due to its presence in the diet, therefore, CA may play a role in the modulation of oxidative processes *in vivo*. In fact, dietary supplementation of CA in rats resulted in a statistically significant increase of alpha-tocopherol both in plasma and lipoprotein (Nardini *et al.*, 1997). While CA was not detectable in plasma under fasting conditions, in postprandial plasma it was present at μmol concentrations, doubling plasma total antioxidant capacity, and lipoproteins from CA-fed rats were more resistant than control to Cu^{2+}-catalyzed oxidation, despite the lack of incorporation of CA in the particles (Nardini *et al.*, 1997).

Additionally, CA is a selective, non-competitive inhibitor for 5-lipoxygenase and therefore for leukotriene biosynthesis (Koshihara *et al.*, 1984). CA and its methyl ester did not inhibit prostaglandin synthase activity at all, but rather stimulate it at higher doses. The biosynthesis of leukotrienes in mouse mast tumor cells was also inhibited completely with CA. Further, platelet aggregation induced by arachidonic acid was inhibited by CA at high doses, while platelet aggregation induced by ADP is not influenced by CA at all (Koshihara *et al.*, 1984).

Apart from this, CA exerts anti-proliferative effects, as it inhibited T-lymphocyte progenitor cells, particularly at lower cell concentrations, an effect which is restored by leukotriene B4, IL-1 and IL-2 (Miller *et al.*, 1989), and altered the colony-forming ability of MCF-7 human breast carcinoma cells (Ahn *et al.*, 1997). CA markedly reduced the IC_{50} of doxorubicin (Dox) in multidrug-resistant MCF-7/Dox cells (Ahn *et al.*, 1997). The level of TGF-β1 in MCF-7/Dox cells was about 3-fold greater than that in MCF-7 cells. In cells pretreated with CA, TGF-β1 and TGF-β2 levels were overexpressed only in MCF-7/Dox cells. These results suggest that CA is potentially a chemosensitizing agent with greater selectivity to drug-resistant MCF-7/Dox cells over parent MCF-7 cells and that the chemosensitizing effect is not mediated by altered drug concentrations in the cells, but may be possibly correlated to the induction of TGF-β isotypes (Ahn *et al.*, 1997).

Topical application of phenolic acids, such as p-coumaric, caffeic, ferulic, gentisic, protocatechuic, syringic and isovanillic acids, was effective against 12-O-tetradecanoylphorbol-13-acetate (TPA)-induced mouse ear oedema (Fernández *et al.*, 1998). Their action is markedly influenced by the inhibition of neutrophil migration into inflamed tissue (Fernández *et al.*, 1998). While the topical application of curcumin together with TPA inhibited the stimulation of [^{3}H]-thymidine uptake into epidermal DNA, CA, FA and chlorogenic acid were less effective as inhibitors of the TPA-dependent stimulation of DNA synthesis (Huang *et al.*, 1988). However, topical application of curcumin, CA, FA and chlorogenic acid together with TPA twice weekly for 20 weeks to mice previously initiated with 7,12-dimethyl benz(a)anthracene (DMBA) inhibited the number of TPA-induced tumors per mouse (Huang *et al.*, 1988).

Reports on the suggested anti-mutagenic and anti-carcinogenic effects of phenolic compounds are conflicting. Ellagic acid, CA, FA, chlorogenic acid and other phenolic compounds such as purpurogallin, quercetin, alizarin and monolactone were reported to inhibit GSH-transferase(s) activity (Das *et al.*, 1984). In a bacterial mutation assay using *Salmonella typhimurium*, CA and its derivative chlorogenic acid had inhibitory effects on the mutagenicity of Trp-P-1 and Glu-P-2 (Yamada and Tomita, 1996). CA completely eliminated the mutagenicity induced by activated Glu-P-2. Some compounds analogous to CA, such as cinnamic acid, coumaric acid, and FA, also significantly decreased the mutagenicity of Glu-P-2 (Yamada and Tomita, 1996). In other experiments, polyphenolic acids such as CA, FA, chlorogenic acid and ellagic acid failed to have such an effect in the *S. typhimurium* strain TA98 as indicator and hepatic S9 mixes as metabolic activating systems, while the plant flavonoids morin, myricetin and quercetin generally inhibited IQ, MeIQ, MeIQx and Trp-P-1 induced mutagenesis in a dose-dependent manner (Alldrick *et al.*, 1986). Using Chinese hamster ovary (CHO K-1) cells, addition of CA and FA significantly increased the frequency of mitomycin C- and UV-induced DNA lesions, as measured by SCE, while X-ray-induced DNA lesions decreased (Sasaki *et al.*, 1989).

Preliminary experiments of our group revealed a significant induction of SCE in 72 h cultured human whole blood cells by the addition of caffeic acid at 0.1 μM (from 3.73 ± 1.94 to 6.04 ± 1.83, $p > 0.001$; Büssing *et al.*, in preparation) Further, genotoxicity of DMBA in B6C3F1 mice was protected by CA, as measured by a 50% decrease of the DMBA-induced micronuclei (Raj *et al.*, 1983). Also, topical application of CA and FA simultaneously with phorbol-12-myristate-13-acetate (PMA) or mezerein resulted in significant protection against DMBA-induced skin tumors in mice (Kaul and Khanduja, 1998). As in rats receiving aminopyrine and nitrite, elevation of serum N-nitrosodimethylamine levels and the serum glutamic pyruvic transaminase levels associated with hepatotoxicity were blocked by these phenolic acids, dietary CA and FA may play a role in the body's defense against carcinogenesis by inhibiting the formation of N-nitroso compounds (Kuenzig *et al.*, 1984).

On the other hand, dietary levels of antioxidants such as the butylated hydroxyanisole, CA, sesamol, 4-methoxyphenol and catechol, known to enhance the incidences of forestomach papillomas and squamous cell carcinomas in F344 rats treated with N-methyl-N′-nitro-N-nitrosoguanidine (MNNG) (Hirose *et al.*, 1987, 1992), slightly increased the incidences of forestomach papillomas, and significantly increased with the antioxidants in combination (Hirose *et al.*, 1998). However, glandular stomach carcinogenesis was not enhanced (Hirose *et al.*, 1987, 1992). With regard to other organs, the incidence of colon tumors was significantly decreased only in the high dose combination group (Hirose *et al.*, 1998).

Other Biologically Active Compounds

Alkaloids from the Korean mistletoe were reported to inhibit the growth of cultured leukemia L1210 and increased the life span of leukemic mice (leukemia P388 in BDF_1 female mice) when applied intraperitoneally (Kwaja *et al.*, 1980). Due to their extreme lability, the structures of these compounds have not been identified. Khwaja *et al.* (1986) hypothesised that the alkaloids may appear as labile glycoconjugates with *Viscum album* proteins and lectins. However, the presence of alkaloids within *Viscum album* is still controversial (see Pfüller, this book).

Actually, less is know about a cytotoxic 5 kDa peptide (NSC 635 089) extracted from a bacterially fermented VA-E by Kuttan and co-workers (1988, 1990, 1992a–c). The tumouricidal activity of this peptide is probably mediated by the induction of NK cells and macrophages (Kuttan and Kuttan 1992b,c, Kuttan, 1993a). Furthermore, the action of a rhamnogalacturonan extracted from VA-E is thought to be mediated by NK-cells and monocytes/macrophages (Hamprecht *et al.*, 1987; Klett and Anderer, 1989; Müller and Anderer, 1990a-c; Zhu *et al.*, 1994), and bridges $CD56^+$ NK cells with tumour cells (Müller and Anderer, 1990b,c). Further, an oligosaccharide isolated from a VA-E induced interferon-γ from $CD4^+$ T cells and TNF-α from monocytes/macrophages (Klett and Anderer, 1989; Müller and Anderer, 1990a).

In a fermented drug extract from mistletoe from pine trees, Stein and Berg (1994, 1996a,b, 1998a) detected an undefined antigen that elicited a strong stimulation of lymphocytes ($CD4^+$ $CD25^+$ T helper cells), monocytes ($CD14^+$ $CD80^+$) with cytokine production (IL-6, TNF-α, and variable Th1- and Th2-cytokine concentrations) in untreated healthy and allergic individuals.

Interactions

Obviously, crude plant extracts from mistletoe contain several distinct components with different biological properties. Interactions of compounds from *Viscum album* are not well described at present (see Stein and Schietzel, this book). Fischer *et al.* (1996b) reported an inhibition of ML-cytotoxicity by the vesicles from *Viscum album*. Furthermore, a synergistic effect was suggested with a mixture of ML and vesicles more effectively increasing uptake of [^{3}H]-thymidine in the DNA of lymphocytes from VA-E-sensitised patients compared to ML and vesicles alone (Fischer, 1997a,b). This effect was suggested to be due to a binding of ML to glycolipids of the chloroplast membranes that form the vesicles. The responding cells were $CD4^+$ T helper cells, but not $CD8^+$ T suppressor/cytotoxic cells (Fischer, 1997a,b).

Preliminary results indicate that the polysaccharides from mistletoe berries induced IL-6 and IFN-γ and proliferation of $CD4^+$ T helper cells (Stein *et al.*, 1999c). Simultaneous addition of ML I or ML III with the polysaccharides may enhance the uptake of the thymidine-analogue BrdU in the DNA of cells as compared to the polysaccharides or ML alone. However, these responses showed strong interindividual differences, as some persons responded with enhanced proliferation, while others showed a suppression of polysaccharide-induced BrdU uptake by the ML, and others did not respond at all.

These results indicate that the immune sytem may respond individually towards distinct stimuli. Although not fully understood, several compounds in *Viscum album* may interact to impact the biological properties of the drug. But this question has to be addressed in further studies.

CLINICAL RELEVANCE OF CANCER TREATMENT WITH MISTLETOE EXTRACTS

Modes of Application

Although treatment of cancer patients with VA-E obviously has effects on immunocompetent cells, such as increased number and activity of lymphocytes, monocytes/macrophages and granulocytes, the clinical relevance of this treatment is controversially discussed. On the other hand, the biological effects of VA-E treatment are highly dependent on the mode of application (Table 4), however, only the subcutaneous application is recommended. One cannot exclude the possibility that the effectiveness of mistletoe therapy in complementary cancer therapy may improve (1) by a differentiated use of VA-E and adapted route of application, and (2) by the combination of conventional treatment strategies with VA-E therapy.

Clinical Studies

Meta-analyses of mistletoe studies

Controlled clinical studies of VA-E in cancer treatment have been reviewed by Kiene (1989, 1996), and Kleijnen and Knipschild (1994). In a meta-analysis of mistletoe studies in bronchus carcinoma, colorectal carcinoma, mamma carcinoma, gastric cancer and femal genital cancer, Kleijnen and Knipschild (1994) reviewed 11 controlled trials in which at least unbiased allocation into the contracted treatment groups was attempted. The principal endpoint was mostly survival time. Four studies reported significant better results with VA-E, while six trials showed a positive trend. The authors stated that methodological quality of these studies was poor. The best designed and performed trial did not find significant differences between the VA-E group and the controls. In this study (Dold *et al.*, 1991), a impressive number of complete remissions and partial remissions were observed both, in VA-E and so called "placebo group" which in fact was treated with a multi-vitamin-preparation. Thus, one cannot exclude the possibility that even in the control group,

Table 4 Summarised effects of VA-E application.

application	*clinical effects*		
	malignant cells	*immune system*	*well-being*
subcutaneous	(↘)	↑ number and activity of immune cells	↑ β-endorphin
intracutaneous*	?	↑ number of of immune cells	↑ pain reduction
intravenous	?	↑ number of juvenile granulocytes	↑ pain reduction
intrapleural	↓ pleurodesis	↑ number of macrophages and eosinophils	palliation
intratumoural	↓ tumour necrosis	?	palliation
oral **	↓ tumour starvation (?)	?	?

* application recommended in the treatment of arthrosis.
** application recommended in the treatment of hypertension.

apart from the multi-vitamines, other treatment strategies were used without the knowledge of the medicals.

Another meta-analysis published by Kiene (1989, 1996) scored 41 trials. Sixteen of these studies were reported to have solidity, nine had questionable solidity, and sixteen had no solidity at all. Twelve of the 16 studies with solidity (involving bronchus carcinoma, colorectal carcinoma, mamma carcinoma, gastric cancer, femal genital cancer, and liver metastases) showed better results for the mistletoe groups in regard of survival time, quality of life, or pleurodesis in cases of malignant pleural effusions.

Mistletoe studies

Cancer patients receiving subcutaneous mistletoe treatment commonly report an increase of their well-being with a feeling of "inner warmth" and improvement of social activity. By all means, the psychological effects of being "non-abandoned" and the recovery of self-determination by injecting the "hopeful drug" to themselves is hardly to be ignored. Anyway, a recent study (Heiny and Beuth, 1994) enrolling 68 patients with breast cancer (TNM stages III-IV) reported a significant release of β-endorphin by subcutaneous application of a VA-E normed on ML I (*Eurixor*) in those patients responding with local skin reactions (redness, swelling) or increased peripheral lymphocyte counts, indicating a modulation of neuroendocrine functions by the treatment.

In a preclinical study with brain tumour patients (e.g. glioblastoma multiforme) (Lenartz *et al.*, 1996), tumour destructive therapy (neurosurgery, perioperative dexamethasone treatment, and local radiotherapy) resulted in a significant decrease of peripheral blood cell counts. As compared to patients receiving standard therapy only, those patients treated with standard therapy and VA-E (*Eurixor*) subcutaneously revealed a significant improvement of peripheral T lymphocyte numbers (i.e. $CD4^+$ T-helper cells and $CD8^+$ T-suppressor/cytotoxic cells) within 12 weeks; however, 3 month later, the T cell numbers did not differ between both groups (Lenartz *et al.*, 1996). Furthermore, VA-E treatment resulted in an improved of quality of life (as measured by Spitzer's standard questionnaire) 6 month after surgery.

In a prospectively randomised clinical study (Heiny *et al.*, 1998) enrolling 79 patients with advanced colorectal cancer treated with standard chemotherapy (5-fluorouracil and folinic acid), the patients receiving chemotherapy and VA-E (*Eurixor*) showed a significant increase of their quality of life (as measured by "Functional Assessment of Cancer Therapy Scale V 3.0" questionnaire) not earlier than 12 weeks after the treatment, indicating that this effect may not due to a placebo effect which should occur in the early phase of the treatment. Furthermore, the duration of severity of mucositis was significantly reduced (Heiny *et al.*, 1998). However, the frequency and length of remission, relapse-free interval or overall survival did not differ between control group and those treated additionally with VA-E.

Similarly, another study involving 20 patients with advanced colorectal cancer receiving standard chemotherapy and 20 patients treated with standard therapy and VA-E (*Helixor*) was also unable to verify significant differences in respect to the number of patients in complete or partial remission (Douwes *et al.*, 1986). Although the response rates of both groups did not differ, the mean survival time of the responder receiving chemotherapy and VA-E was higher as compared to the responder receiving chemotherapy alone (Douwes *et al.*, 1986), indicating that VA-E may have a beneficial effect by improving the quality of life.

Although survival time and survival rate did not significantly differ between the treatment groups in a randomised and controlled study on the treatment of patients with non-small cell lung carcinoma published by Dold *et al.* (1991), the well-being of cancer patients significantly increased in the VA-E group (*Iscador*) as compared to the controls.

In a prospective comparative study (Gutsch *et al.*, 1988), patients were treated after radical mastectomy by different adjuvant therapies: 177 breast cancer patients were treated with conventional chemotherapy (CMF – cyclophosphamide, methotrexate, 5-fluorouracil), 192 patients with subcutaneously applied VA-E (*Helixor*), and 274 patients without chemotherapy or VA-E therapy. Here, the 5 year survival rate of CMF (67.7%) or VA-E (69.1%) treated patients was significantly higher as compared to the patients treated by mastectomy alone (59.7%). The survival rates of all patients treated with CMF or VA-E did not differ significantly. As compared to CMF or controls, however, the survival rate of VA-E treated patients was higher especially in those patients with more than 4 tumour-positive lymph nodes.

Intrapleural administration of mistletoe extracts is reported to result in pleurodesis in cancer patients with malignant pleural effusions (Salzer, 1977; Böck and Salzer, 1980; Böck, 1983a,b; Stumpf and Schietzel, 1994). In a recent study (Stumpf and Schietzel, 1994), 20 cancer patients with malignant pleural effusions were treated intrapleurally with VA-E. Regarding pleurodesis, the overall response rate was 72%, displaying only 1.2% side effects of WHO classification I. The decline of tumour cells in the effusion liquid correlated negatively with the number of instillations. The efficacy was suggested to be due to the cytostatic and immunomodulatory properties of the intrapleurally applied VA-E. Salzer and Popp observed an increase of eosinophil granulocytes, $CD4^+$ T helper cells, and NK cells in the effusions (Salzer, 1986; Salzer and Popp, 1990), while others described a transient increase of macrophages and eosinophils, and a constant increase of $CD8^+$ T cells (Stumpf and Büssing, 1997). Thus, VA-E-mediated pleurodesis might be due to a stimulation of antitumour immunity rather than mechanical sclerosis.

After intratumoural injections of VA-E, Koch (1993) described in animals local inflammation and subsequent degeneration of epithelial carcinoma within 7 to 10 d. Regarding this application in cancer patients, only a few case reports are published (Drees *et al.*, 1996; Stumpf *et al.*, 1997a; Mathes, 1997; Scheffler *et al.*, 1996). These reports describe effective reduction of tumour volume. To clarify whether tumour regression by an intratumoural injection of VA-E will be of benefit for the patients, controlled clinical studies especially in colorectal cancer patients with liver dissemination are required.

In conclusion, an increase of life quality and well-being was one of the most consistent effects of subcutaneous treatment with VA-E, while an improvement of survival (reported preferentially in older reports) may depend on several aspects such as individual responsiveness, tumour status, drug concentration, and route of application. Intrapleural and intralesional injection of potent VA-E may have beneficial effects as the tumour cells may be attacked directly, whether by the toxic proteins from *Viscum album* and/or by VA-E-activated immune cells.

CONCLUSIONS

Experimental research, however, was initiated within the last century and suggested that the VA-E may have beneficial effects as a cancer treatment. Also, treatment of arthrosis with intracutaneously applied VA-E extracts seems to be effective in several cases. Due to the lack of convincing studies, the benefit of VA-E treatment for hypertension and arteriosclerosis remains unclear.

The claims of pharmacologists to define the main biologically relevant component present in VA-E is difficult because several biologically active components are present in the drug used as a cancer treatment (Figure 6). It is obvious that VA-E contains potent cytotoxic proteins such as the ML and viscotoxins which induce apoptosis and accidental cell death, respectively. The mechanisms of cytotoxicity are known in several details, but several aspects remain to be clarified. Apart from an immunomodulation, these cytotoxic effects might be of relevance in the treatment of cancer patients. The clinical benefit must be examined carefully because the effects depend on the mode of application. However, introduction of VA-E normed or standardised on single components suggested to exert the main biological effects of the drug, obviously did not improve the quality of clinical results as compared to whole plant extracts standardised on other criteria. Encouraging results of cytotoxicity to cultured and implanted tumour cells are only reproducible in humans by the direct application of VA-E into the tumour or intracavitally. Preliminary results indicate that the intra- or peritumoural application of VA-E might be more effective in tumour reduction as compared to a subcutaneous one. Subcutaneous and intravenous application are recommended to modulate the immune system. The possible impact of the DNA stabilising properties of VA-E requires further research. An improvement of life quality and well-being is one of the most consistent effects of subcutaneous treatment with VA-E. However, the last and final chapters of the mistletoe story remain to be written, and thus, more rigorous scientific investigations are encouraged.

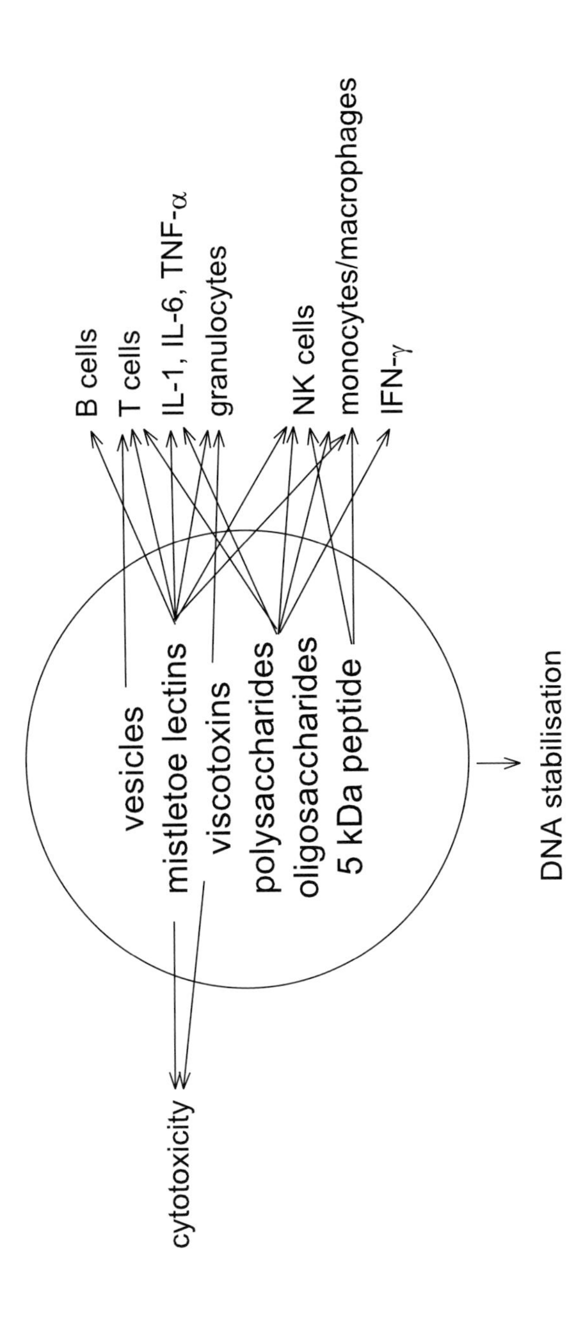

Figure 6 Immunological effects of *Viscum album* components. While the immunomodulating and the cytotoxic effects of *Viscum album* were clearly attributed to several compounds, as indicated by the arrows, the components responsible for the DNA stabilising properties of VA-E remain to be defined.

REFERENCES

Acatrinei, G.N. (1966) (Polyploide und Aneuploide in Zwiebelwurzeln unter Einfluß physiologisch-aktiver Substancen des Extraktes aus der Mistel (*Viscum album* L.)) (russian) *Genetika*, **4**, 97–104.

Acatrinei, G.N. (1967) (Changes in the mechanism of cell division induced by physiologically active substances in a plant extract) (russian) *Fisol. rast.*, **14**, 271–275.

Acatrinei, G.N. (1969) (Effect of physiologically active substances from mistletoe (*Viscum album*) extract on the nuclear division of onion cells; binuclear cells, synchronous and asynchronous bimitoses) (russian) *Genetika*, **5**, 170–174.

Ahn, C.H., Choi, W.C., and Kong, J.Y. (1997) Chemosensitizing activity of caffeic acid in multidrug-resistant MCF-7/Dox human breast carcinoma cells. *Anticancer Research*, **17**, 1913–1917.

Alldrick, A.J., Flynn, J., and Rowland, I.R. (1986) Effects of plant-derived flavonoids and polyphenolic acids on the activity of mutagens from cooked food. *Mutation Research*, **163**, 225–232.

Alvi, N.K., Rizvi, R.Y., and Hadi, S.M. (1986) Interaction of quercetin with DNA. *Biosci Rep*, **6**, 861–868.

Anderson, D., Basaran, N., Dobrzy[nacute]ska, M.M., Basaran, A.A., and Yu, T.W. (1997) Modulating effects of flavonoids on food mutagens in human blood and sperm samples in the comet assay. *Teratogenesis, Carinogenesis and Mutagenesis*, **17**, 45–58.

Angerhofer, C.K., Shier, W.T., and Vernon, L.P. (1990) Phospholipase activation in the cytotoxic mechanism of thionin purified from nuts of *Pyrularia pubera. Toxicon*, **28**, 547–557.

Antony, S., Kuttan, R., and Kuttan, G. (1999) Inhibition of lung metastasis by adoptive immunotherapy using iscador. *Immunological Investigations*, **28**, 1–8.

Arends, M.J., McGregor, A.H., and Wyllie, A.H. (1994) Apoptosis is inversely related to necrosis and determines the growth in tumors bearing contitutively expressed *myc*, *ras*, and *HPV* oncogenes. *Journal of Pathology*, **144**, 1045–1057.

Avila, M.A., Velasco, J.A., Cansado, J., and Notario, V. (1994) Quercetin mediates the down-regulation of mutant p53 in the human breast cancer cell line MDA-MB468. *Cancer Research*, **54**, 2424–2428.

Bantel, H., Engels, I.H., Voelter, W., Schulze-Osthoff, K., and Wesselborg, S. (1999) Mistletoe lectins activates caspase-8/FLICE independently of death receptor signaling and enhances anticancer drug-induced apoptosis. *Cancer Research*, **59**, 2083–2090.

Barbieri, L., Valbonesi, P., Bonora, E., Gorini, P., Bolognesi, A. and Stirpe, F. (1997) Polynucleotide:adenosine glycosidase activity of ribosome-inactivating proteins: effect on DNA, RNA and poly(A). *Nucleic Acids Research*, **25**, 518–522.

Barotto, N.N., López, C.B., Eynard, A.R., Fernández Zapico, M.E., and Valentich, M.A. (1998) Quercetin enhances pretumorous lesions in the NMU model of rat pancreatic carcinogenesis. *Cancer Letters*, **129**, 1–6.

Basaran, A.A., Yu, T.W., Plewa, M.J., and Anderson, D. (1996) An investigation of some turkish herbal medicines in *Salmonella typhimurium* and in the COMET assay in human lymphocytes. *Teratogenesis, Carinogenesis and Mutagenesis*, **16**, 125–138.

Benson, S.C., Olsnes, C., Pihl, A., Skorre, J., and Abraham, A. (1975) On the mechanism of protein synthesis inhibition by abrin and ricin. *European Journal of Biochemistry*, **59**, 573

Berger, M., and Schmähl, D. (1983) Studies on tumour-inhibiting efficacy of Iscador in experimental animal tumours. *Journal of Cancer Research and Clinical Oncology*, **105**, 262–265.

Beuth, J., Ko, H.L., Gabius, H.J., and Pulverer, G. (1991) Influence of treatment with the immunomodulatory effective dose of β-galactoside-specific lectin from mistletoe on tumour colonization in BALB/c-mice for two experimental model systems. *In vivo*, 5, 29–32.

Beuth, J., Ko, H.L., Gabius, H.J., Burrichter, H., Oette, K., and Pulverer, G. (1992) Behaviour of lymphocyte subsets and expression of activation markers in response to immunotherapy with galactoside-specific lectin from mistletoe in breast cancer patients. *Clinical Investigator*, **70**, 658–661.

Beuth, J., Ko, H.L., Tunggal, L., Steuer, M.K., Geisel, J., Jeljaszewicz, J., *et al.* (1993a) Thymocyte proliferation and maturation in response to galactoside-specific mistletoe lectin-1. *In vivo*, 7, 407–410.

Beuth, J., Ko, H.L., Tunggal, L., Geisel, J., and Pulverer, G. (1993b) Vergleichende Untersuchungen zur immunaktiven Wirkung von Galaktosid-spezifischem Mistellektin. Reinsubstanz gegen standardisierten Extrakt. *Arzneimittel-Forschung/Drug Research*, **43**, 166–169.

Beuth, J., Ko, H.L., Tunggal, L., Buss, G., Jeljaszewicz, J., Steuer, M.K., *et al.* (1994a) Immunaktive Wirkung von Mistellektin-1 in Abhängigkeit von der Dosierung. *Arzneimittel-Forschung/Drug Research*, **44**, 1255–1258.

Beuth, J., Ko, H.L., Tunggal, L., Buss, G., Jeljaszewicz, J., Steuer, M.K., *et al.* (1994b) Einfluß von wäßrigen, auf Mistellektin-1 standardisierten Mistelextrakten auf die In-vitro-(Tumor)Zellproliferation. *Deutsche Zeitschrift für Onkologie*, **26**, 1–6.

Beuth, J., Stoffel, B., Ko, H.L., Jeljaszewicz, J., and Pulverer, G. (1995) Immunomodulatory ability of galactoside-specific lectin standardized and depleted mistletoe extract. *Arzneimittel-Forschung/Drug Research*, **45**, 1240–1242.

Beuth, J., Stoffel, B., Samtleben, R., Staak, O., Ko, H.L., Pulverer, G., *et al.* (1996) Modulating activity of mistletoe lectins 1 and 2 on the lymphatic system in BALB/c mice. *Phytomedicine*, **2**, 239–273.

Beuth, J. (1997) Clinical relevance of immunoactive mistletoe lectin-1. *Anticancer Drugs*, **8** (Suppl 1), S53-S55.

Bloksma, N., van Dijk, H., Korst, P., and Willers, J.M. (1979) Cellular and humoral adjuvant activity of a mistletoe extract. *Immunobiology*, **156**, 309–319.

Bloksma, N., Schmiermann, P., de Reuver, M., van Dijk, H., and Willers, J.M. (1982) Stimulation of humoral and cellular immunity by *Viscum* preparations. *Planta Medica*, **4**, 221–227.

Böcher, E., Stumpf, C., Büssing, A., Schietzel, M. (1996) Prospektive Bewertung der Toxizität hochdosierter *Viscum album* L.-Infusionen bei Patienten mit progredientem Malignomen. *Zeitschrift für Onkologie*, **28**, 97–106.

Böck, D., and Salzer, G.: Morphologischer Nachweis einer Wirksamkeit der Iscadorbehandlung maligner Pleuraergüsse und ihre klinischen Ergebnisse. *Krebsgeschehen*, **12**, 49–53.

Böck, D. (1983a) Neue zytomorphologische Ergebnisse bei lokaler Behandlung des karzinomatösen Pleuraergusses. *Krebsgeschehen*, **15**, 33–34.

Böck, D. (1983b): Lokalbehandlung der Pleurakarzinose: Elektronenmikroskopische Befunde. *Krebsgeschehen*, **15**, 35–39.

Bräunig, B., Dorn, M., Adler, C., and Knick, E. (1993) Blutdruckenkende Wirkung dreier Mistel-Zubereitungen bei Patienten mit leichter und mittlerer Hypertonie. *Ärztezeitschrift für Naturheilverfahren*, **34**, 46–52.

Büssing, A., Azhari, A., Ostendorp, H., Lehnert, A., and Schweizer, K. (1994) *Viscum album* L. extracts reduce sister chromatid exchanges in cultured peripheral blood mononuclear cells. *European Journal of Cancer*, **30A**, 1836–1841.

Büssing, A., Lehnert, A., Schink, M., Mertens, R., and Schweizer, K. (1995a) Effect of *Viscum album* L. on rapidly proliferating amniotic fluid cells: Sister chromatid exchange frequency and proliferation index. *Arzneimittel-Forschung/Drug Research*, **45**, 81–83.

Büssing, A., Ostendorp, H., and Schweizer, K. (1995b) In vitro-Effekte von *Viscum album* (L.)-Präparationen auf kultivierte Leukämie-Zellen und mononukleäre Zellen des peripheren Blutes. *Tumourdiagnostik und Therapie*, **16**, 49–53.

Büssing, A., Regnery, A., and Schweizer, K. (1995c) Effects of *Viscum album* L. on cyclophosphamide-treated peripheral blood mononuclear cells in vitro: Sister chromatid exchanges and activation/proliferation marker expression. *Cancer Letters*, **94**, 199–205.

Büssing, A., Jungmann, H., Suzart, K., and Schweizer, K. (1996a) Suppression of sister chromatid exchange-inducing DNA lesions in cultured peripheral blood mononuclear cells by *Viscum album* L. *Journal of Experimental and Clinical Cancer Research*, **15**, 103–114.

Büssing, A., Jurin, M., Zarkovic, N., Azhari, T., and Schweizer, K. (1996b) DNA-stabilisierende Wirkungen von *Viscum album* L. – Sind Mistelextrakte als Adjuvans während der konventionellen Chemotherapie indiziert? *Forschende Komplementärmedizin*, **3**, 244–248.

Büssing, A., Suzart, K., Bergmann, J., Pfüller, U., Schietzel, M., and Schweizer, K. (1996c) Induction of apoptosis in human lymphocytes treated with *Viscum album* L. is mediated by the misteltoe lectins. *Cancer Letters*, **99**, 59–72.

Büssing, A., Suzart, K., Schweizer, K., and Schietzel, M. (1996d) Killing und Inflammation – Über die Apoptose-induzierende Potenz von *Viscum album* L.-Extrakten. *Zeitschrift für Onkologie*, **28**, 2–9.

Büssing, A. (1996e) Induction of apoptosis by the mistletoe lectins. A review on the mechanisms of cytotoxicity mediated by *Viscum album* L. *Apoptosis*, **1**, 25–32.

Büssing, A., Stumpf, C., Stumpf, R.T., Wutte, H., and Schietzel, M. (1996f) Therapiebegleitende Untersuchung immunologischer Parameter bei Tumour-Patienten nach hochdosierter intravenöser Applikation von *Viscum album* L.-Extrakten. *Zeitschrift für Onkologie*, **28**, 54–59.

Büssing, A., Suzart, K., and Schweizer, K. (1997) Differences in the apoptosis-inducing properties of *Viscum album* L. extracts. *Anticancer Drugs*, **8** (Suppl 1), S9-S14.

Büssing, A., Stein, G.M., and Pfüller, U. (1998a) Selective killing of $CD8^{+}$ cells with a "memory" phenotype ($CD62L^{lo}$) by the N-acetyl-D-galactosamine-specific lectin from *Viscum album* L. *Cell Death and Differentiation*, **5**, 231–240.

Büssing, A., Multani, A.S., Pathak, S., Pfüller, U., and Schietzel, M. (1998b) Induction of apoptosis by the N-acetyl-galactosamine-specific toxic lectin from *Viscum album* L. is associated with a decrease of nuclear p53 and Bcl-2 proteins and induction of telomeric associations. *Cancer Letters*, **130**, 57–68.

Büssing, A., Schaller, G., and Pfüller, U. (1998c) Generation of reactive oxygen intermediates by the thionins from *Viscum album* L. *Anticancer Research*, **18**, 4291–4296.

Büssing, A., and Schietzel, M. (1999) Apoptosis-inducing properties of *Viscum album* L. extracts from different host trees correlate with their content of toxic mistletoe lectins. *Anticancer Research*, **19**, 23–28.

Büssing, A., Wagner, M., Wagner, B., Stein, G.M., Schietzel, M., Schaller, G., *et al.* (1999b) Induction of mitochondrial Apo2.7 molecules and generation of reactive oxygen-intermediates in cultured lymphocytes by the toxic proteins from *Viscum album* L. *Cancer Letters*, **139**, 79–88.

Büssing, A., Stein, G.M., Wagner, M., Wagner, B., Schaller, G., Pfüller, U., *et al.* (1999c) Accidental cell death and generation of reactive oxygen intermediates in human lymphocytes induced by thionins from *Viscum album* L. *European Journal of Biochemistry*, **262**, 79–87.

Büssing, A., Stein, G.M., Stumpf, C., and Schietzel, M. (1999d) Release of interleukin-6 in cultured B-chronic lymphocytic leukaemia cells is associated with both, activation and cell death *via* apoptosis. *Anticancer Research*, **19**, 3953–3960.

Büssing, A., Stein, G.M., Pfüller, U., and Schietzel, M. (1999e) Induction of Fas ligand (CD95L) by the toxic mistletoe lectins in human lymphocytes. *Anticancer Research*, **19**, 1785–1790.

Büssing, A., Vervecken, W., Wagner, M., Wagner, B., Pfüller, U., and Schietzel, M (1999f) Expression of mitochondrial Apo2.7 molecules and caspase-3 activation in human lymphocytes treated with the ribosome-inhibiting mistletoe lectins and the cell membrane permeabilizing viscotoxins. *Cytometry*, **37**, 131–139.

Büssing, A., Rosenberger, A., Stumpf, C., and Schietzel M. (1999g) Verlauf lymphozytärer Subpopulationen bei Tumorpatienten nach subkutaner Applikation von Mistelextrakten. *Forschende Komplementärmedizin*, **6**, 196–204.

Büssing, A. Stein, G.M., Pfüller, U., and Schietzel, M (1999h) Differential binding of toxic lectins from *Viscum album* L., ML I and ML III, to human lymphocytes. *Anticancer Research*, **19**, 5095–5100.

Büssing, A. (1999) Biologische Wirkungen der Mistel. *Zeitschrift für Onkologie*, **31**, 35–43.

Calomme, M., Pieters, L., Vlietinck, A., and Vanden Berghe, D. (1996) Inhibition of bacterial mutagenesis by Citrus flavonoids. *Planta Medica*, **62**, 222–226.

Caria, H., Chaveca, T., Laires, A., and Rueff, J. (1995) Genotoxicity of quercetin in the micronucleus assay in mouse bone marrow erythrocytes, human lymphocytes, V79 cell line and identification of kinetochore-containing (CREST staining) micronuclei in human lymphocytes. *Mutation Research*, **343**, 85–94.

Chernov, V.A. (1954) (Die Anwendung von Mistelpräparaten in der experimentellen Chemotherapie von Tumouren) (russian). Trud. Akad. med. Nauk. SSSR, *Vopr. Onkol.*, 7, 139–150.

Chernov, V.A. (1955) (Die Wirkung von Mistelpräparaten auf die Metastasierung von Kaninchen Brown-Pearce Tumouren) (russian). *Vopr. Onkol.*, **1**, 38–48.

Crawford, A.C. (1905) Pharmacological notes on two American plants. *American Journal of Pharmacy*, **77**, 493–494.

Crawford, A.C. (1911) The pressor action of an American mistletoe. *Journal of the American Medical Association*, **57**, 865–868.

Crawford, A.C., and Watanabe, W.K. (1914) Parahydroxyphenylethylamine, a pressor compound in an American mistletoe. *Journal of Biological Chemistry*, **19**, 303.

Csokay, B., Prajda, N., Weber, G., and Olah, E. (1997) Molecular mechanisms in the antiproliferative action of quercetin. *Life Science*, **60**, 2157–2163.

Czeczot, H., Tudek, B., Kusztelak, J., Szymczyk, T., Dobrowolska, B., Glinkowska, G., *et al.* (1990) Isolation and studies of the mutagenic activity in the Ames test of flavonoids naturally occurring in medical herbs. *Mutation Research*, **240**, 209–216.

Das, M., Bickers, D.R., and Mukhtar, H. (1984) Plant phenols as in vitro inhibitors of glutathione S-transferase(s). *Biochemical and Biophysical Research Communications*, **120**, 427–433.

Dietrich, J.B., Ribéreau-Gayon, G., Jung, M.L., Franz, H., Beck, J.P., and Anton, R. (1992) Identity of the N-terminal sequences of the three A chains of mistletoe (*Viscum album* L.) lectins: homology with ricin-like plant toxins and single chain ribosome-inhibiting proteins. *Anticancer Drugs*, **3**, 507–511.

Dold, U., Edler, L., Mäurer, H.C., Müller-Wening, D., Sakellariou, B., Trendelenburg, F., *et al.* (1991) *Krebszusatztherapie beim fortgeschrittenen nicht-kleinzelligen Bronchialkarzinom. Multizentrische kontrollierte Studie zur Prüfung der Wirksamkeit von Iscador nd Polyerga.* Georg Thieme Verlag, Stuttgart, New York.

Doser, C., Doser, M., Hülsen, H., and Mechelke, F. (1989) Influence of carbohydrates on the cytotoxicity of an aqueous mistletoe drug and of purified lectins tested on human T-leukemia cells. *Arzneimittel-Forschung/Drug Research*, 39, 647–651.

Douwes, F.R., Wolfrum, D.I., and Migeod, F. (1986) Results of a prospective randomized study on chemotherapy versus chemotherapy plus "biological response modifier" in metastatic colorectal carcinoma. *Krebsgeschehen*, **6**, 155–164.

Drees, M., Berger, D.P., Dengler, W.A., and Fiebig, G.H. (1996) Direct cytotoxic effect of preparations used as unconventional methods in cancer therapy in human tumour xenografts in the clonogenic assay and in nude mice. In W. Arnold, P. Köpf-Maier, and B. Micheel, (eds.), *Immunodeficient animals. Models for cancer research. Contributions in Oncology*, Basel, **51**, 115–122.

Drüen, A. (1943) Ein Beitrag zur Arthrosebehandlung mit Mistelextrakten. *Deutsche Medizinische Wochenschrift*, **69**, 249–251.

Dunnick, J.K. and Hailey, J.R. (1992) Toxicity and carcinogenicity studies of quercetin, a natural component of foods. *Fundam Appl Toxicol*, **19**, 423–431.

Duthie, S.J., Johnson, W., and Dobson, V.L. (1997) The effect of dietary flavonoids on DNA damage (strand breaks and oxidised pyrimdines) and growth in human cells. *Mutation Research*, **390**, 141–151.

Elsner, W. (1940) Neuzeitliche Behandlung der Osteoarthrosen und der Spondyosis deformans mit Mistelextrakten. *Zentralblatt für Chirurgie*, **67**, 1104–1112.

Ebster, H., and Jarisch, A. (1929) Pharmakologische Untersuchungen über die Mistel. I. Mitteilung: Die Herzwirkung. *Archiv für experimentelle Pathologie und Pharmakologie*, **83**, 297–311.

Enders, A. (1940) Die Pharmakologie des herzwirksamen Stoffes der Mistel. *Archiv für experimentelle Pathologie und Pharmakologie*, **196**, 328–342

Endo, Y., Mitsui, K., Motizuki, M., and Tsurugi, K. (1987) The mechanism of action of ricin and related toxic lectins on eukaryotic ribosomes. The site and the characteristics of the modification in the 28S ribosomal RNA caused by the toxins. *Journal of Biological Chemistry*, **262**, 5908–5912.

Endo, Y., Tsurugi, K., and Franz, H. (1988) The site of action of the A-chain of mistletoe lectin I on eukaryotic ribosomes – the RNA N-glycosidase activity of the protein. *FEBS Letters*, **231**, 378–380.

Endo, Y. and Tsurugi, K. (1988) The RNA *N*-glycosidase activity of ricin A-chain. The characteristics of the enzymatic activity of ricin A-chain with ribosomes and with rRNA. *Journal of Biological Chemistry*, **263**, 8735–8739.

Evans, J., Wang, Y., Shaw, K.P., and Vernon, L.P. (1989) Cellular responses to *Pyrularia* thionin are mediated by Ca^{2+} influx and phospholipase A_2 activation and are inhibited by thionin tyrosin iodination. *Proceedings of the National Academy of Science USA*, **86**, 5849–5853.

Ewen, S.W.B., Bardocz, S., Grant, G., Pryme, I.F., and Pusztai, A. (1998) The effects of PHA and mistletoe lectin binding to epithelium of rat and mouse gut. In S. Bardocz, U. Pfüller, A. Pusztai, (eds.), *COST 98. Effects of antinutrients on the nutritional value of legume diets*, Vol. 5., Office for Official Publications of the European Communites, Luxembourg, pp. 221–225.

Fernández, M.A., Sáenz, M.T., and García, M.D. (1998) Anti-inflammatory activity in rats and mice of phenolic acids isolated from Scrophularia frutescens. *Journal of Pharmaceutical Pharmacology*, **50**, 1183–1186.

Ferriola, P.C., Cody, V., and Middleton, E. Jr. (1989) Protein kinase C inhibition by plant flavonoids. Kinetic mechanisms and structure-activity relationships. *Biochem Pharmacol*, **38**, 1617–1624.

Fischer, S., Scheffler, A., and Kabelitz, D. (1996a) Activation of human $\gamma\delta$ T-cells by heat-treated mistletoe plant extracts. *Immunology Letters*, **52**, 69–72.

Fischer, S., Scheffler, A., and Kabelitz, D. (1996b) Reaktivität von T-Lymphozyten gegenüber Mistelinhaltsstoffen. In R. Scheer, H. Becker, P.A. Berg, (eds), *Grundlagen der Misteltherapie. Aktueller Stand der Forschung und klinische Anwendung*. Hippokrates Verlag, Stuttgart, pp. 213–223.

Fischer, S., Scheffler, A., and Kabelitz, D. (1997a) Stimulation of the specific immune system by mistletoe extracts. *Anticancer Drugs*, **8** (Suppl 1), S33-S37.

Fischer, S., Scheffler, A., and Kabelitz, D. (1997b) Oligoclonal *in vitro* response of CD4 T cells to vesicles of mistletoe extracts in mistletoe treated cancer patients. *Cancer Immunology Immunotherapy*, **44**, 150–156.

Franz, H., Haustein, B., Luther, P., Kuropka, U., and Kindt, A. (1977) Isolierung und Charakterisierung von Inhaltsstoffen der Mistel (*Viscum album* L.). *Acta Biol Med Germ*, **36**, 113–117

Franz, H., Ziska, P., and Kindt, A. (1981) Isolation and properties of three lectins from mistletoe (*Viscum album* L.). *Biochemical Journal*, **195**, 481–484.

Franz, H. (1986) Mistletoe lectins and their A and B chains. *Oncology*, **43** (Suppl 1), 23–34.

Franz, H. (1989) Viscaceae lectins. *Advances in lectin research*, **2**, 28–59.

Fujita, M., Nagai, M., Murata, M., Kawakami, K., Irino, S., and Takahara, J. (1997) Synergistic cytotoxic effect of quercetin and heat treatment in a lymphoid cell line (OZ) with low HSP70 expression. *Leukocyte Research*, **21**, 139–145.

Fukunaga, T., Nishiya, K., Kajikawa, I., Takeya, K., and Itokawa, H. (1989) Studies on the constituents of Japanese mistletoes from different host trees, and their antimicrobial and hypotensive properties. *Chem. Pharm. Bull.*, **37**, 1543–1546.

Furutani, M., Kashiwagi, K., Ito, K., Endo, Y., and Igarashi, K. (1992) Comparison of the modes of action of a Vero toxin (a Shiga-like toxin) from *Escherichia coli*, of ricin, and of sarcin. *Archives of Biochemistry and Biophysics*, **293**, 140–146.

Gao, Y., Herndon, J.M., Zhang, H., Griffith, T.S., and Ferguson, T.A. (1998) Antiinflammatory effects of CD95 ligand (FasL)-induced apoptosis. *Journal of Experimental Medicine*, **188**, 887–896.

Gabius, H.J., Gabius, S., Joshi, S., Koch, B., Schroeder, M., Manzke, W.M., *et al.* (1994) From ill-defined extracts to the immunomodulatory lectin: Will there be a reason for oncological application of mistletoe? *Planta Medica*, **60**, 2–7.

Gaultier, M.R. (1907) Action hypotensive de léxtrait aqueux de gui. *La Semaine Médicale*, **43**, 513.

Gaultier, R. (1910) Etudes physiologiques sur le Gui (*Viscum album*). *Arch Internat. de Pharmacodynam et de Thérapie*, **20**, 96–116.

Göckeritz, W., Körner, I.J., Kopp, J., Bergmann, J., Pfüller, K., Eifler, R., *et al.* (1994) Mistletoe lectins: Comparative studies on cytotoxicity, receptor binding and their effect on the cytosolic calcium content in human lymphocytes. In E. van Driessche, J. Fischer, S. Beeckmans and T.C. Bog-Hansen, (eds.), *Lectins: Biology, Biochemistry, Clinical Biochemistry*, Vol. 10, Textop, Hellerup, pp. 345–354.

Goetsch, W. (1930) Viscysatum zur Vorbereitung und Unterstützung einer Kur mit natürlichen kohlensauren Mineralbädern. *Fortschritte der Medizin*, **48**, 992

Gorter, R.W., val Wely, M., Stoss, M., and Wollina, U. (1998) Subcutaneous infiltrates induced by injection of mistletoe extracts (Iscador). *American Journal of Therapeutics*, **5**, 181–187.

Graf, E, (1992) Antioxidant potential of ferulic acid. *Free Radical Biology & Medicine*, **13**, 435–448.

Gray, A.M., and Flatt, P.R. (1999) Insulin-secreting activity of the traditional antidiabetic plant *Viscum album* (mistletoe). *Journal of Endocrinology*, **160**, 409–414.

Green, D.R. and Martin, S.J. (1995) The killer and the executioner: how apoptosis controls malignancy. *Current Opinion Immunology*, **7**, 694–703.

Green, D.R. and Reed, J.C. (1998) Mitochondria and Apoptosis. *Science*, **281**, 1309–1312.

Griffiths, G.D., Leek, M.D., and Gee, D.J. (1987) The toxic plant proteins ricin and abrin induce apoptotic changes in mammalian lymphoid tissues and intestine. *Journal of Pathology*, **151**, 221–229.

Grisar, G. (1969) Arthrosis deformans – ein häufiges Problem in der täglichen Praxis. *Zeitschrift für Allgemeinmedizin/Der Landarzt*, **45**, 1585–1589.

Groh, H. (1965) Die Behandlung der Kniegelenksarthrose. *Münchner Medizinische Wochenschrift*, **107**, 2180.2183.

Gutsch, J., Berger, H., Scholz, G., and Denck, H. (1988) Prospektive Studie beim radikal operierten Mammakarzinom mit Polychemotherapie, Helixor und unbehandelter Kontroll. *Deutsche Zeitschrift für Onkologie*, **4**, 94–101.

Hagiwara, M., Inoue, S., Tanaka, T., Nunoki, K., Ito, M., and Hidaka, H. (1988) Differential effects of flavonoids as inhibitors of tyrosine protein kinases and serine/threonine protein kinases. *Biochemical Pharmacology*, **37**, 2987–2992.

Hahne, M., Renno, T., Schroeter, M., Irmler, M., French, L., Bornard, T., *et al.* (1996) Activated B cells express functional Fas ligand. *European Journal of Immunology*, **26**, 721–724.

Hajto, T. (1986) Immunomodulatory effects of iscador: A *Viscum album* preparation. *Oncology*, **43** (Suppl 1), 51–65.

Hajto,T., Hostanska, K., and Gabius, H.J. (1989) Modulatory potency of the β-galactoside-specific lectin from mistletoe extract (Iscador) on the host defense system *in vivo* in rabbits and patients. *Cancer Research*, **49**, 4803–4808.

Hajto, T., Hostanska, K., Frey, K., Rordorf, C., and Gabius, H.J. (1990) Increased secretion of tumour necrosis factor α, interleukin 1, and interleukin 6 by human mononuclear cells exposed to β-galactoside-specific lectin from clinically applied mistletoe extract. *Cancer Research*, **50**, 3322–3326.

Hajto, T., Hostanska, K., Fischer, J., and Lentzen, H. (1996) Investigations of cellular parameters to establish the response of a biomodulator: galactoside-specific lectin from *Viscum album* plant extract. *Phytomedicine*, **3**, 129–137.

Ham, S.S., Kang, S.T., Choi, K.P., Park, W.B., and Lee, D.S. (1998) Antimutagenic effect of Korean mistletoe extracts. *J Korean Soc Food Sci Nutr*, **27**, 359–365.

Hanzlik, P.J., and French, W.O. (1924) The pharmacology of *Phoradendron flavescens* (American mistletoe). *The Journal of Pharmacology and Experimental Therapy*, **23**, 269–306.

Hamprecht, K., Handgretinger, R., Voetsch, W., and Anderer, F.A. (1989) Mediation of human NK-activity by components in extracts of *Viscum album*. *International Journal of Immunopharmacology*, **9**, 199–209.

Hardigree, A.A., and Epler, J.L. (1978) Comparative mutagenesis of plant flavonoids in microbial systems. *Mutation Research*, **58**, 231–239.

Hatcher, J.F., and Bryan, G.T. (1985) Factors affecting the mutagenc activity of quercetin for *Salmonella typhimurium* TA98: metal ions, antioxidants and pH. *Mutation Research*, **148**, 13–23.

Havas, L.J. (1937) Effects of Colchicin and *Viscum album* preparations upon phytocarcinomata caused by *B. tumefaciens*. *Nature*, **139**, 371–372

Havas, L.J. (1939) Growth of induced plant tumours. *Nature*, **143**, 789–791.

Healy, K.C. (1995) Telomere dynamics and telomerase activation in tumour progression: prospects for prognosis and therapy. *Oncology Research*, **7**, 121–130.

Heiny, B.M., and Beuth, J. (1994) Misteltoe extract standardized for the galactoside-specific lectin (ML-1) induces β-endorphin release and immunopotentiation in breast cancer patients. *Anticancer Research*, **14**, 1339–1342.

Heiny, B.M., Albrecht, V., and Beuth, J. (1998) Lebensqualitätsstabilisierung durch Mistellektin-1 normierten Extrakt beim fortgeschrittenen kolorektalen Karzinom. *Onkologe*, 4 (Suppl 1), 35–39.

Henze, C., and Ludwig, W. (1937) Über die Herzwirkung der Mistel. *Archiv für experimentelle Pathologie und Pharmakologie*, **187**, 694–705.

Hirose, M., Fukushima, S., Sakata, T., Inui, M., and Ito, N. (1983) Effect of quercetin on two-stage carcinogenesis of the rat urinary bladder. *Cancer Letters*, **21**, 23–27.

Hirose, M., Masuda, A., Imaida, K., Kagawa, M., Tsuda, H., and Ito, N. (1987) Induction of forestomach lesions in rats by oral administrations of naturally occurring antioxidants for 4 weeks. *Japanese Journal of Cancer Research*, **78**, 317–321.

Hirose, M., Kawabe, M., Shibata, M., Takahashi, S., Okazaki, S., and Ito, N. (1992) Influence of caffeic acid and other o-dihydroxybenzene derivatives on N-methyl-N′-nitro-N-nitrosoguanidine-initiated rat forestomach carcinogenesis. *Carcinogenesis*, **13**, 1825–1828.

Hirose, M., Takesada, Y., Tanaka, H., Tamano, S., Kato, T., and Shirai, T. (1998) Carcinogenicity of antioxidants BHA, caffeic acid, sesamol, 4-methoxyphenol and catechol at low doses, either alone or in combination, and modulation of their effects in a rat medium-term multi-organ carcinogenesis model. *Carcinogenesis*, **19**, 207–212.

van der Hoeven, J.C., Bruggeman, I.M., and Debets, F.M. (1984) Genotoxicity of quercetin in cultured mammalian cells. *Mutation Research*, **136**, 9–21.

Hoffmann-Axthelm, W., and Zellner, R. (1954) Über die konservativeBehandlung chronischer Arthropathien des Kiefergelenkes. *Deutsche zahnärztliche Zeitschrift*, 9, 1036–1042.

Hostanska, K., Hajto, T., Spagnoli, G.C., Fischer, J., Lentzen, H., and Herrmann, R. (1996) A plant lectin derived from *Viscum album* induces cytokine gene expression and protein production in cultures of human peripheral blood mononuclear cells. *Natural Immunity*, **14**, 295–304.

Hostanska, K., Hajto, T., Weber, K., Fischer, J., Lentzen, H., Sütterlin, B., *et al.* (1996–1997) A natural immunity-activating plant lectin, *Viscum album* agglutinin-I, induces apotosis in human lymphocytes, monocytes, monocytic THP-1 cells and murine thymocytes. *Natural Immunity*, **15**, 295–311.

Huang, M.T., Smart, R.C., Wong, C.Q., and Conney, A.H. (1988) Inhibitory effect of curcumin, chlorogenic acid, caffeic acid, and ferulic acid on tumor promotion in mouse skin by 12-O-tetradecanoylphorbol-13-acetate. *Cancer Research*, **48**, 5941–5946.

Hülsen, H., and Mechelke, F. (1982) The influence of a mistletoe preparation on suspension cell cultures of human leukemia and human myeloma cells. *Arzneimittel-Forschung/Drug Research*, **32**, 1126–1127.

Hülsen, H., Doser, C., and Mechelke, F. (1986) Differences in the in vitro effectiveness of preparations produced from mistletoesmistletoe of various host trees. *Arzneimittel-Forschung/Drug Research*, **36**, 433–436.

Hülsen, H., Kron, R., and Mechelke, F. (1989) Influence of *Viscum album* preparations on the natural killer cell-mediated cytotoxicity of peripheral blood. *Naturwissenschaften*, **76**, 530–531.

Hülsen, H., Jilg, S., and Mechelke, F. (1992) Einfluß von Mistelpräparaten auf die *In-vitro*-Aktivität die natürlichen Killer-Zellen von Krebspatienten. *Therapeutikon*, **6**, 585–588.

Hülsen, H., and Born, U. (1993) Einfluß von Mistelpräparaten auf die In-vitro-Aktivität die natürlichen Killer-Zellen von Krebspatienten. Teil 2. *Therapeutikon*, **7**, 434–439.

Hughes, J.N., Lindsay, C.D., and Griffiths, G.D. (1996) Morphology of ricin and abrin exposed endothelial cells is consistent with apoptotic cell death. *Hum. Exp. Toxicol.*, **15**, 443–451.

Irmler, B. (1989) Viscum album als Therapeutikum bei Erkrankungen der Gelenke. *Natura-Med*, **4**, 638–641.

Ishikawa, M., Oikawa, T., Hosokawa, M., Hamada, J., Morikawa, K., and Kobayashi, H. (1985) Enhancing effect of quercetin on 3-methylcholanthrene carcinogenesis in C57Bl/6 mice. *Neoplasma*, **32**, 435–441.

Iwahashi, H., Ishii, T., Sugata, R., and Kido, R. (1990) The effects of caffeic acid and its related catechols on hydroxyl radical formation by 3-hydroxyanthranilic acid, ferric chloride, and hydrogen peroxide. *Archives of Biochemistry and Biophysics*, **276**, 242–247.

Janssen, O., Scheffler, A., and Kabelitz, D. (1993) In vitro effects of mistletoe extracts and mistletoe lectins. Cytotoxicity towards tumour cells due to the induction of programmed cell death (apoptosis), *Arzneimittel-Forschung/Drug Research*, **43**, 1221–1227.

Jarisch, A., and Henze, C. (1937) Über Blutdrucksenkung durch chemische Erregung depressorischer Nerven. *Archiv für experimentelle Pathologie und Pharmakologie*, **187**, 106–730

Jarisch, A. (1938) Die blutdrucksenkende Wirkung der Mistel. *Wiener Klinische Wochenschrift*, **37**, 1032–1035.

Joller, P.W., Menrad, J.M., Schwarz, T., Pfüller, U., Parnham, M.J., Weyhenmeyer, R., *et al.* (1996) Stimulation of cytokine production via a special standardized mistletoe preparation in an in vitro human skin bioassay. *Arzneimittel-Forschung/Drug Research*, **46**, 649–653.

Jung, M.L., Baudino, S., Ribéreau-Gayon, G., and Beck, J.P. (1990) Characterization of cytotoxic proteins from mistletoe (*Viscum album* L.). *Cancer Letters*, **51**, 103–108.

Jurin, M., Zarkovic, N., Hrzenjak, M., and Ilic, Z. (1993) Antitumourous and immunomodulatory effects of the *Viscum album* L. preparation Isorel. *Oncology*, **50**, 393–398.

Kang, T.B. and Liang, N.C. (1997) Studies on the inhibitory effects of quercetin on the growth of HL-60 leukemia cells. *Biochem Pharmacol*, **54**, 1013–1018.

Kaul, A., and Khanduja, K.L. (1998) Polyphenols inhibit promotional phase of tumorigenesis: relevance of superoxide radicals. *Nutr Cancer*, **32**, 81–85.

Khwaja, T.A., Varven, J.C., Pentecost, S., and Pande, H, (1980a) Isolation and biologically active alkaloids from Korean misteltoe, *Viscum album, coloratum. Experientia*, **36**, 599–560.

Khwaja, T.A., Dias, C.B., Papoian, T., and Pentecost, S. (1980b) Studies on cytotoxic and immunologic effects of *Viscum album* (mistletoe). *Proceedings of the American Association of Cancer Research*, **22**, 253.

Khwaja, T.A., Dias, C.B., and Pentecost, S. (1986) Recent studies on the anticancer activities of mistletoe (*Viscum album*) and its alkaloids. *Oncology*, **43** (Suppl 1), 42–50.

Kiene, H. (1989) Klinische Studien zur Misteltherapie karzinomatöser Erkrankungen. Eine Übersicht. *Therapeutikon*, **3**, 347–353.

Kiene, H. (1996) Beurteilung klinischer Studien zur Misteltherapie. In R. Scheer, H. Becker, and P.A.Berg, (eds.*), Grundlagen der Misteltherapie. Aktueller Stand der Forschung und klinische Anwendung*, Hippokrates Verlag, Stuttgart, pp. 484–496.

Kienholz, E (1990). Behandlung von Arthrosen mit intracutaner Mistelinjektion. Erfahrungen mit der Arthrosetherapie bei 319 Patienten. *Natura-Med*, **4**, 201–202.

Kleijnen, J., and Knipschild, P. (1994) Mistletoe treatment for cancer. Review of controlled trials in humans. *Phytomedicine*, **1**, 253–260.

Klett, C.Y., and Anderer, F.A. (1989) Activation of natural killer cell cytotoxicity of human blood monocytes by a low molecular weight componenet from *Viscum album* extract. *Arzneimittel-Forschung/Drug Research*, **39**, 1580–1585.

Klug (1906) Viscolan, eine neue Salbengrundlage. *Deutsche Medizinische Wochenschrift*, **51**, 2071–2072.

Koch, F.E. (1938a) Experimentelle Untersuchungen über entzündungs- und nekroseerzeugende Wirkungen von *Viscum album. Zschr. ges. exp. Med.*, **103**, 740–749.

Koch, F.E. (1938b) Experimentelle Untersuchungen über lokale Beeinflussung von Impfgeschwulsten. *Zeitschrift für Krebsforschung*, **47**, 325–335.

Kochi, S.K. and Collier, R.J. (1993) DNA fragmentation and cytolysis in U937 cells treated with diphtheria toxin or other inhibitors of protein synthesis. *Experimental Cell Research*, **208**, 296–302.

Köteles, G.J., Kubasova, T., Hurna, E., Horváth, G., and Pfüller, U. (1998) Cellular and cytogenetic approaches in testing toxic and safe concentrations of mistletoe lectins. In S. Bardocz, U. Pfüller, A. Pusztai, (eds), *COST 98. Effects of antinutrients on the nutritional value of legume diets,* Vol. 5, Luxembourg: Office for Official Publications of the European Communites, pp. 81–86.

Koshihara, Y., Neichi, T., Murota, S., Lao, A., Fujimoto, Y., and Tatsuno, T. (1984) Caffeic acid is a selective inhibitor for leukotriene biosynthesis. *Biochimica et Biophysica Acta*, **792**, 92–97.

Konopa, J., Woynarowski, J.M., and Lewandowska-Gumieniak, M. (1980) Isolation of Viscotoxins. Cytotoxic basic polypeptides from *Viscum album* L. *Hoppe-Seyler's Zeitschrift für Physiologische Chemie*, **361**, 1525–1533.

Kovacs, E., Hajto, T., and Hostanska, K. (1991) Improvement of DNA repair in lymphocytes of breast cancer patients treated with *Viscum album* extract (Iscador). *European Jounal of Cancer*, **27**, 1672–1676.

Kovacs, E., Kuehn, J.J., Werner, M., and Hoffmann, J. (1995) Effect of Iscador on DNA repair after radiation or cyclophosphamide. Correlation with IFN-γ production. *Onkologie*, **18**, S2, 651.

Kovacs, E., Kuehn, J.J., Werner, M., and Hoffmann, J. (1996) Die Wirkung einer Behandlung mit *Viscum album* (Iscador) auf die DNA-Reparatur der Lymphozyten bei Karzinompatienten. *Forschende Komplementärmedizin*, **3** (Suppl 1), V19, 18.

Kroemer, G., Zamzami, N., and Susin, S.A. (1997) Mitochodrial control of apoptosis. *Immunology today*, **18**, 44–51.

Kuban, G. (1991) Behandlung der Gonarthrose und der *Periarthrosis humeroscapularis* mit Plenosol-Quaddelung. *Naturheilpraxis mit Naturmedizin*, **44**, 782–785.

Kubasova, T., Pfüller, U., Bojtor, I., Köteles, G.J. (1998) Modulation of immune response by mistletoe lectin I as detected on tumor model *in vivo*. In S. Bardocz, U. Pfüller, A. Pusztai, (eds.), *COST 98. Effects of antinutrients on the nutritional value of legume diets*, Vol. 5., Office for Official Publications of the European Communites, Luxembourg, pp. 202–207.

Kuenzig, W., Chau, J., Norkus, E., Holowaschenko, H., Newmark, H., Mergens, W., *et al.* (1984) Caffeic and ferulic acid as blockers of nitrosamine formation. *Carcinogenesis*, **5**, 309–313.

Kunze, E., Schukz, H., Ahrens, H., and Gabius, H.J. (1997) Lack of an antitumoral effect of immunomodulatory galactoside-specific mistletoe lectin on N-methyl-N-nitrosourea-induced urinary bladder carcinogensis in rats. *Exp. Toxic Pathol.*, **49**, 167–180.

Kunze, E., Schukz, H., and Gabius, H.J. (1998) Inability of galactoside-specific mistletoe lectin to inhibit N-methyl-N-nitrosourea-induced tumor development ihn the urinary bladder of rats and to mediate a local cellular immune esponse after long-term adminisration. *Journal of Cancer Research and Clinical Oncology*, **124**, 73–87.

Kuo, S.M. (1996) Antiproliferative potency of structurally distinct dietary flavonoids on human colon cancer cells. *Cancer Letters*, **110**, 41–48.

Kuttan, G., Vasudevan, D.M., and Kuttan, R. (1988) Isolation and identification of a tumour reducing component from mistletoe extract (Iscador). *Cancer Letters*, **41**, 307–314.

Kuttan, G., Vasuddevan, D.M., and Kuttan, R. (1990) Effect of a preparation from *Viscum album* on tumour development in vitro and in mice. *Journal of Ethnopharmacology*, **29**, 35–41.

Kuttan, G., Vasudevan, D.M., and Kuttan R. (1992a) Tumour reducing activity of an isolated active ingredient from mistletoe extract and its possible mechanism of action. *Journal of Cancer Research and Clinical Oncology*, **11**, 7–12.

Kuttan, G., and Kuttan, R. (1992b) Immunomodulatory activity of a peptide isolated from *Viscum album* extract (NSC 635 089). *Immunological Investigations*, **21**, 285–296.

Kuttan, G., and Kuttan, R. (1992c) Immunological mechanism of action of the tumour reducing peptide from mistletoe extract (NSC 635089) cellular proliferation. *Cancer Letters*, **66**, 123–130.

Kuttan, G. (1993a) Tumouricidal activity of mouse peritoneal macrophages treated with *Viscum album* extract. *Immunological Investigations*, **22**, 431–440.

Kuttan, G., and Kuttan, R. (1993b) Reduction of leukopenia in mice by "*viscum album*" administration during radiation and chemotherapy. *Tumouri*, **79**, 74–76.

Kuttan, G., Menon, L.G., and Kuttan, R. (1996) Prevention of 20-methylcholanthrene-induced sarcoma by a mistletoe extract, Iscador. *Carcinogenesis*, **17**, 1107–1109.

Kuttan, G., Menon, L.G., Anthony, S., and Kuttan, R. (1997) Anticarcinogenic and antimetastatic activity of Iscador. *Anticancer Drugs*, **8** (Suppl 1), S15-S16.

Larocca, L.M., Teofili, L., Maggiano, N., Piantelli, M., Ranelletti, F.O., and Leone, G. (1996) Quercetin and the growth of leukemic progenitors. *Leukaemia and Lymphoma*, **23**, 49–53.

Legel, C. (1942) Behandlung der Arthrosis deformans mit dem Mistelextrakt Plenosol. Zur baldigen Wiederherstellung der Arbeitsfähigkeit. *Fortschritte in der Therapie*, **18**, 184–189.

Lenartz, D., Stoffel, B., Menzel, J., and Beuth, J. (1996) Immunoprotective activity of the galactoside-specific lectin from mistletoe after tumour destructive therapy in glioma patients. *Anticancer Research*, **16**, 3799–3802.

Lenartz, D., Andermahr, J., Plum, G., Menzel, J., and Beuth, J. (1998) Efficiency of treatment with galactoside-specific lectin from mistletoe against rat glioma. *Anticancer Research*, **18**, 1011–1014.

Luther, P., Theise, H., Chatterjee, B., Karduck, D., and Uhlenbruck, G. (1980) The lectin from *Viscum album* L. – Isolation, characterization, properties and structure. *International Journal of Biochemistry*, 11, 429–435.

MacGregor, J.T., and Jurd, L. (1978) Mutagenicity of plant flavonoids: structural requirements for mutagenic activity in Salmonella typhimurium. *Mutation Research*, **54**, 297–309.

Männel, D.N., Becker, H., Gundt, A., Kist, A., and Franz, H. (1991) Induction of tumor necrosis factor expression by a lectin from *Viscum album. Cancer Immunology Immunotherapy*, **33**, 177–182.

Martin, S.J. (1993) Apoptosis: suicide, execution or murder? *Trends in Cell Biology*, **3**, 141–144.

Mathes, H. (1997) Intraläsionale Mistelinjektionen in Lebermetastasen bei kolorektalem Karzinom und in das primäre Hepatozelluläre Karzinom (HCC) (abstract). *Der Merkurstab*, **50** (Sonderheft Juni), 41.

Mattausch, F. (1938) Über blutdruck- undkreislaufsteuernde Wirkungen der Mistel. *Wiener Medizinische Wochenschrift*, **45**, 1175–1177.

Mengs, U., Clare, C.B., Poiley, J.A. (1997) Genotoxicity testing of an aqueous mistletoe extract *in vitro*. *Arzneimittel-Forschung/Drug Research*, **47**, 316–319.

Mengs, U. (1998) Toxicity of an aqueous mistletoe extract: acute and subchronic toxicity in rats, genotoxicity *in vitro*. In S. Bardocz, U. Pfüller, A. Pusztai, (eds.), *COST 98. Effects of antinutrients on the nutritional value of legume diets,* Vol. 5, Luxembourg: Office for Official Publications of the European Communites, pp. 77–80.

Mengs, U., Weber, K., Schwarz, T., Hajto, T., Hostanska, K., and Lentzen, H. (1998) Effects of the standardized mistletoe preparation Lektinol on granulopoiesis and pulmonary metastases in mice. In S. Bardocz, U. Pfüller, A. Pusztai, (eds), *COST 98. Effects of antinutrients on the nutritional value of legume diets,* Vol. 5, Luxembourg: Office for Official Publications of the European Communites, pp. 194–201.

Metzner, G., Franz, H., Kindt, A., Fahlbusch, B., and Süss, J. (1985) The *in vitro* activity of lectin I from mistletoe (ML I) and its isolated A and B chains on functions of macrophages and polymorphonuclear cells. *Immunobiology*, **169**, 461–471.

Metzner, G., Franz, H., Kindt, A., Schuman, I., and Fahlbusch, B. (1987) Effects of lectin I from mistletoe (ML I) and its isolated A and B chains on human mononuclear cells: mitogenic activity and lymphokine release. *Pharmazie*, **42**, 337–340.

Miller, A.M., Elfenbein, G.J., and Barth, K.C. (1989) Regulation of T-lymphopoiesis by arachidonic acid metabolites. *Experimental Hematology*, **17**, 198–202.

Möckel, B., Schwarz, T., Zinke, H., Eck, J., Langer, M., and Lentzen, H. (1997) Effects of mistletoe lectin I on human blood cell lines and peripheral blood cells. Cytotoxicity, apoptosis and induction of cytokines. *Arzneimittel-Forschung/Drug Research*, **47**, 1145–1151.

Montanaro, I., Sperti, S., Mattioli, A., Testoni, G., and Stirpe, F. (1975) Inhibition by ricin of protein synthesis *in vitro*. *Journal of Biochemistry*, **146**, 127–131.

Müller, J.A. (1932) Zur Kenntnis der Inhaltsstoffe der nordischen Mistel, *Viscum album* L. *Archiv der Pharmazie*, **270**, 449–476.

Müller, G. (1962) Plenosol-Therapie in der Allgemeinpraxis. *Der Landarzt*, **38**, 622–623.

Müller, Th., and Müller, I. (1978a) Zur Möglichkeit der Früherkennung der *Arthrosis coxae* und deren Behandlung. *Erfahrungsheilkunde*, **27**, 120–121.

Müller, Th., and Müller, I. (1978b) Früherkennung der *Arthrosis coxae* und deren Behandlung. *Erfahrungsheilkunde*, **27**, 364–365.

Müller, E.A., and Anderer, F.A. (1990a) A *Viscum album* oligosaccharide activating human natural cytotoxicity is an interferon γ inducer. *Cancer Immunology Immunotherapy*, **32**, 221–227.

Müller, E.A., and Anderer, F.A. (1990b) Chemical specificity of effector cell/tumour cell bridging by a *Viscum album* rhamnogalacturonan enhancing cytotoxicity of human NK cells. *Immunopharmacology*, **19**, 69–77.

Müller, E.A., and Anderer, F.A.(1990c) Synergistic action of a plant rhamnogalacturonan enhancing antitumour cytotoxicity of human natural killer and lymphokine-activated killer cells: chemical specificity of target cell recognition. *Cancer Research*, **50**, 3646–3651.

Multani, A.S., Hopwood, V.L., and Pathak, S. (1996) A modified fluorescence in situ hybridization (FISH) technique. *Anticancer Research*, **16**, 3435–3438.

Nagata, S., and Golstein, P. (1995) The Fas death factor. *Science*, **267**, 1449–1456.

Nanba, T. (1980) *Genshokuwakanyakuzukan*, Hoikusya, Tokyo, pp. 172.

Nardini, M., Natella, F., Gentili, V., Di Felice, M., and Scaccini, C. (1997) Effect of caffeic acid dietary supplementation on the antioxidant defense system in rat: an *in vivo* study. *Archives of Biochemistry and Biophysics*, **342**, 157–160.

Nardini, M., Pisu, P., Gentili, V., Natella, F., Di Felice, M., Piccolella, E., and Scaccini, C. (1998) Effect of caffeic acid on tert-butyl hydroperoxide-induced oxidative stress in U937. *Free Radical Biology & Medicine*, **25**, 1098–1105.

Narimanov, A.A., Popova, O.I., and Muraviova, D.D. (1992) *Viscum album* L. polysaccharides used to change the sensitivity of mice to γ-radiation. *Radiobiologiia*, **32**, 868–872.

Nienhaus, J., and Leroi, R. (1970) Tumourhemmung und Thymusstimulation durch Mistelpräparate. *Elemente der Naturwissenscahften*, **13**, 45–54.

Nienhaus, J., Stoll, M., and Vester, F. (1970) Thymus stimulation and cancer prophylaxis by *Viscum* proteins. *Experientia*, **26**, 523–525.

Nishioka, H, Imoto, M., Sawa, T., Hamada, M., Naganawa, H., Takeuchi, T., and Umezawa, K. (1989) Screening og phosphatidylinositol kinase inhibitors from Streptomyces. *J. Antibiot.* **42**, 823–825.

Obatomi D.K., Bikomo, E.O., and Temple, V.J. (1994) Anti-diabetic properties of the African mistletoe in streptozotocin-induced diabetic rats. *Journal of Ethnopharmacology*, **43**, 13–17.

Oda, T., Komatsu, N., and Muramatsu, T. (1997) Cell lysis by ricin D and ricin E in various cell lines. *Biosci. Biotechnol. Biochem.*, **61**, 291–297.

Ogawa, S., Hirayama, T., Nohara, M., Tokuda, M., Hirai, K., and Fukui, S. (1985) The effect of quercetin on the mutagenicity of 2-acetylaminofluorene and benzo[alpha]pyrene in *Salmonella typhimurium* strains. *Mutation Research*, **142**, 103–107.

Ogawa, S., Hirayama, T., Sumida, Y., Tokuda, M., Hirai, K., and Fukui, S. (1987a) Enhancement of the mutagenicity of 2-acetylaminofluorene by flavonoids and the structural requirements. *Mutation Research*, **190**, 107–112.

Ogawa, S., Hirayama, T., Fujioka, Y., Ozasa, S., Tokuda, M., Hirai, K., *et al.* (1987b) Mutagenicity modulating effect of quercetin on aromatic amines and acetamines. *Mutation Research*, **192**, 37–46.

Ogawa, S., Hirayama, T., Tokuda, M., Hirai, K., and Fukui, S. (1987c) Mutagenicity-enhancing effect of quercetin on the active metabolites of 2-acetylaminofluorene with mammalian metabolic activation systems. *Mutation Research*, **192**, 241–246.

Oliveira, N.G., Rodrigues, A.S., Chaveca, T., and Rueff, J. (1997) Induction of an adaptive response to quercetinm, mitomycin C and hydrogen peroxide by low doses of quercetin in V79 Chinese hamster cells. *Mutagenesis*, **12**, 457–462.

Olsnes, S., Stirpe, F., Sansvig, K., and Pihl, A. (1982) Isolation and characterization of Viscumin, a toxic lectin from *Viscum album* L. (mistletoe). *Journal of Biological Chemistry*, **257**, 13263–13270.

Orlowski, P. (1932) Über die blutdruckherabsetzende Wirkung der Mistel. *Wiener Medizinische Wochenschrift*, **33**, 1069–1070.

Osario e Castro, V.R., van Kuiken, B., and Vernon, L.P. (1989) Action of thionin isolated from *Pyrularia pubera* on human erythrocytes. *Toxicon*, **27**, 501–510.

Osario e Castro, V.R., Ashwood, E.R., Wood, S.G., and Vernon, L.P. (1990) Hemolysis of erythrocytes and fluorescence polarization changes elicited by peptide toxins, aliphatic alcohols, related glycols and benzylidene derivatives. *Biochimica et Biophysica Acta*, **1029**, 252–258.

Pathak, S., Dave, B.J., and Gagos, S.H. (1994a) Chromosome alterations in cancer development and apoptosis. *In Vivo*, **8**, 843–850.

Pathak, S., Risin, S., Brown, N.M., and Berry, K.K. (1994b) Telomeric association of chromosomes is an early manifestation of programmed cell death. *International Journal of Oncology*, **4**, 323–328.

Peters, G. (1957) Übersichten Insulin-Ersatzmittel pflanzlichen Ursprungs. *Deutsche Medizinische Wochenschrift*, **82**, 320–322.

Pic, and Bonnamour (1923) *Phytothérapie*, Paris.

Plaumann, B., Fritsche, M., Rimpler, H., Brandner, G., and Hess, R.D. (1996) Flavonoids activate wild-type p53. *Oncogene*, **13**, 1605–1614.

Pora, A., Pop, E., Roska, D., and Radu, A. (1957) Der Einfluß der Wirtspflanze auf den Gehalt an hypotensiven und herzwirksamen Prinzipien der Mistel (*Viscum album* L.). *Pharmazie*, **12**, 528–538.

Pryme, I.F., Pusztai, A., Bardocz, S., and Ewen, S.W.B. (1998a) The induction of gut hyperplasia by phytohaemagglutinin in the diet and limitation of tumour growth. *Histology and Histopatholgy*, **13**, 575–583.

Pryme, I.F., Bardocz, S., Grant, G., Pusztai, A., and Pfüller, U. (1998) The plant lectins PHA and ML-1 suppress the growth of a lymphosarcoma tumour in mice. In S. Bardocz, U. Pfüller, A. Pusztai, (eds.), *COST 98. Effects of antinutrients on the nutritional value of legume diets*, Vol. 5., Office for Official Publications of the European Communites, Luxembourg, pp. 215–220.

Pusztai, A., Grant, G., Gelencsér, E., Ewen, S.W.B., Pfüller, U., Eifler, R., *et al.* (1998) Effects of orally administered mistletoe (type-2 RIP) lectin on growth, body composition, small intestinal structure, and insulin levels in young rats. *Nutritional Biochemistry*, **9**, 31–36.

Raj, A.S., Heddle, J.A., Newmark, H.L., and Katz, M. (1983) Caffeic acid as an inhibitor of DMBA-induced chromosomal breakage in mice assessed by bone-marrow micronucleus test. *Mutation Research*, **124**, 247–253.

Reis, H. (1986) Zur Behandlung degenerativer Skeletterkrankungen. *Erfahrungsheilkunde*, **35**, 2–8.

Ribéreau-Gayon, G., Jung, M.L., Baudino, S., Sallé, G., and Beck, J.P. (1986a) Effects of mistletoe (*Viscum album* L.) extracts on cultured tumour cells. *Experientia*, **42**, 594–599.

Ribéreau-Gayon, G., Jung, M.L., Di Scala, D., and Beck, J.P. (1986b) Comparison of the effects of fermented and unfermented mistletoe preparations on cultured tumour cells. *Oncology*, **43** (Suppl 1), 35–41.

Ribéreau-Gayon, H., Jung, M.L., Beck, J.P. and Anton, R. (1995) Effect of fetal calf serum on the cytotoxic activity of mistletoe (*Viscum album* L.) lectins in cell culture. *Phytotherapy Research*, **9**, 336–339.

Ribéreau-Gayon, G., Dumont, S., Müller, C., Jung, M.L., Poindron, P., and Anton, R. (1996) Mistletoe lectins I, II, and III induce the production of cytokines by cultured human monocytes. *Cancer Letters*, **109**, 33–38.

Ribéreau-Gayon, G., Jung, M.L., Frantz, M., and Anton, R. (1997) Modulation of the cytotoxicity and enhancement of cytokine release induced by *Viscum album* L. extracts or mistletoe lectins. *Anticancer Drugs*, **8** (Suppl 1), S3-S8.

Ripperger (1937) *Grundlagen zur praktischen Pflanzenheilkunde*, Stuttgart.

Roderfeld (1950) Praktische Erfahrungen über die Behandlung von Arthrosen und Polyarthritis mit dem Mistelpräparat Plenosol. *Deutsches Gesundheitswesen*, **5**, 953–954.

Rosell, S., and Samuelsson, G. (1966) Effect of mistletoe viscotoxin and phoratoxin on blood circulation. *Toxicon*, **4**, 107–110.

Rueff, J., Laires, A., Borba, H., Chaveca, T., Gomes, M.I., and Halpern, M. (1986) Genetic toxicology of flavonoids: the role of metabolic conditions in the induction of reverse mutation, SOS functions and sister-chromatid exchanges. *Mutagenesis*, **1**, 179–183.

Sahu, R.K., Basu, R., and Sharma, A. (1981) Genetic toxicological of some plant flavonoids by the micronucleus test. *Mutation Research*, **89**, 69–74.

Sahu, S.C., and Gray, G.C. (1997) Pro-oxidant activity of flavonoids: effects on glutathione and glutathione S-transferase in isolated rat liver nuclei. *Cancer Letters*, **104**, 193–196.

Salzer, G. Die lokale Behandlung karzinomatöser Pleuraergüsse mit dem Mistelpräprat Iscador. *Österreichische Zeitschrift für Onkologie*, **4**, 13–14.

Salzer, G. (1986) Pleura Carcinosis. Cytomorphological findings with the mistletoe preparation Iscador and other pharmaceuticals. *Oncology*, **43**, 66–70.

Salzer, G., and Popp, W. (1990) Die lokale Iscadorbehandlung der Pleurakarzinose. In Jungi, Senn, (eds.), *Krebs und Alternativmedizin II*. Springer-Verlag, Heidelberg, pp. 70–83.

Samuelsson, G. (1959) Phytochemical and pharmacological studies on *Viscum album* L. III. Isolation of a hypotensive substance: γ-aminobutyric acid.. *Svensk Farmaceutisk Tidskrift*, **62**, 169–190.

Samuelsson, G. (1961) Phytochemical and pharmacological studies on *Viscum album* L. V. Further improvements in the isolation methods for Viscotoxin. Studies on Viscotoxin from *Viscum album* growing on Tillia cordata mill. *Svensk Farmaceutisk Tidskrift.*, **65**, 481–494.

Samuelsson, G. and Pettersson, B. (1970) Separation of viscotoxins from the European mistletoe, *Viscum album* L. (Loranthaceae) by chromatography on sulfoethyl sephadex. *Acta Chemica Scandinavica*, **24**, 2751–2756.

Samulesson, G. (1974) Mistletoe toxins. *System. Zool.*, **22**, 566–569.

Sargiacomo, M. and Hughes, R.C. (1982) Interaction of ricin-sensitive and ricin-resistant cell lines with other carbohydrate-binding toxins. *FEBS Letters*, **141**, 14–18.

Sasaki, Y.F., Imanishi, H., Ohta, T., and Shirasu, Y. (1989) Modifying effects of components of plant essence on the induction of sister-chromatid exchanges in cultured Chinese hamster ovary cells. *Mutation Research*, **226**, 103–110.

Sauviat, M.P., Berton, J., and Pater, C. (1985) Effect of phoratoxin B on the electrical and mechanical activities of rat papillary muscle. *Acta Pharmac. Sin.*, **6**, 91–92.

Schaller, G., Urech, K., and Giannattasio, M. (1996) Cytotoxicity of different viscotoxins and extracts from European subspecies of *Viscum album* L. *Phytotherapy Research*, **10**, 473–477.

Scheffler, A., Mast, H., Fischer, S., and Metelmann, H.R. (1996) Komplette Remission eines Mundhöhlenkarzinoms nach alleiniger Mistelbehandlung. In R. Scheer, H. Becker, P.A. Berg, (eds.), *Grundlagen der Misteltherapie. Akueller Stand der Forschung und klinische Anwendung*. Hippokrates Verlag, Stuttgart, pp. 453–464.

Schimmel (1971) Die Arthrosebehandlung in der Praxis. *Zeitschrift für Allgemeinmedizin/Der Landarzt*, **47**, 657–661.

Schmidtke-Schrezenmeier, G., Reck, R., and Gerster, G. (1992) Behandlung der nichtaktivierten Gonarthrose. Besserung durch ein Phytotherapeutikum. *Therapiewoche*, **42**, 1322–1325.

Schrader, G. and Apel, K. (1991) Isolation and characterization of cDNAs encoding viscotoxins of mistletoe (*Viscum album*). *European Journal of Biochemistry*, **198**, 549–553.

Schultze, J.L., Stettin, A., and Berg, P.A. (1991) Demonstration of specifically sensitized lymphocytes from patients treated with an aqueous mistletoe extract (*Viscum album* L.) *Klinische Wochenschrift*, **69**, 397–403.

Schwartz, L.M. and Osborn, B.A. (1993) Programmed cell death, apoptosis and killer genes. *Immunology today*, **14**, 582–590.

Seeger, P.A.G. (1965a-c) Über die Wirkung von Mistelextrakten (Iscador und Plenosol) auf die Vitalität und Verimpfbarkeitvon Tumourzellen des Ehrlichschen Ascitescarcinoms der Maus, zugleich ein Beitrag zur Bedeutung der lymphocytären Geschwulstabwehr als erfolgreichstem Wirkungsfaktor von Viscumextrakten. *Erfahrungsheilkunde*, **14**, 149–174, 253–272, 301–317.

Selawry, O.E., Schwartz, M.R., and Haar, H. (1959) Tumour-inhibitory activity of products of Loranthacae (mistletoe) (abstract) *Proceedings of the American Association of Cancer Research*, **3**, 62–63.

Selawry, O.S., Vester, F., Mai, W., and Schwartz, M.R. (1961) Zur Kenntnis der Inhaltsstoffe von *Viscum album*, II. Tumourhemmende Inhaltsstoffe. *Hoppe-Seyler's Zeitschrifft für physiologische Chemie*, **324**, 262–281.

Sharma, H.W., Maltese, J.Y., Zhu, X., Kaiser, H.E., and Narayanan, R. (1996) Telomeres, telomerase and cancer: is the magic bullet real? *Anticancer Research*, **16**, 511–515.

Sickel, K. (1971) Plenosol^R in der Therapie arthrotischer Veränderungen. *Zeitschrift für Allgemeinmedizin/Der Landarzt*, **47**, 1667–1669.

Sommer, K.H. (1957) Die Behandlung der sogenannten Verbrauchsarthrosen mit Plenosol. *Ther. Gegenw.*, **96**, 216–217.

Staak, J.O., Stoffel, B., Wagner, H., Pulverer, G., and Beuth, J.: In-vitro-Zytotoxizität der Viscum album-Agglutinine I und II. *Zeitschrift für Onkologie*, **30**, 29–33.

Stein, G. and Berg, P.A.(1994) Non-lectin component in a fermented extract from *Viscum album* L. grown on pines induced proliferation of lymphocytes from healthy and allergic individuals *in vitro*. *European Journal of Clinical Pharmacology*, **47**, 33–38.

Stein, G.M., and Berg, P.A. (1996a) Evaluation of the stimulatory activity of a fermented mistletoe lectin I-free mistletoe extract on T-helper cells and monocytes in healthy individuals in vitro. *Arzneimittel-Forschung/Drug Research*, **46**, 635–639.

Stein, G.M., Meink, H., Durst, J., and Berg, P.A. (1996b) Release of cytokines by a fermented lectin-1 (ML-1) free mistletoe extract reflects diffferences in the reactivity of PBMC in healthy and allergic individuals and tumour patients. *European Journal of Clinical Pharamacoogy*, **151**, 247–252.

Stein, G.M., and Berg, P.A. (1997a) Mistletoe extract-induced effects on immunocompetent cells: *in vitro* studies. *Anticancer Drugs*, **8** (Suppl 1), S39-S42.

Stein, G.M., Stettin, A., Schultze, J., and Berg, P.A. (1997b) Induction of anti-mistletoe lectin antibodies in relation to different mistletoe extracts. A short review. *Anticancer Drugs*, **8** (Suppl 1), S57-S59.

Stein, G.M., Berg, P.A. (1998a) Flow cytometric analyses of the specific activation of peripheral blood mononuclear cells from healthy donors after in vtro timulation with a fermented mistletoe extract and mistletoe lectins. *European Journal of Cancer*, **34**, 1105–1110.

Stein, G.M. and Berg, P.A. (1998b) Modulation of cellular and humoral immune responses during exposure of healthy individuals to an aqueous mistletoe extract. *European Journal of Medical Research*, **3**, 307–314.

Stein, G.M., Henn, W., von Laue, H.B., and Berg, P.A. (1998a) Modulation of cellular and humoral immune responses of tumor patients by mistletoe therapy. *European Journal of Medical Research*, **3**, 194–202.

Stein, G.M., von Laue, H.B., Henn, W., and Berg, P.A. (1998b) Human anti-mistletoe lectin antibodies. In S. Bardocz, U. Pfüller, A. Pusztai, (eds.), *COST 98. Effects of antinutrients on the nutritional value of legume diets,* Vol. 5, Luxembourg: Office for Official Publications of the European Communites, pp. 168–175.

Stein, G.M., Schaller, G., Pfüller, U., Wagner, M., Wagner, B., Schietzel, M., *et al.* (1999a) Characterisation of granulocyte stimulation by thionins from European mistletoe and from wheat. *Biochimica et Biophysica Acta*, **1426**, 80–90.

Stein, G.M., Schaller, G., Pfüller, U., Schietzel, M., and Büssing, A. (1999b) Thionins from *Viscum album* L.: influence of viscotoxins on the activation of granulocytes. *Anticancer Research*, **19**, 1037–1042.

Stein, G.M., Edlund, U., Pfüller, U., Büssing, A., and Schietzel, M. (1999c) Influence of polysaccharides from *Viscum album* L. on human lymphocytes, monocytes and granulocytes *in vitro*. *Anticancer Research*, **19**, 3907–3919

Stettin, A., Schultze, J.L., Stechemesser, E., and Berg, P.A. (1990) Anti-mistletoe-antibodies are produced in patients during therapy with an aqueous mistletoe extract derived from *Viscum album* L. and neutralize lectin-induced cytotoxicity in vitro. *Klinische Wochenschrift*, **68**, 896–900.

Stirpe, F., Legg, R.F., Onyon, L.J., Ziska, P., and Franz, H. (1980) Inhibition of protein synthesis by a toxic lectin from *Viscum album* L. (mistletoe). *Biochemical Journal*, **190**, 843–845.

Stirpe, F., Sandvig, K., Olsnes, S., and Pihl, A. (1982) Action of Viscumin, a toxic lectin from mistletoe, on cells in culture. *Journal of Biological Chemistry*, **257**, 13271–13277.

Stirpe, F.. Barbieri, L., Giulia, M., Soria, M., and Lappi, D.A. (1992) Ribosome-inactivating proteins from plants: present status and future prospects. *Bio/Technology*, **10**, 405–412.

Stoffel, B., Krämer, K., Mayer, H., and Beuth, J. (1997) Immunomodulating efficacy of combined administration of galactose-specific lectin standardized mistletoe extract and sodium selenite in BALB/c-mice. *Anticancer Research*, **17**, 1893–1896.

Strauss, O. (1931) Über ein neues Hypotonikum. *Wiener Medizinische Wochenschrift*, **24**, 831.

Stumpf, C., and Schietzel, M. (1994) Intrapleurale Instillation eines Extraktes aus *Viscum album* (L.) zur Behandlung maligner Pleuraergüsse. *Tumourdiagnostik und Therapie*, **15**, 57–62.

Stumpf, C., Ramirez-Martinez, S., Becher, A., Stein, G.M., Büssing, A., and Schietzel, M. (1997a) Intratumourale Mistelapplikation bei stenosierendem Rezidiv eines Cardia-Carcinoms. *Erfahrungsheilkunde*, Sonderausgabe August,: 509–513.

Stumpf, C., and Büssing, A. (1997b) Stimulation of antitumour immunity by intrapleural instillation of a *Viscum album* L. extracts. *Anticancer Drugs*, **8** (Suppl 1 1), S23–S26.

Suda, T., Okazaki, T., Naito, Y., Yokota, T., Arai, N., Ozaki, S., *et al.* (1995) Expression of the Fas ligand in cells of T cell lineage. *Journal of Immunology*, **154**, 3806–3813.

Suzuki, S., Takada, T., Sugawara, Y., Muto, T., and Kominami R (1991) Quercetin induces recombinational mutations in cultured cells as detected by DNA fingerprinting. *Japanese Journal of Cancer Research*, **82**, 1061–1064.

Swanston-Flatt, S.K., Day, C., Bailey, C.J., and Flatt, P.R. (1989) Evaluation of traditional plant treatments for diabetes: studies in streptozotocin diabetic mice. *Acta Diabetol. Lat.*, **26**, 51–55.

Teeter, M.M., Ma, X.Q., Rao, U., and Whitlow, M. (1990) Crystal structure of a protein-toxin α1-purothionin at 2.5 Å and a comparison with predicted models. *Proteins Structure Function Genet.* **8**, 118–132.

Timoshenko, A.V. and Gabius, H.J. (1993) Efficient induction of superoxide release from human neutrophils by the galactoside-specific lectin from *Viscum album*. *Biol Chem Hoppe Seyler*, **374**, 237–243.

Timoshenko, A.V. and Gabius, H.J. (1995) Influence of the galactoside-specific lectin from *Viscum album* and its subunits on cell aggregation and selected intracellular parameters of rat thymocytes. *Planta Medica*, **61**, 130–133.

Urech, K., Schaller, G., Ziska, P., and Giannattasio, M. (1995) Comparative study on the cytotoxic effect of viscotoxin and mistletoe lectin on tumour cells in culture. *Phytotheraoy Research*, **9**, 49–55.

Vasold, A., and Händdel, F. (1954) Über die kombinierte intracutane und intravenöse Anwendung von Plenosol bei dgenerativen Gelenkerkrrankungen. *Therapie der Gegenwart*, **93**, 456–458.

Vehmeyer, K., Hajto, T., Hostanska, K., Könemann, S., Löser, H., Saller, R., *et al.* (1998) Lectin-induced increase in clonogenic growth of haematopoietic progenitor cells. *European Journal of Haematology*, **60**, 16–20.

Vernon, L.P. (1996) Pyrularia thionin. Physical properties, binding to phospholipid bilayers and cellular responses. In B.R. Singh and A.T. Tu, (eds.), *Natural Toxins II*. Plenum Press, New York, pp. 279–291.

Vervecken, W., Kleff, S., Pfüller, U., and Büssing, A. (2000) Induction of apoptosis by mistletoe lectin I and its subunits. No evidence for cytotoxic effects caused by isolated subunits. *The International Journal of Biochemistry & Cell Biology*, **32**, 317–326.

Vester, F., and Nienhaus, J. (1965) Cancerostatic protein components from *Viscum album*. *Experientia*, **21** (Suppl 4), 197–199.

Vester, F., Schweiger, A., Seel, A., and Stoll, M. (1968) Die Hemmwirkung basischer Proteine aus *Viscum album* auf die RNA-Synthese in Yoshida-Ascites. *Hoppe-Seyler's Zeitschrift für Physiologische Chemie*, **349**, 865–866.

Weber, G., Shen, F., Prajda, N., Yang, H., Li, W., Yeh, A., *et al.* (1997) Regulation of the signal transduction program by drugs. *Adv Enzyme Regul*, **37**, 35–55.

Weber, K., Mengs, U., Schwarz, T., Hajto, T., Hostanska, K., Allen, T.R., *et al.* (1998) Effects of a standardized mistletoe peparation on metastatic B16 melanoma colonization in murine lungs. *Arzneimittel-Forschung/Drug Research*, **48**, 497–502.

Wei, Y.Q., Zhao, X., Kariya, Y., Fukata, H., Teshigawara, K., and Uchida, A. (1994) Induction of apoptosis by quercetin: involvement of heat shock protein. *Cancer Research*, **54**, 4952–4957.

Wenzel-Seifert, K., Lentzen, H., and Seifert, R. (1997) In U-937 promonocytes, mistletoe lectin I increases basal $[Ca^{2+}]_i$, enhances histamine H_1- and complement C5a-receptor mediated rises in $[Ca^{2+}]_i$, and induces cell deah. *Naunyn-Schmidebergs's Arch Pharmacol.*, **355**, 190–197.

Winterfeld, K., and Kronenthaler, A. (1942) Zur Chemie des blutdrucksenkenden Bestandteiles der Mistel (*Viscum album*). III. Mitteilung. *Archiv der Pharmazie*, **280**, 103–115.

Winterfeld, K., and Bijl, L.H. (1948) Viscotoxin, ein neuer Inhaltsstoff der Mistel (*Viscum album* L). *Liebigs Ann.*, **561**, 107–115.

Winterfeld, K., and Rink, M. (1948) Über die Konstitution des Viscotoxins. *Liebigs Ann.*, **561**, 186–193.

Winterfeld, K., and Leiner, M. (1956) Zur Kenntnis des Viscotoxins. *Archiv der Pharmazie*, **289**, 358–364.

Woynarowski, J.M. and Konopa, J. (1980) Interaction between DNA and Viscotoxins. Cytotoxic basic polypeptides from *Viscum album* L. *Hoppe-Seyler's Zeitschrift für Physiologische Chemie*, **361**, 1535–1545.

Yamada, J., and Tomita, Y. (1996) Antimutagenic activity of caffeic acid and related compounds. *Biosci Biotechnol Biochem*, **60**, 328–329.

Yokoo, T., and Kitamura, M. (1997) Unexpected protection of glomerular mesangial cells from oxidant-triggered apoptosis by bioflavonoid quercetin. *American Journal of Physiology*, **273** (2 Pt 2), F206–212.

Yoshida, M., Yamamoto, M., and Nikaido, T. (1992) Quercetin arrests human leukemic T-cells in late G1 phase of the cell cycle. *Cancer Research*, **52**, 6676–6681.

Zarkovic, N., Trbojevic-Cepe, M., Ilic, Z., Hrzenjak, M., Grainca, S., and Jurin, M. (1995) Comparison of the effects of high and low concentrations of the separated *Viscum album* L. lectins and of the plain mistletoe plant preparation (Isorel) on the growth of normal and tumour cells *in vitro*. *Periodicum Biologorum*, **97**, 61–65.

Zarkovic, N., Kalisnik, T., Loncaric, I., Borovic, S., Mang, S., Kissel, D., *et al.* (1998) Comparison of the effects of Viscum album lectin ML-1 and fresh plant extract (Isorel) on the cell growth in vitro and tumorigenicity of melanoma B16F10. *Cancer Biotherapy & Radiopharmaceuticals*, **13**, 121–131.

Zell, J., Ecker, R., Batz, H., and Vestweber, A.M. (1993) Misteltherapie bei Gon- und Coxarthrose. Segment- oder Immunterapie? – Eine Pilotstudie. *Heilkunst*, **106**, 23–26.

Zhu, B.T., Ezell, E.L., and Liehr, J.G. (1994) Catechol-O-methyltransferase-catalyzed rapid O-methylation of mutagenic flavonoids. Metabolic inactivation as a possible reason for their lack of carcinogenicity *in vivo*. *Journal of Biological Chemistry*, **269**, 292–299.

Zhu, H.G., Zollner, T.M., Klein-Franke, A., and Anderer, F.A. (1994) Enhancement of MHC-unrestricted cytotoxic activity of human CD56+ CD3- natural killer (NK) cells and CD3+ T cells by rhamnogalacturonan: target cell specificity and activity against NK-insensitive targets. *Journal of Cancer Resarch and Clinical Oncology*, **120**, 383–388.

Zipf, H.F. (1950) Untersuchungen über den Herzwirksamen Mistelstoff Viscotoxin. *Archiv für experimentelle Pathologie und Pharmakologie*, **209**, 165–180.

Ziska, P., Franz, H., and Kindt, A. (1978) The lectin from Viscum album L. purification by biospecific affinity chromatography. *Experientia*, **34**, 123–124

Zschiesche, W. (1966) Die Wirkung von Iscador auf die Phagozytoseaktivität des dreticulo-histiozytären Systems. *Monatsbericht der Deutschen Akademie der Wissenschaften*, **8**, 750.

10. TOXICOLOGY OF MISTLETOE EXTRACTS AND THEIR COMPONENTS

GERBURG M. STEIN

Krebsforschung Herdecke, Communal Hospital, University Witten/Herdecke, Herdecke, Germany

INTRODUCTION

Plants of the Viscaceae and related families are widely distributed all over the world. Within these families, European (*Viscum album* L.) and American mistletoe (*Phoradendron* species) are the most common species used for medicinal purposes, and are also very popular for decoration.

For development of remedies in general, knowledge of toxicology is essential to ensure drug safety. Apart from the allergic potential (Berg and Stein, this book), the following chapter reviews and critically discusses toxicological data with respect to oral and parenteral exposure towards parts of the plant and plant extracts.

ORAL EXPOSURE

Exposure to the Whole Plant or Parts of the Plant

Toxicity of oral mistletoe exposure is controversially discussed. Already in 1952, Winterfeld stated that oral application of powdered *Viscum album* extracts (VA-E) or drops were well tolerated and did never induce toxic reactions (Winterfeld, 1952). In general, oral uptake of parts of the plant material from *Viscum album* seems to be relatively less toxic (Frohne and Pfänder, 1987), although it induced nausea and bloody diarrhoea in few individuals (Haupt, 1995). Baker (1985) described similar symptoms especially after ingestion of the berries: nausea, vomiting, explosive diarrhoea, hypertension followed by shock. Frohne and Pfänder (1982) reported 57 cases of ingestion of European mistletoe but did not observe severe toxicity.

Also internet search for mistletoe related toxicity revealed no entry for *Viscum*, *Loranthus* or *Phoradendron* as poisonous drugs in different data bases such as Canadian Poisonous plants information system (Munro, 1996). In addition, no fatal effects were observed for American mistletoe *Phoradendron* as documented by the American Association of Poison Control Centers (Litovitz and Veltri, 1985, 1995; Veltri and Litovitz, 1984).

Although detailed data are lacking, Arena (1979) mentioned a fatality after ingestion of tea brewed from American mistletoe berries, and also two other authors mentioned death of adults and children after eating of mistletoe berries (Adkins, 1983; Weinberger, 1984). Since no conclusive data were presented and even the plant genus was not characterised, these reports, however, remain very obscure.

Krenzelok *et al.* (1997) summarised accidental or intentional exposures (n = 1,754) to American mistletoe. Ingestion occurred mainly to children (92.1%), but no fatalities were observed and 99.2% had an outcome without morbidity. Toxicity from *Phoradendron* species has been already evaluated by Hall *et al.* (1986). They did not observe severe effects. Also Spiller *et al.* (1996) dealt with accidental ingestion of *Phoradendron*. Out of 92 cases, 6 gastrointestinal upset, two mild drowsiness, one eye irritation, one ataxia and one seizure were reported, while neither arrhythmia nor cardiovascular changes were observed.

Exposure to Plant Extracts

Especially in Germany, oral uptake of mistletoe preparations is used for antihypertensive treatment. Influence on blood pressure and cardiac effects were shown to be related to cholin, acetylcholin and viscotoxins (Büssing, this book; Rosell and Samuelsson, 1966; Samuelsson, 1974; Winterfeld, 1952; Winterfeld and Dörle, 1942; Winterfeld and Kronenthaler, 1942).

Fermented VA-E (*Iscador* Quercus, Mali) were found to be non-toxic at high concentrations after oral application into mice (Nienhaus and Leroi, 1970). 2 ml of the extract per mouse (corresponding to 20 mg of the fresh plant), i.e. 1000 mg/kg, were well tolerated and revealed no toxic symptoms.

Effects of oral application of purified *Viscum album* components on the intestine and the immune system were performed recently, and are reviewed by Büssing (this book).

The most startling and debated report on toxic effects of *Viscum album* on the liver came from Harvey and Colin-Jones (1981) who reported a case of clinically proven hepatitis with slight inflammatory-cell infiltration due to oral application of tablets containing a mixture of 5 different plant extracts. This hepatitis reappeared after re-exposition with the remedy two years later and was further proven by oral challenge test. Main constituents of the remedy were kelp, mistletoe (*Viscum album*), motherwort (*Leonurus cardiaca*), skullcap (*Scutellaria* spec.) and wild lettuce. Since mistletoe contains highly cytotoxic components (mistletoe lectins and viscotoxins in addition to phenylethylamine, tyramine, acetylcholine etc.), mistletoe was suggested to be the responsible drug. However, no further proof for the correlation to this drug was given and thus, this conclusion was discussed in the following years by other authors (Anderson and Phillipson, 1982; Fletcher Hyde, 1981; Stirpe, 1983), since especially the latter molecules have not been reported to induce adverse effects on the liver and are inactive after oral application (Farnsworth and Loub, 1981). In addition, another report on plant extract induced hepatotoxicity described 4 case reports on liver injury after ingestion of tablets containing skullcap and valerian but without *Viscum album* (MacGregor *et al.*, 1989), and also Buajordet and Bodd (1992) reported about liver damage due to skullcap ingestion. From these data, it was suggested that in the report from Harvey and Colin-Jones hepatitis was rather due to skullcap than to *Viscum album* (Galeazzi, 1992; MacGregor *et al.*, 1989).

Weeks and Proper (1989) also reported a case of chronic active hepatitis after ingestion of a herbal remedy containing mistletoe, skullcap, valerian and other plants. However, there are no conclusive data on which herb might be responsible.

In contrast, anti-hepatotoxic action of a variety of plants was studied by Yang *et al.* (1987) and mistletoe from Formosa (*Loranthus parasiticus*) was found to protect liver damage in two experimental systems (CCl_4- and GalN-induced cytotoxicity).

PARENTERAL APPLICATION

Application of *Viscum album* Extracts

In vivo toxicity of VA-E (*Iscador*) and a purified protein fraction was determined in the mouse system (Varese mice) (Nienhaus and Leroi, 1970). Since oral ingestion of high amounts were not toxic, further application routes were studied. Different forms of toxicity occurred: rapid death of animals at the LD_{100} within 5 minutes with all animals exerting convulsions but no signs of organ toxicity when > 2 ml/mouse were applied subcutaneously (s.c.) or intraperitoneally (i.p.), or slower death within 0.5–12 h with animals showing irregularly spasms but also no signs of organ toxicity (LD_{100} s.c. 2 ml/mouse, i.p. 0.75–1 ml/mouse depending upon the extract used). Within the LD_{50} dose (s.c. 1.4–6 ml/mouse, i.p. 0.5–6 ml/mouse), organ alterations in the surviving animals became visible after a few days: reduction of weight of spleen and thymus. Long term toxicity was observed only with the purified but not well characterised proteins with death after 3–4 days which was associated with atrophy of the liver and other metabolic organs and disintegration of the thymus. In contrast, spleen and thymus were enlarged between 30 and 100% and 30–200%, respectively, using concentrations of 5–10% of the LD_{50} of these proteins. In all cases, intravenous (i.v.) and intracardiac (i.c.) application was more toxic to the mice than s.c., i.p., or intramuscular (i.m.) exposure.

LD_{50} of a VA-E (*Iscador*) was also determined by Rentea and colleagues (1981). They found different levels for distinct animals after i.p. injection: in CD-1 outbred albino mice LD_{50} was 700 mg/kg (*Iscador* Mali), in C57/BL6 mice it was 348 mg/kg and in Sprague-Dwaley rats it was 378 mg/kg. Animals injected with lethal doses developed haemorrhagic peritonitis and died with tonic and clonic seizures. Lower doses did not induce peritonitis. These data are comparable to those described by Nienhaus and Leroi (1970). Also, Hajto and Hostanska (1986) estimated the LD_{50} of VA-E (*Iscador* Mali) in mice being 168 mg/kg corresponding to the weight of fresh plant. Luther *et al.* (1986) measured a LD_{50} of 500 mg/kg (mouse) after i.v. and of 1200 mg/kg (mouse) after s.c. application of VA-E (*Iscador* Quercus).

Acute and subchronic toxicity of VA-E was further investigated by Mengs (1998). Rats received 25 or 100 mg/kg of a VA-E (*Lektinol*) i.v. corresponding to 10 or 40 μg mistletoe lectin (ML) per kg bodyweight (bw). In the high dose group, all animals died within 5 minutes and showed sedation, dyspnoea, ataxia, exophthalmus, and spasms. In the low dose group, all animals survived and showed sedation and dyspnoea. Subchronic toxicity of doses of 0.2, 1.5 or 5 mg/kg was determined in rats, which were exposed to VA-E i.v. for 4 weeks. No relevant organ toxicity was found according to the parameters measured: food consumption, body weight gain, clinical signs, ophthalmoscopy, haematology, urine analysis, necropsy, organ weight, histology.

There are few reports on the *in vivo* application of high doses of VA-E in human. In breast cancer patients, high dose therapy with the VA-E *Abnobaviscum* Fraxini (45 mg s.c., once weekly for 4 months) revealed no hepatic or renal side effects (Mahfouz *et al.,* 1998). Infusion of high amounts of VA-E (mean dose: 600 mg *Helixor*) in patients with progressive cancer and distant metastasis or primary tumours, which could not be treated by surgery, no signs of hepatotoxic, haematologic or nephrotoxic effects were observed (Böcher *et al.*, 1996).

Application of Purified Components

For the different mistletoe components, *in vivo* toxicological data are available, although in many publications, the application route was not mentioned. Vester *et al.* (1968a,b) isolated different fractions from *Viscum album*, the so-called "Vester proteins". For some of these fractions, toxicity was determined in animals. Also, Winterfeld and Dörle (1942) and Koch (1938) described the toxicity of distinct *Viscum album* fractions or crude extracts. However, most of these fractions are not clearly characterised and thus, these data are of little value today.

For the purified viscotoxins, the lethal dose on rabbits was 0.8 mg/kg bw (Winterfeld and Bijl, 1948). After i.p. application of viscotoxins, LD_{50} in mice was 0.5 mg/kg and after i.v. application in cats, LD_{50} was 0.1 mg/kg. Sublethal doses led to hypotension, bradycardia, and negative inotropic effects (Samuelsson, 1974).

First attempts to define toxicity of the ML were made by Lutsik (1975). The ML are type-II ribosome inactivating proteins (Endo *et al.,* 1988; Sweeney *et al.*, 1993). For the "*Viscum album* lectin" (ML-1), LD_{50} was estimated to be 80 μg/kg bw in mice. According to Luther *et al.* (1986), LD_{50} (mice) of ML-1 (*Viscum album* agglutinin, VAA I) was 28 mg/kg, and of ML-3 (VAA II) 49 mg/kg. However, application route, the source of these data and the activity of the lectin remain unclear. It is well established that different methods used for determination of the activity of the ML revealed no comparable results (Tröger, 1992). Thus, published data on the toxicological effects varied considerably. In contrast to the high LD_{50} values, Franz and coworkers revealed much lower values after i.p. injection: LD_{50} of ML-1 was 28 μg/kg, of ML-2 1.5 μg/kg and of ML-3 55 μg/kg (Franz, 1986; Franz *et al.*, 1981).

In rats, viscumin (ML-1) at an i.p. dose of 100 μg/kg was lethal within 1 day, while 10 μg/kg was lethal within 3–4 days (Stirpe *et al.*, 1980). These mice expressed toxic effects with lesions like ascites, haemorrhages in the pancreas and congested intestine, similar to ricin poisoning.

Although cytotoxicity of the ML against various established cell lines *in vitro* has been described in a lot of publications (Franz, 1989, 1993; Janssen *et al.*, 1998;

Konopa *et al.*, 1980; Kopp *et al.*, 1993; Ribéreau-Gayon *et al.*, 1986; Urech *et al.*, 1995), the cause of their *in vivo* cytotoxicity remains unclear. Intraperitoneal injection of lethal doses of ML-1 into mice revealed no drastic alterations in histological investigations and thus, Franz (1993) concluded that animal death was not related to a generalised cell injury, while the release of cytokines like TNF-α could help explain animal death. Probably, haemagglutinating activity of the lectins may also contribute to the toxic effect. I.p. injection of 0.3–1.2 μg/mouse revealed a clear cut response in the liver with decrease of glycogen levels, an increase of thiamine phosphatase in the hepatocytes, and changes in the activity of alkaline phosphatase in the sinusoidal endothelial cells. However, they did not explain the death of the animals, especially because typical signs of toxicity like lipid accumulation and altered lipid metabolism in the liver, were lacking (Gossrau and Franz, 1990; Franz, 1991).

GENOTOXICITY AND MUTAGENICITY

Cytogenetic damage and mutagenicity were studied by Büssing *et al.* (1994), who investigated the rate of sister chromatid exchanges (SCE) in phytohemagglutinin-stimulated PBMC from healthy controls after VA-E (*Helixor* A, *Helixor* P) incubation *in vitro*. These SCE are considered to be sensitive and specific indicators for cytogenetic damage and mutagenicity (Allen and Latt, 1976; Perry and Evans, 1975). The VA-E decreased the number of SCE and thus, protects DNA damage (Büssing *et al.*, 1994). In addition, cyclophosphamide-induced DNA lesions were reduced when PBMC were incubated with VA-E (*Helixor* A) (Büssing *et al.*, 1995b). Also, SCE of rapidly proliferating amniotic fluid cells decreased in the presence of VA-E (*Iscador* P), however, only at very high concentrations (Büssing *et al.*, 1995a). Yet, it is still unclear, which component is responsible for these protective effects. ML-1 seems to be of minor importance for this DNA protection (Büssing *et al.*, 1998). DNA repair might contribute to these effects (Kovacs *et al.* 1991).

Findings of Köteles *et al.* (1998) using purified ML-1 are in contrast to the data on the whole plant extracts. In human lymphocytes treated with ML-1 at 7.4–7400 ng/ml, they observed an increase of micronuclei (MN) formation, which are recognised as cellular damage of cytogenetic nature. It remains to be seen whether this effect may be associated with apoptosis, which is known to be induced by the ML in these concentrations (Büssing *et al.*, 1996). Interestingly, irradiation of human whole blood cells revealed a reduction of radiation-associated micronuclei forming at non-toxic ML-1 concentrations. Comparison of the effects of the three ML in irradiated human lung carcinoma cells (Calu-1 cell line) revealed that ML-1 and ML-2 at lower concentrations reduced MN formations, while ML-3 treatment lead to an increase. In addition, in irradiated rabbits, micronuclei carrying lymphocytes were reduced in ML-1 treated animals in contrast to the untreated rabbits. This might be due to a rapid killing of damaged cells by ML-1, since ML-1 was applicated after irradiation of the animals. It remains, however, to be shown, which mechanisms are responsible for the increase of lung metastases (Lewis lung tumour cells) in irradiated mice due to s.c. treatment with ML-1 at different doses

(Kubasova *et al.*, 1998). This finding was obtained in mice irradiated before and after application of tumour cells.

Genotoxicity of an aqueous VA-E (*Lektinol*) was studied *in vitro* by Mengs *et al.* (Mengs *et al.*, 1997; Mengs, 1998). In the different test systems, such as bacterial mutation assay, mammalian cell gene mutation assay, *in vitro* cytogenetic analysis, and cell transformation assay, they did not find gene mutations, chromosomal aberrations or signs of carcinogenicity. In addition, AMES test with other aqueous VA-E (*Helixor* A, M, P) also revealed no mutagenic potency (Table 1). AMES test was performed using *Salmonella typhimurium* strains TA 98, TA 100 in the presence of rat liver microsomes. Strains TA 1535, TA 1537 and TA 1538 were also tested and revealed similar results (data not shown).

Schimmer *et al.* (1994) tested *Viscum* tinctura, an alcoholic extract from *Viscum* but did not find a mutagenic effect using the AMES test. In addition, using Korean mistletoe in different bacterial short-term test, such as AMES test, SOS spot test and other test systems, Ham *et al.* (1998) did not observe a mutagenic but an anti-mutagenic effect of VA-E against chemically induced mutations.

Basaran and co-workers (1996) investigated the genotoxic potential of an aqueous extract from berries from *Viscum album* without definition of the host tree. They bought the material from a herbalist market. Mutagenicity was studied in the AMES test and DNA damage with the COMET assay. Although the exact amounts of VA-E remain unclear (40 and 80 or 400 and 800 μg for the AMES test and 40 or 400 μg for the COMET assay), no mutagenic effects was detected in the AMES test. In con-

Table 1 Mutagenicity (AMES) test of aqueous VA-E (*Helixor*) in the presence of rat liver microsomes*.

VA-E [μl]	*Helixor A*		*Helixor M*		*Helixor P*	
	TA 98	*TA 100*	*TA 98*	*TA 100*	*TA 98*	*TA 100*
0	33 ± 3	99 ± 16	37 ± 6	103 ± 10	52 ± 4	103 ± 5
0.32	39 ± 7	110 ± 21	40 ± 4	97 ± 4	49 ± 5	102 ± 12
1.6	34 ± 4	94 ± 8	38 ± 6	106 ± 19	56 ± 11	103 ± 3
8	28 ± 4	109 ± 6	24 ± 5	95 ± 7	48 ± 5	93 ± 5
40	37 ± 4	120 ± 18	35 ± 8	95 ± 8	42 ± 3	100 ± 8
200	36 ± 8	100 ± 11	29 ± 5	98 ± 5	39 ± 2	97 ± 6
positive control**	382 ± 49	401 ± 22	382 ± 49	401 ± 22	382 ± 49	401 ± 22

* AMES test was performed by Labor L+S GmbH (Bad Bocklet, Germany) using *Salmonella typhimurium* strains TA 98, TA 100. Only concentrations of the different VA-E were tested which were not toxic, ranging from 0.32 to 200 μl of the respective extract (50 mg/ml, corresponding to the weight of fresh plant). Given are mean values ± standard deviation of triplicates. A second independent set of experiments revealed similar results although values were higher. Helixor A/M/P: extract derived from *Viscum album* grown on fir trees/apple trees/pines.

** positive control was 2-aminoanthracene.

trast, the COMET assay revealed a significant increase of DNA damage by strand breakage. There is doubt, whether the latter effect is of relevance for genotoxicity/mutagenicity since the so-called DNA ladder is a typical marker for apoptosis, and VA-E are well known to induce apoptosis in different cell types (Büssing *et al.*, 1996, Janssen *et al.*, 1993).

From all these data, there are few hints for a cytogenetic damage but rather for protective effects especially of *Viscum album* components.

PROTECTION AGAINST ML CYTOTOXICITY

In vitro, it was shown that anti-ML-1 antibodies neutralised ML-1 mediated cytotoxic effects towards human PBMC and inhibited cytokine release (Ribéreau-Gayon *et al.*, 1996; Schultze *et al.*, 1991; Stein *et al.*, 1998, 1999b; Stettin *et al.*, 1990). Also, VA-E (*Iscador* Mali) mediated inhibition of protein synthesis was neutralised by an antiserum towards ML-1 (Holtskog *et al.*, 1988). In addition, serum components such as haptoglobin and transferrin, showed similar effects (Ribéreau-Gayon *et al.*, 1995, 1997). It is possible that these molecules, the antibodies and the serum components, protect the patient from the direct toxic effects especially of the ML. Yet, it is unclear, whether specific antibodies towards the viscotoxins are also produced during therapy with VA-E, because anti-ML-1 antibodies were shown to cross-react with the viscotoxins in the Western Blot (Stein *et al.*, 1998, 1999b).

In vivo, anti-ML antibodies produced by rabbits neutralise lethal doses of certain protein fractions derived from *Viscum album* (LD_{100} = 10 μg) in a dilution of 1:50 (Vester *et al.*, 1968). In these experiments, even tenfold amounts of this dose, i.e. 100 μg were tolerated without any toxicity when applicated with the serum of the immunised rabbits. Protective effects not only against systemic but also cutaneous (local) toxicity was achieved by passive immunisation of guinea pigs using the rabbit antiserum. Furthermore, in the immunised rabbits, the antigen was applicated into the conjunctiva but was without effect regarding toxic reactivity and sensitisation. These data suggest that the antibodies produced during immunisation, which may correspond to the initial period of the therapy in patients, may neutralise the antigen (ML-1) mediated effects. However, detailed experiments addressing this question are lacking and thus, the role of these anti-ML antibodies during therapy of cancer patients still remains unclear.

CONCLUDING REMARKS

Reflecting toxicological data of VA-E and the frequent application of VA-E, when applied s.c. in the therapeutic concentrations, are safe remedies. There is no evidence for toxic effects in humans even after application of high doses, and adverse effects occur very seldom (Berg and Stein, this book; Stein and Berg, 1999a), an observation, which is of great importance especially for the long term treatment of the patients. Although the extracts contain highly cytotoxic proteins, serum components and production of anti-ML antibodies may protect the patients from cytotoxic effects. In this respect it is noteworthy, that according to the anthroposophic therapeutic recommendations subcutaneous application starts with very low doses, which increase during further therapy, a therapeutic application scheme, which resembles classical immunisation.

In addition, one has to assume that VA-E contain a variety of distinct proteins and other components, which may differ in their composition and their concentration between the commercial available preparations. Especially these different proteins imply the potential, like for all applicated non-self proteins, to induce a specific immune response.

REFERENCES

Adkins, W.S. (1983) *Bulletin*. The Intermountain Regional Poison Control Center, **1**, 1–7.

Allen, J.W. and Latt, S.A. (1976) Analysis of sister chromatide exchange formation *in vivo* in mouse spermatogonia as a new test system for environmental mutagens. *Nature*, **260**, 449–451.

Anderson, L.A and Phillipson, J.D. (1982) Mistletoe – the magic herb. *Pharmaceutical J.*, **228**, 437–439.

Arena, J.M. (1979). Poisoning. In C.C. Thomas, (ed.), *Toxicology, symptoms, treatments*, Springfield Illinois, 4th edition, pp. 191–192, 474, 548–554, 739–741.

Baker, M.D. (1985) Holiday hazards. *Pediatric Emergency Care*, **1**, 210–214.

Basaran, A.A., Yu, T.W., Plewa, M.J., and Anderson, D. (1996) An investigation of some turkish herbal medicines in *Salmonella typhimurium* and in the COMET assay in human lymphocytes. *Teratogenesis, Carcinogenesis and Mutagenesis*, **16**, 125–138.

Böcher, E., Stumpf, C., Büssing, A., and Schietzel, M. (1996) Prospektive Bewertung der Toxizität hochdosierter *Viscum album* L.- Infusionen bei Patienten mit progredienten Malignomen. *Zeitschrift für Onkologie*, **28**, 97–106.

Buajordet, I. and Bodd, E. (1992) Scullcap-leverskade. *Tidsskr. Nor. Lægeforen*, **15**, 2006.

Büssing, A. (1996) Induction of apoptosis by mistletoe lectins. A review on the mechanisms of cytotoxicity mediated by *Viscum album* L. *Apoptosis*, **1**, 25–32.

Büssing, A., Azhari, T., Ostendorp, H., Lehnert, A., and Schweizer, K. (1994) *Viscum album* L. extracts reduce sister chromatid exchanges in cultured peripheral blood mononuclear cells. *Eur. J. Cancer*, **30A**, 1836–1841.

Büssing, A., Lehnert, A., Schink, M., Mertens, R., and Schweizer, K. (1995a) Effects of *Viscum album* L. on rapidly proliferating amniotic fluid cells. Sister chromatid exchange frequency and proliferation index. *Arzneimittel Forsch./Drug Res.*, **45**, 81–83.

Büssing, A., Multani, A.S., Pathak, S., Pfüller, U., and Schietzel, M. (1998) Induction of apoptosis by the N-acetyl-galactosamine-specific toxic lectin from *Viscum album* L. is associated with a decrease of nuclear p53 and Bcl-2 proteins and induction of telomeric associations. *Cancer Lett.*, **130**, 57–68.

Büssing, A., Regenery, A., and Schweizer, K. (1995b) Effects of *Viscum album* L. on cyclophosphamide-treated peripheral blood mononuclear cells *in vitro*: sister chromatid exchanges and activation/proliferation marker expression. *Cancer Lett.*, **94**, 199–205.

Endo, Y., Tsurugi, K., and Franz, H. (1988) The site of action of the A-chain of mistletoe lectin I on eucaryotic ribosomes. *FEBS Lett.*, **231**, 378–380.

Farnsworth, N.R. and Loub, W.D. (1981) Mistletoe hepatitis. *Brit. Med. J.*, **283**, 1058.

Fletcher Hyde, F. (1981) Mistletoe hepatitis. *Brit. Med. J.*, **282**, 739.

Franz, H. (1991) Mistletoe lectins (2). *Advances in Lectin Research*, **4**, 33–50.
Franz, H. (1986) Mistletoe lectins and their A and B chains. *Oncology*, **43** suppl 1, 23–34.
Franz, H. (1993) The *in vivo* toxicity of toxic lectins is a complex phenomenon. In E. Van Driessche, H. Franz, S. Beeckmans, U. Pfüller, A. Kallikorm, and T.C. Bøg-Hansen, (eds.), *Lectins: Biology, Biochemistry, Clinical Biochemistry*. Vol. 8, Textop, Hellerup, Denmark, pp. 5–9.
Franz, H. (1989) Viscaceae lectins. *Advances in Lectin Research*, **2**, 28–59.
Franz, H., Ziska, P., and Kindt, A. (1981) Isolation and properties of three lectins from mistletoe (*Viscum album* L.). *Biochem. J.*, **195**, 481–484.
Frohne, D. and Pfänder, H.J. (1982) *Viscum album*. In *Giftpflanzen*. Wissenschaftliche Verlagsgesellschaft, Stuttgart, pp. 155–156 (cited according to: Luther, P, Becker, H. (1986) Die Mistel. Botanik, Lektine, medizinische Anwendung, VEB Verlag Volk und Gesundheit).
Frohne, D. and Pfänder, H.J. (1987) *Viscum album*. In *Giftpflanzen*. Wissenschaftliche Verlagsgesellschaft, Stuttgart, pp. 155–156.
Galeazzi, R.L. (1989) Unerwünschte Wirkungen von Naturheilmitteln. *Therapeut. Umschau*, **49**, 86–92.
Gossrau, R. and Franz, H. (1990) Histochemical response of mice to mistletoe lectin I (ML I). *Histochemistry*, **94**, 531–537.
Hajto, T. and Hostanska, K. (1986) An investigation of the ability of *Viscum album* activated granulocytes to regulate natural killer cells *in vivo*. *Clin. Tri. J.*, **23**, 345–358.
Hall, A.H., Spoerke, D.G., and Rumack, B.H. (1986) Assessing mistletoe toxicity. *Annals of Emergency Medicine*, **15**, 1320–1323.
Ham, S.S., Kang, S.T., Choi, K.P., Park, W.B., and Lee, D.S. (1998) Antimutagenic effect of Korean mistletoe extracts. *J. Korean. Soc. Food Sci. Nutr.*, **27**, 359–365.
Harvey, J. and Colin-Jones, D.G. (1981) Mistletoe hepatitis. *Brit. Med. J.*, **282**, 186–187.
Haupt, H. (1995) Giftige und weniger giftige Pflanzen. *Kinderkrankenschwester*, **14**, 479–480.
Holtskog, R., Sandvig, K., and Olsnes, S. (1988) Characterization of a toxic lectin in Iscador, a mistletoe preparation with alleged cancerostatic properties. *Oncol.*, **45**, 172–179.
Janssen, O., Fischer, S., Fiebig, H.H., Scheffler, A., and Kabelitz, D. (1998) Cytotoxicity of mistletoe extracts and mistletoe lectins towards tumour cells due to the induction of apoptosis. In S. Bardocz, U. Pfüller and A. Pusztai, (eds.), *COST 98. Effects of antinutritients on the nutritional value of legume diets*, European Commission, Luxembourg Vol. 5, pp. 157–163.
Janssen, O., Scheffler, A., and Kabelitz, D. (1993) *In vitro* effects of mistletoe extracts and mistletoe lectins. *Arzneimittel-Forsch./Drug Res.*, **43**, 1221–1227.
Koch, F.E. (1938) Experimentelle Untersuchungen über entzündungs- und nekroseerzeugende Wirkung von *Viscum album*. *Z. ges. exp. Medizin*, **103**, 740–749.
Köteles, G.J., Kubasova, T., Hurna, E., Horváth, G., and Pfüller, U. (1998) Cellular and cytogenetic approaches in testing toxic and safe concentrations of mistletoe lectins. In S. Bardocz, U. Pfüller and A. Pusztai, (eds.), *COST 98. Effects of antinutritients on the nutritional value of legume diets*, European Commission, Luxembourg Vol. 5, pp. 81–86.
Konopa, J., Woynarowski, J.M., and Lewandowska-Gumieniak, M. (1980) Isolation of viscotoxins. Cytotoxic basic polypeptides from *Viscum album* L. *Hoppe-Seyler Z. Physiol. Chem.*, **361**, 1525–1533.
Kopp, I., Koerner, I.J., Pfüller, U., Göckeritz, W., Eifler, R., Pfüller, K., and Franz, H. (1993) Effects of mistletoe lectins I, II and III on normal and malignant cells. In E. van Driessche, H. Franz, S. Beeckmans, U. Pfüller, A. Kallikom, and T.C. Bøg-Hansen, (eds.), *Lectins: Biology, Biochemistry, Clinical Biochemistry*, Textop, Hellerup, Denmark, Vol. 8, pp. 41–47.

Kovacs, E., Hajto, T., and Hostanska, K. (1991) Improvement of DNA repair in lymphocytes of breast cancer patients treated with *Viscum album* extract (*Iscador*). *Eur. J. Cancer* **27**, 1672–1676.

Krenzelok, E.P., Jacobsen, T.D., and Aronis, J. (1997) American mistletoe exposures. *Am. J. Emerg. Med.*, **15**, 516–520.

Kubasova, T., Pfüller, U., Bojtor, I., and Köteles, G.J. (1998) Modulation of immune response by mistletoe lectin I as detected on tumour model *in vivo*. In S. Bardocz, U. Pfüller and A. Pusztai, (eds.), *COST 98. Effects of antinutritients on the nutritional value of legume diets*, European Commission, Luxembourg, Vol. 5, pp. 202–207.

Litovitz T.L., Felberg L., Soloway R.A., Ford M., and Geller, R. (1995) 1994 Annual report of the American Association of Poison Control centers toxic exposure surveillance system. *Am. J. Emerg. Med.*, **13**, 551–597.

Litovitz, T. and Veltri, J.C. (1985) 1984 Annual report of the American Association of poison control centers national data collection system. *Am. J. Emerg. Med.*, **3**, 423–450.

Luther, P., Uhlenbruck, G., Reutgen, H., Samtleben, R., Sehrt, I., and Ribéreau-Gayon, G. (1986) Are lectins from *Viscum album* interesting tools in lung diseases? *Z. Erkrank. Atm. org.*, **166**, 247–256.

Lutsik M.D. (1975) [Die Antitumoreigenschaften des PHA der Mistel (*Viscum album*).] *Dokl. Akad. Nauk. Ukrainsk. SSR* Ser. B, **6**, 541–544.

MacGregor, F.B., Abernethy, V.E., Dahabra, S., Cobden, I., and Hayes, P.C. (1989) Hepatotoxicity of herbal remedies. *Br. Med. J.*, **299**, 1156–1157.

Mahfouz, M.M., Ghaleb, H.A., Zawawy, A., and Scheffler, A. (1998) Significant tumour reduction, improvement of pain and quality of life and normalisation of sleeping patterns of cancer patients treated with a high dose of mistletoe. *Annals of Oncol.*, **9**, Suppl 2, 129.

Mengs, U. (1998) Toxicity of an aqueous mistletoe extract: acute and subchronic toxicity in rats, genotoxicity *in vitro*. In S. Bardocz, U. Pfüller and A. Pusztai, (eds.), *COST 98. Effects of antinutritients on the nutritional value of legume diets*, European Commission, Luxembourg, Vol. 5, pp. 77–80.

Mengs, U., Clare, C.B., and Poiley, J.A. (1997) Genotoxicity testing of an aqueous mistletoe extract *in vitro*. *Arzneimittel-Forsch./Drug Res.*, **47**, 316–319.

Munro, D.B. (1996) Canadian poisonous plants information system. Biol. Resource Program, Eastern Cereal and Oilseed Res. Centre, Res. Branch, Ottawa, Ontario K1 A OC6, Canada (internet database)

Nienhaus, J. and Leroi, R. (1970) Tumorhemmung und Thymusstimulation durch Mistelpräparate. *Elemente der Naturw.*, **13**, 45–54.

Perry, P. and Evans, H.J. (1975) Cytological detection of mutagen-carcinogen exposure by sister chromatid exchange. *Nature,* **258**, 121–125.

Rentea, R., Lyon, E., and Hunter, R. (1981) Biologic properties of Iscador: a *Viscum album* preparation. *Lab. Invest.*, **44**, 43–48.

Ribéreau-Gayon, G., Dumont, S., Muller, C., Jung, M.L., Poindron, P., and Anton, R. (1996) Mistletoe lectins I, II and III induce the production of cytokines by cultured human monocytes. *Cancer Lett.*, **109**, 33–38.

Ribéreau-Gayon, G., Jung, M.L., Beck, J.P., Anton, R. (1995) Effect of fetal calf serum on the cytotoxic activity of mistletoe (*Viscum album* L.) in cell culture. *Phytotherapy Res.*, **9**, 336–339.

Ribéreau-Gayon, G., Jung, M.L., Di Scala, D., and Beck, J.P. (1986) Comparison of the effects of fermented and unfermented mistletoe preparations on cultured tumour cells. *Oncology*, **43**, suppl 1, 35–41.

Ribéreau-Gayon, G., Jung, M.L., Frantz, M., Anton, R. (1997) Modulation of the cytotoxicity and enhancement of cytokine release induced by *Viscum album* L. extracts or mistletoe lectins. *Anti-Cancer Drugs*, **8** suppl. 1, S3–S8.

Rosell, S. and Samuelsson, G. (1966) Effect of mistletoe viscotoxin and phoratoxin on blood circulation. *Toxicon*, **4**, 107–110.

Samuelsson, G. (1974) Mistletoe toxins. *Syst. Zool.*, **22**, 566–569.

Schimmer, O., Krüger, A., Paulini, H., and Haefele, F. (1994) An evaluation of 55 commercial plant extracts in the Ames mutagenicity test. *Pharmazie,* **49**, 448–451.

Schultze, J.L., Stettin, A., and Berg, P.A. (1991) Demonstration of specifically sensitised lymphocytes in patients treated with an aqueous mistletoe extract (*Viscum album* L). *Klin. Wochenschr.*, **69**, 397–403.

Spiller, H.A., Willias, D.B., Gorman, S.E., and Sanftleben, J. (1996) Retrospective study of mistletoe ingestion. *Clin. Toxicol.*, **34**, 405–408.

Stein, G.M. and Berg, P.A. (1999a) Characterisation of immunological reactivity of patients with adverse effects during treatment with an aqueous mistletoe extract. *Eur. J. Med. Res.*, **4**, 169–177

Stein, G.M., Pfüller, U., and Berg, P.A. (1999b) Recognition of different antigens of mistletoe extracts by anti-mistletoe lectin antibodies. *Cancer Lett.*, **135**, 165–170.

Stein, G.M., von Laue, H.B., Henn, W., and Berg, P.A. (1998) Human anti-mistletoe lectin antibodies. In S. Bardocz, U. Pfüller and A. Pusztai, (eds.), *COST 98. Effects of antinutritients on the nutritional value of legume diets,* European Community, Luxembourg, Vol. 5, pp. 168–175.

Stettin, A., Schultze, J.L., Stechemesser, E., and Berg, P.A. (1990) Anti-mistletoe lectin antibodies are produced in patients during therapy with an aqueous mistletoe extract derived from *Viscum album* L. and neutralise lectin-induced cytotoxicity *in vitro*. *Klin. Wochenschr.*, **68**, 896–900.

Stirpe, F. (1983) Mistletoe toxicity. *Lancet,* **I**, 295.

Stirpe, F., Legg, R.F., Onyon, L.J., and Ziska, P. (1980) Inhibition of protein synthesis by a toxic lectin from *Viscum album* L. (mistletoe). *Biochem. J.*, **190**, 843–845.

Sweeney, E.C., Palmer, R.A., and Pfüller, U. (1993) Crystallisation of the ribosome inactivating protein ML-1 from *Viscum album* (mistletoe) complexed with β-D-galactose. *J. Mol. Biol.*, **234**, 1279–1281.

Tröger, W. (1992) Nachweismethoden von Mistellektinen. *Der Merkurstab*, **6**, 456–460.

Urech, K., Schaller, G., Ziska, P., and Giannattasio, M. (1995) Comparative study on the cytotoxic effect of viscotoxin and mistletoe lectin on tumour cells in culture. *Phytother. Res.*, **9**, 49–55.

Veltri, J.C. and Litovitz, T. (1984) 1983 Annual report of the American Association of poison control centres national data collection system. *Am. J. Emerg. Med.*, **2**, 420–443.

Vester, F., Bohne, L., and El-Fouly, M. (1968) Zur Kenntnis der Inhaltsstoffe von *Viscum album*, IV. *Hoppe Seyler's Z. Physiol. Chem.*, **349**, 495–511.

Vester, F., Sell, A., Stoll, M., and Müller, J.M. (1968) Zur Kenntnis der Inhaltsstoffe von *Viscum album* III. Isolierung und Reinigung cancerostatischer Proteinfraktionen. *Hoppe-Seyler's Z. Physiol. Chem.*, **349**, 125–147.

Weeks, G.R. and Proper, J.S. (1989) Herbal Medicines – Gaps in our knowledge. *Aust. J. Hosp. Pharm.*, **19**, 155–157.

Weinberger, H.L. (1984) Preventing those holiday poisonings. *Contemp. Pediatr.*, **14**, 49–54.

Winterfeld, K. (1952) Die Wirkstoffe der Mistel (*Viscum album* L.). *Pharm. Ztg.*, **88**, 573–574.

Winterfeld, K., and Bijl, L.H. (1948) Viscotoxin, ein neuer Inhaltsstoff der Mistel (*Viscum album* L.) *Liebigs Ann.*, **561**, 107–115.
Winterfeld, K. and Dörle, E. (1942) Über die Wirkstoffe der Mistel (*Viscum album* L.). (2. Mitteilung) *Arch. Pharmaz.*, **280**, 23–36.
Winterfeld, K. and Kronenthaler, A. (1942) Zur Chemie des blutdrucksenkenden Bestandteiles der Mistel (*Viscum album*). *Arch. Pharmaz.*, **280**, 103–115.
Yang, L.L., Yen, K.Y., Kiso, Y., and Hikino, H. (1987) Antihepatotoxic actions of Formosan plant drugs. *J. Ethnopharmacol.*, **19**, 103–110.

11. ADVERSE EFFECTS DURING THERAPY WITH MISTLETOE EXTRACTS

GERBURG M. STEIN[1] AND PETER A. BERG[2]

[1]*Krebsforschung Herdecke, Communal Hospital, University Witten/Herdecke, Herdecke, Germany,* [2]*Medical Clinic, Department of Internal Medicine II, University of Tübingen, Otfried-Müller-Strasse 10, 72076 Tübingen, Germany*

INTRODUCTION

Extracts from European mistletoe (*Viscum album* L.) are very popular for adjuvant treatment of neoplastic disorders especially in the German speaking areas (Grothey *et al.*, 1998; Weis *et al.*, 1998). Although surgery, radio- and chemotherapy are the therapy of choice for patients especially with epithelial tumours, these therapeutic strategies are not always successful, and this may be the reason why quite a few patients ask for alternative regimens (Eisenberg, 1997; Grothey *et al.*, 1998; Weis *et al.*, 1998). With respect to this concept, the immunodefence mechanisms are altered in cancer patients (Coeugniet and Kühnast, 1987; Hara *et al.*, 1992; Elsässer-Beile *et al.*, 1993a,b; Zielinski *et al.*, 1990), and, therefore, substances which exert an effect on cells involved in the tumour defence are of interest. Thus, different extracts of animal (thymic petides) or plant origin (Lentinan from *Lentinula edodes*, an edible Shitake mushroom used in Japan, and extracts from *Echinacea purpurea*) are used. Especially extracts from mistletoe have been shown in *in vitro* and *in vivo* investigations to stimulate immunocompetent cells, to exert cytotoxic activity against different cell populations, especially tumour cells and to be of benefit in the tumour defence in animal models (for review see Büssing, this book).

Although *Viscum album* extracts (VA-E) are well tolerated in general, there are some reports describing side effects during therapy. This chapter reviews these reports (Table 1); unfortunately most of these reports provide very few data to characterise the kind of mechanism. Added in this review are own data on patients with supposed adverse effects gained during a period of about 6 years. These immunological examinations were pursued in co-operation with a manufacturer of aqueous mistletoe extracts (*Helixor* Heilmittel, Rosenfeld, Germany).

REVIEW OF THE CASE REPORTS ON TOLERABILITY AND ADVERSE EFFECTS OF MISTLETOE EXTRACTS

Tolerability of Mistletoe Treatment

Extracts from European mistletoe have been reported to be well tolerated in several studies (Schlodder, 1996). Even intravenous or subcutaneous (i.v./s.c.) application of

Table 1 Reports on adverse effects after VA-E exposure.

Adverse effect	*Treatment regimen*	*Mistletoe as causative agent?*	*Reference*
Well documented reports			
Anaphylactic reaction	Only VA-E injection	likely	Pichler and Angeli, 1991
Appearance of NHL nodules in application sites of VA-E	VA-E injection	likely	Hagenah *et al.*, 1998
Less well documented reports			
Anaphylactic reaction	Only VA-E injection	likely	Friess *et al.*, 1996
	VA-E, no further data available	unclear	Anonymous, 1998
Acute asthmatic reaction	VA-E injection, no further data available	unclear	Anonymous, 1998
Allergic colitis	VA-E injection	most likely	Ottenjann, 1992
Allergic rhinitis	Exposure to tea dust from mistletoe	most likely	Seidemann, 1984
Generalised erythema with necrotic areas, high fever	VA-E injection in combination with not further specified complementary treatments	unclear	Lange-Wantzin *et al.*, 1983
Sarcoidosis	VA-E injection, further treatment with *Echinacea*, *Baptisia*, *Thuja*, Roxatidine, Influenza vaccination	possible	Zürner, 1992
Urticaria, Quincke's edema	VA-E injection, cyclophosphamide	most likely	Lange-Wantzin *et al.*, 1983

high doses were shown to be without serious complications (Böcher *et al.*, 1996; Mahfouz *et al.* 1998). Mild but not severe side effects were reported during different studies. While no signs for liver or kidney injury or haematotoxicity were observed after the application of VA-E, the main side effects were strong local reactions, temporary forming of nodules at the injection site, "flu-like symptoms", slight fever and eosinophilia (Böcher *et al.*, 1996; Boie and Gutsch, 1980; Boie *et al.*, 1981; Dold *et al.*, 1991; Douwes and Wolfrum, 1986; Finelli and Limberg, 1998; Gutsch *et al.*, 1988; Hajto *et al.*, 1991; Hassauer *et al.*, 1979; Kjaer, 1989; Schlodder, 1996).

In addition, tolerability of an unfermented mistletoe extract (*Viscum album* QuFrF) was studied in healthy controls and HIV patients in a phase I/II study and mainly confirmed previous observations (Stoss *et al.*, 1999; Stoss *et al.* in press; van Wely *et al.*, 1999). Interestingly, exacerbation of gingivitis or herpes simplex infections were observed mainly in the HIV patients.

Adverse Effects of Mistletoe Exposure

Adverse effects induced by subcutaneous mistletoe exposure are relatively rare and only few reports are available. Schlodder (1996) reviewed common side effects during subcutaneous therapy with these plant extracts. Mainly inflammatory application site reactions were observed, seldom are allergic or pseudo-allergic reactivity like urticaria, swelling of the regional lymph nodes, phlebitis. Strong local reactions depend upon the concentration of the drug, and dose reduction mostly led to tolerability of the application. Anaphylactic reactions were reported only in rare cases.

Anaphylactic reactions

One of the few reports on adverse effects during mistletoe exposure dealt with an anaphylactic reaction after subcutaneous application of *Iscador* Quercus c. Arg. (Pichler and Angeli, 1991). Because therapy was initially well tolerated by a patient with disseminating seminom with only small application site reactions, therapy was continued after 3 weeks. Increase of the injected dose led to an anaphylactic reaction after half an hour with nausea, generalised itchiness and collapse with blood pressure at the lower level of the normal range. Different skin tests remained negative in contrast to the lymphocyte transformation test (LTT), which showed a stimulation index of 58 in the presence of the offending extract. Since immediate skin tests were negative, the authors concluded that the reaction was induced by a complement-mediated specific IgG response and not by a type 1 hypersensitivity reaction. The LTT *per se* gives evidence for the presence of sensitised lymphocytes, as also observed in patients without side effects (Schultze *et al.*, 1991; Stein and Berg, 1998a; Stein *et al.*, 1998b), but gives no answer with respect to the immunological effector reactions.

There are two other reports on an anaphylactic reaction during therapy with VA-E, although no detailed data are provided. The first concerns a patient with pancreatic cancer (Friess *et al.*, 1996) and the other one describes a patient with chronic obstructive pulmonary disease (COPD) being also allergic to a variety of substances (Anonymous, 1998).

Appearance of NHL nodules in previous application sites of VA-E

The most recent report on adverse effects deals with the detection of lymphoma cells of a Non-Hodgkin-lymphoma (NHL) at the VA-E (*Iscador* Qu c Hg) injection site (Hagenah *et al.*, 1998). The patient suffered for several years from a progredient invasive NHL first located at cervical lymph nodes at the Waldeyer's tonsilar rings later followed by infiltration of eye lids. After first local irradiation, patient underwent different chemotherapy regimes when leukaemic cells were poured out. Within chemotherapy, he received an extract from mistletoe grown on oak trees. 5 weeks later, i.e. after 14 VA-E injections, several new subcutaneous nodules appeared only at the application sites, which histologically were shown to be proliferating lymphoma cells. 6 weeks later, patient died because of pneumonia.

Although it is very difficult to explain the provided data, it seems unlikely that VA-E stimulated the ongoing leucaemic process. Subcutaneous nodules might rather be an expression of a strong application site reaction. Unfortunately, the authors did not provide detailed data on the composition of the leukocyte counts of 3190/μl. Since the patient suffered from a chemotherapy-induced T-cell lymphopenia, one may speculate that the chemotherapy-insensitive lymphoma cells are the only population being reactive to *Viscum album* antigens.

Other adverse effects, less well documented

Skin reactions: Subcutaneous application of an aqueous VA-E (*Helixor*) was suggested to be responsible for allergic reactions of two cancer patients (Lange Wantzin *et al.*, 1983). One patient with chronic myeloic leukaemia (CML) receiving busulfane therapy developed a generalised erythema with large necrotic areas and a high fever reaction and diarrhoea. Unfortunately, in this patient therapy was continued in such a way that the extract was injected into the inflamed skin. Other factors could also be responsible in view of the fact that the patient received further complementary therapies. In another case, a patient with breast cancer was treated with VA-E (*Helixor* M) and, two days after the beginning of cyclophosphamide (CP) therapy, he developed urticaria and Quincke's edema (Lange Wantzin *et al.*, 1983). Skin test towards VA-E was positive. However, it is known that in patients without adverse effects VA-E also may induce a positive skin test reaction (Cougniet and Kühnast, 1987). Since application of CP was well tolerated later, VA-E co-application might have been responsible.

Allergic rhinitis: Allergic rhinitis induced by mistletoe tea was observed in an atopic patient who exposed himself by handling mistletoe professionally (Seidemann, 1984). Allergic reactivity was proven by positive skin (Prick) and RAST tests and re-exposition.

Acute asthmatic reaction: There was one report on an acute asthmatic reaction, which started shortly after subcutaneous injection of VA-E in a colon carcinoma patient suffering from chronic obstructive pulmonary disease (COPD). However, no detailed data were presented and, therefore, no firm conclusions can be drawn (Anonymous, 1998).

Allergic colitis: Haemorrhagic colitis with mucous and bloody diarrhoea was reported to be due to subcutaneous VA-E (*Iscador*) therapy (Ottenjann, 1992). In this patient, re-exposure to VA-E led to re-appearance of the symptoms. Histologically, colitis was confirmed and necrotic areas and an infiltration of eosinophiles was found in the terminal ileum. In addition, blood eosinophilia (22%), a shift to the juvenile forms of neutrophils, as well as an exanthema could be observed. After cessation of the therapy, symptoms ceased slowly. One could speculate that probably a Th2 (see below) related response with the release of IL-5 was responsible for the eosinophilic reaction.

Sarcoidosis: Zürner (1992) described a patient with disseminated sarcoma of the uterus receiving VA-E therapy (*Helixor* Mali). After 14 days she developed erythema nodosum-like skin reactions, with signs for the Löfgren's syndrome. Biopsy of the skin revealed an epitheloid-cell granulomatosis. The syndrome is suggested to be mediated by yet unknown immune reactions and sarcoidosis was also found to occur, although very rarely, in sarcoma patients (Brincker, 1986). At the time of treatment, the patient also received roxatidine (H_2-receptor antagonist) and another immunostimulating remedy containing extracts from *Echinacea*, *Baptisia*, and *Thuja*. In addition, she was vaccinated against influenza. Immunological reactivity of this patient was studied in our laboratory. Lymphocyte transformation test (incorporation of [^{3}H]-thymidin into lymphocytes to detect specific proliferation of the cells) revealed a very strong reactivity with a stimulation index of 140 induced by the mistletoe extract. Low levels of anti-mistletoe lectin-1 (ML-1) antibodies of the IgG and IgA type were found, as it is known to occur during therapy (Stein *et al.*, 1997, 1998a,b,c; Stettin *et al.* 1990).

CHARACTERISATION OF IMMUNOLOGICAL REACTIVITY OF PATIENTS WITH SUPPOSED SIDE EFFECTS DURING THERAPY WITH AN AQUEOUS MISTLETOE EXTRACT

From 1991 to September 1996, immunological reactivity of 43 cases of adverse effects supposed to be related to therapy with an aqueous VA-E (*Helixor*) were studied (Stein and Berg, 1999) in order to define the type of reactivity. According to the following criteria, patients were divided into two groups: patients with proven adverse effects (group 1; n = 34, mean age of 52 ± 12.9 y, mean duration of mistletoe therapy 4.3 ± 7.3 months) had a close relationship between VA-E exposure and onset of symptoms, which had to resemble symptoms observed with classical allergens. The second group consisted of patients with unproven, i.e. most unlikely relationship between clinical manifestation and VA-E therapy (group 2; n = 9, mean age of 56 ± 7.5 y, mean duration of mistletoe therapy 11.4 ± 19.4 months). Symptoms were not related to VA-E therapy because of exposure to other remedies or alterations of the primary disease. Patients suffered from different tumours as reported elsewhere (Stein and Berg, 1999). Manifestations were either of local or of systemic nature (Table 2).

Immunological reactivity of these patients was compared to the reactivity of tumour patients treated with the same VA-E for more than 2 years (n = 14, mean

Table 2 Supposed adverse effects during VA-E therapy in the 44 patients[1].

adverse effects	*group 1 (n = 34)*	*group 2 (n = 9)*
exclusively local reactions: redness, swelling, itching, exanthema	14	2
systemic reactions: generalised urticaria, Quincke's edema, rhinitis, conjunctivitis, high fever, severe myalgic reactions, anaphylactic reaction	20	7[2]

[1] Group 1: patients with proven relation between adverse effects and mistletoe application, group 2: patients with unproven relation. A combination of local and systemic symptoms was seen in some of the patients.
[2] one case of anaemia (see patients and materials and methods).

age of 60.9 ± 9.5 y, mean duration of mistletoe therapy 5.9 ± 3.4 years). These patients, however, did not exert adverse effects due to VA-E exposure.

Immunological reactivity was assessed by means of proliferation ([^{3}H]-thymidine incorporation) and cytokine release of VA-E (*Helixor*)-stimulated peripheral blood mononuclear cells (PBMC). Antibody production against a component present in many VA-E, i.e. mistletoe lectin-1 (ML-1), was detected in the sera of the patients by ELISA.

Proliferation Studies

Although a strong individual variation in the cellular *in vitro* immune response became obvious, proliferation of PBMC from patients of group 1 differed significantly from that of group 2 patients and the VA-E-treated tumour patients without adverse effects (TTP; Figure 1). Immune response induced by the control mitogen pokeweed mitogen (PWM) did not stimulate different reactivities between the groups of patients. Length of therapy of the patients was not responsible for the increased cellular response (data not shown).

Interestingly, cellular reactivity of patients with severe local manifestations was strongly enhanced as compared to the reactivity of patients suffering from systemic adverse effects (Stein and Berg, 1999). This strong cellular reactivity might reflect a delayed type hypersensitivity (DTH) reaction.

Detection of anti-ML-1 Antibodies and Role of anti-ML-1 Antibodies of the IgE Type during VA-E Treatment

Specific humoral immune response of the patients with supposed adverse effects during VA-E therapy was detected by the presence of anti-ML-1 antibodies in the sera of the patients following the ELISA technique as described (Stein *et al.*, 1994; Stettin *et al.*, 1990). It is well known that anti-ML-1 antibodies are produced during

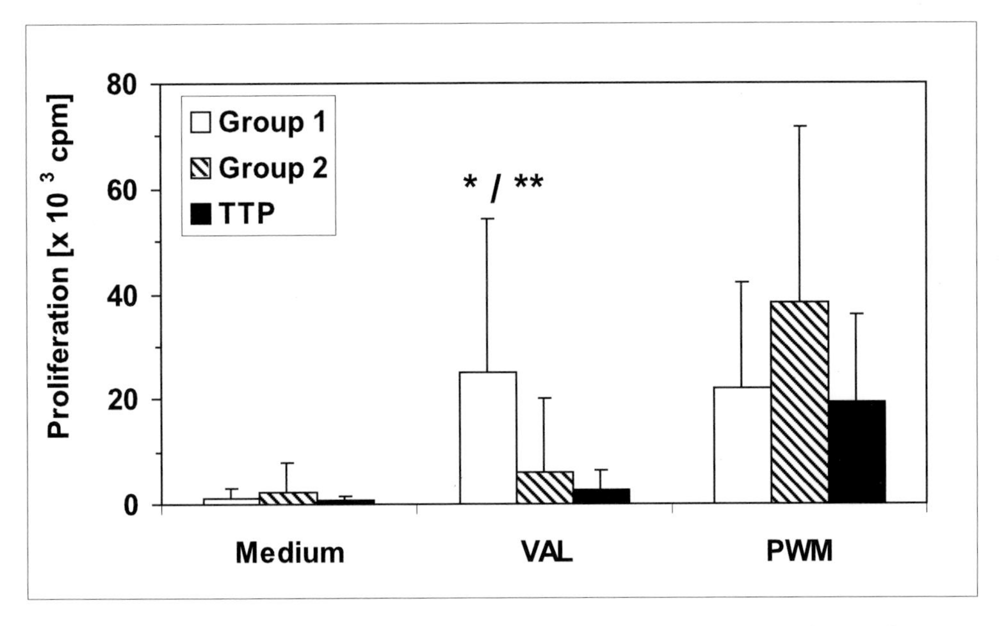

Figure 1 Proliferation of PBMC from the different groups of patients and controls stimulated with VA-E (*Helixor*) *in vitro*. Proliferation was measured by [^{3}H]-thymidine incorporation after 7 days. Given are mean ± standard deviation of each group. Group 1: patients with proven relation between adverse effects and mistletoe application (n = 34); group 2: patients with unproven relation ($n = 9$); TTP: VA-E (*Helixor*) treated tumour patients without adverse effects ($n = 14$). * $p < 0.05$ as compared to group 2, ** $p < 0.01$ as compared to TTP (Man-Whitney U-statistics).

therapy with VA-E in a dose-dependent manner (Stettin *et al.*, 1990). In that study, antibodies were mainly of the IgG type, while IgA and IgM antibodies were only infrequently found and no IgE antibodies were detectable. With respect to IgG, IgA and IgM antibodies (see Figure 2), in our study, there were no significant differences between the patients of group 1 and the treated tumour patients without adverse effects. In contrast to the findings from Stettin and colleagues, in 15 patients of group 1, an IgE response was present. As demonstrated in Figure 2, the IgE response from group 1 patients differed significantly from those patients treated with VA-E for more than 2 years (TTP). There was, however, no correlation of the anti-ML-1 antibody response of the IgE type with the clinical manifestation.

In contrast to patients receiving a vesicular mistletoe preparation (*Abnobaviscum*), anti-ML-1 antibodies of the IgE type were not induced in patients receiving an aqueous VA-E (*Helixor*; Figure 3) (Stein *et al.* 1998c). Thus, the production of anti-ML-1 antibodies of the IgE type during therapy with VA-E in patients without adverse effects seems to depend upon the composition and the kind of the applied VA-E (Stein *et al.*, 1998c, Stettin *et al.*, 1990). The anti-ML-1 antibodies of the IgE type which had been stimulated by the aqueous but not vesicular extract showed a close relationship to adverse effects induced during this kind of therapy.

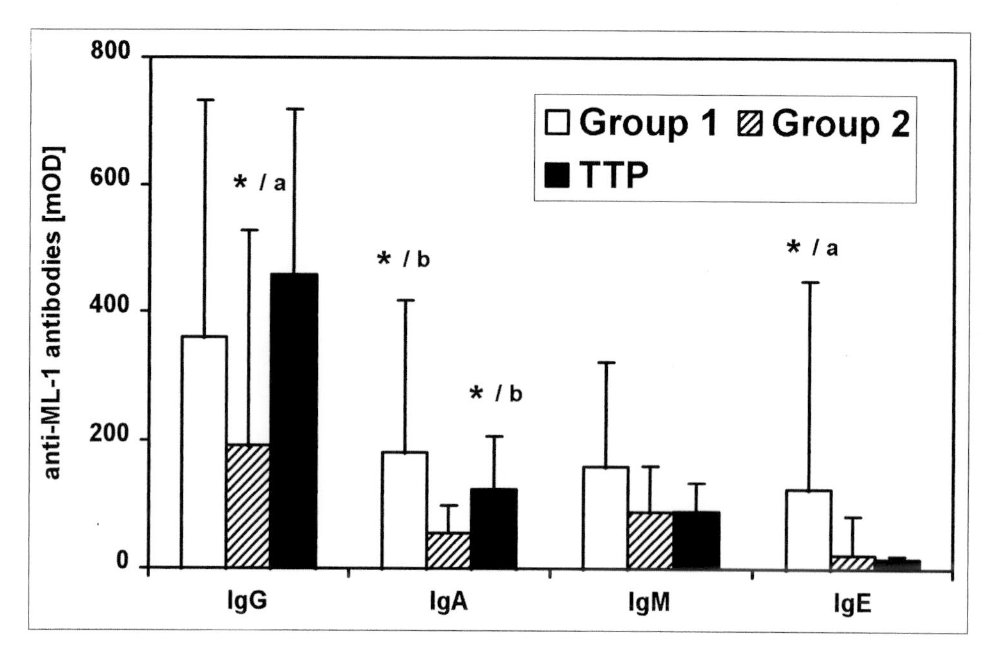

Figure 2 Anti-ML-1 antibodies in the sera from VA-E treated patients of the different groups as measured by ELISA. Given are mean ± standard deviation of each group. Group 1: patients with proven relation between adverse effects and mistletoe application, n = 34; group 2: patients with unproven relation, n = 9; TTP: VA-E (*Helixor*) treated tumour patients without adverse effects, n = 14. */a: $p < 0.05$ as compared to TTP, */b: $p < 0.05$ as compared to group 2 (Man-Whitney U-statistics).

Comparison of Humoral and Cellular Immune Reactivity

As shown in Figure 4, cellular immune responses were observed most frequently in the patients of group 1, while the humoral immune response dominated in the patients of group 2 and VA-E-treated patients without adverse effects. In group 2, 44% of the patients did not show any reactivity. Since there was no relevant correlation between cellular and humoral immune responses (Stein and Berg, 1999), distinct antigens may be responsible for these two types of reactivity.

Analysis of the Cytokine Pattern Induced by the Mistletoe Extract

Different T-cell populations have been shown to be responsible for either cellular (Th1) or humoral (Th2) reactivity (Mosman and Coffman, 1989) and are characterised by the cytokines, which they secrete: Th1 cells produce IFN-γ and IL-2, while Th2 cells are characterised by IL-4 and IL-5 (Howard *et al.*, 1993). IL-6 and TNF-α are the respective mediators released mainly by monocytes/macrophages (Durum and Oppenheim, 1993).

Classical allergic reactions towards pollen or dust mite antigens have been associated with a strong Th2 response. To evaluate, whether in the patients with VA-E

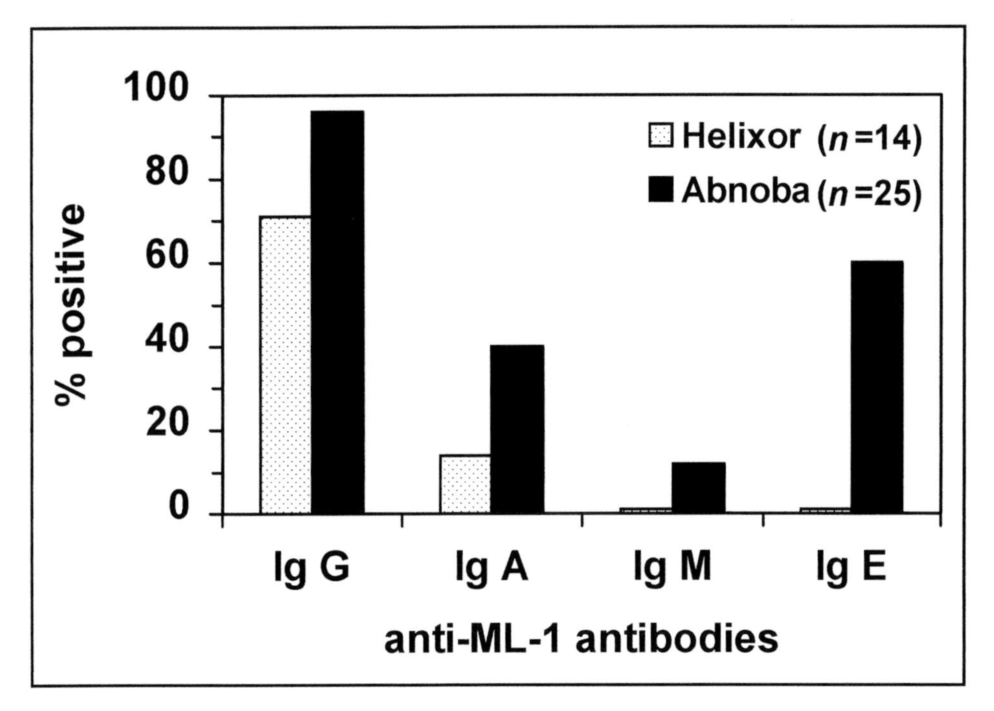

Figure 3 Production of anti-ML-1 antibodies of the different classes in tumour patients after treatment with VA-E for more than two years without exerting adverse effects. Antibodies were measured by ELISA (Stein *et al.*, 1998c).

induced adverse effects also a specific cytokine pattern was present with respect to the Th1/Th2 cytokines, supernatants of the cell cultures stimulated with the *in vivo* applicated VA-E for 6 days *in vitro* were analysed for these cytokines. While in most supernatants cytokines of the monocyte/macrophage lineage (IL-6/TNF-α) were found (data not shown), T-cell-specific cytokines were expressed independent from the adverse effects and did not allow a specification of the type of allergic reactivity (Table 3).

Conclusions from the Immunological Investigations

From the immunological investigations on the patients with supposed adverse effects the following conclusions may be drawn:

- Adverse effects are very rare and may be characterised as a hyperresponsiveness towards a foreign antigen being exposed to the skin-associated immune system.
- The pronounced cellular immune response towards VA-E (*Helixor*) most likely corresponds to a classical delayed type hypersensitivity (DTH) reaction.
- Adverse effects consist predominantly of pronounced application site reactions or systemic reactions like urticaria.

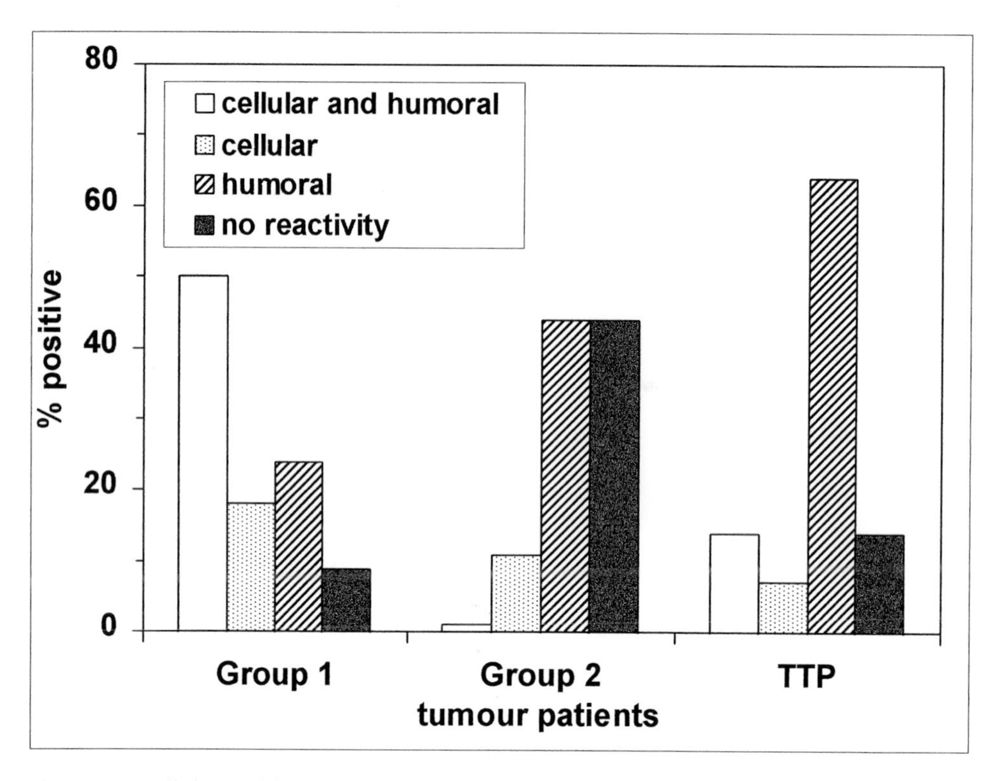

Figure 4 Cellular and humoral immune responses of the patients of the different groups. Cellular immune response was detected by proliferation induced by VA-E and humoral reactivity was assessed by ELISA (anti-ML-1 antibodies). Group 1: patients with proven relation between adverse effects and mistletoe application (n = 34); group 2: patients with unproven relation (n = 9); TTP: VA-E (*Helixor*) treated tumour patients without adverse effects (n = 14).

Table 3 Summary of the VA-E (*Helixor*)-induced Th1 or Th2 cytokine patterns in the supernatants of PBMC from patients of the different groups: number (%) positive.

VA-E (μg/ml)	*Group 1 (n = 23)*	*Group 2 (n = 6)*	*TTP (n = 14)*
0	8 (34)	0	3 (21)
100	5 (21)	1 (17)	5 (36)
1,000	9 (39)	1 (17)	6 (43)
10,000	10 (43)	0	5 (36)

Cytokines were measured in the supernatants of 6 day cultured PBMC by ELISA. For the cytokine pattern, the difference in the cytokine release between the unstimulated control cell culture and the VA-E stimulated cells was of relevance independent of the pre-existing pattern. Th1: IFN-γ/IL-2, Th2: IL4/IL-5. In most cases of Th1 or Th2 positive reactivity pattern there was also secretion of IL-6/TNF-α. Group 1: patients with proven relation between adverse effects and mistletoe application, group 2: patients with unproven relation, TTP: VA-E-treated tumour patients without adverse effects.

- A clear-cut correlation of the adverse effects with respect to the Th1 and Th2 immune response was not found.

However, the presence of anti-ML-1 antibodies of the IgE type could indicate that type I allergic reactions can play a role in some of the observed side effects.

CONCLUDING REMARKS

Although adverse effects in general and allergic reactions in particular were only rarely observed, application of VA-E may induce side effects, which in most instances are harmless. Hyperresponsiveness towards a foreign antigen may be genetically determined and individual factors as e.g. the underlying disease may influence the clinical manifestation of the adverse reactions. These factors certainly have to be considered in patients exerting local or systemic reactions in the course of VA-E therapy.

It is suggested that both, pseudo-allergic reactions induced *via* secretion of inflammatory cytokines (IL-1, IL-6) as well as migration of DTH cells at the injection site are involved in the pathogenesis. It is noteworthy that reduction of the initial doses causing the adverse effects in many cases was not associated with any further symptoms, which rather excludes the generation of memory cells during treatment with VA-E in these cases. However, it cannot be excluded that in some cases memory cells towards different epitopes were generated because in some patients re-exposure with VA-E again led to adverse reactions independently of the dosage.

From all these data it seems, quite possible that pseudo-allergic reactions as well as DTH reactions are involved in the pathogenesis of the adverse effects.

REFERENCES

Anonymous (1998) Allergische reacties op *Iscador*. *Geneesmiddelenbulletin*, **32**, 97.

Böcher, E., Stumpf, C., Büssing, A., and Schietzel, M. (1996) Prospektive Bewertung der Toxizität hochdosierter *Viscum album* L.- Infusionen bei Patienten mit progredienten Malignomen. *Zeitschrift für Onkologie*, **28**, 97–106.

Boie, D. and Gutsch, J. (1980) Helixor bei Kolon- und Rektumkarzinom. *Krebsgeschehen*, **23**, 65–76.

Boie, D., Gutsch, J., and Burkhardt, R. (1981) Die Behandlung von Lebermetastasen verschiedener Primärtumoren mit Helixor. *Therapiewoche*, **31**, 1865–1869.

Brincker, H. (1986) Sarcoid reactions in malignant tumours. *Cancer Treat. Rev.*, **13**, 147–56.

Coeugniet, E. and Kühnast, R. (1987) Hautreaktion nach intrakutaner Injektion eines *Viscum album* Extraktes. *Therapiewoche*, **37**, 626–623.

Dold, U., Edler, L., Mäurere, H.C., Mueller-Werning, D., Sakellariou, B., Trendelenburg, F., *et al.* (1991) *Krebszusatztherapie beim fortgeschrittenen nicht-kleinzelligen Bronchialkarzinom*, Thieme Verlag Stuttgart, New York.

Douwes, F.R. and Wolfrum, D.I. (1986) Prospektive randomisierte Studie zur adjuvanten Therapie kolorektaler Karzinome mit einem "Biological Response Modifier" (BRM). *Therapiewoche*, **36**, 116–122.

Durum, S.K. and Oppenheim, J.J. (1993) Pro-inflammatory cytokines and immunity. In W.E. Paul, (ed.), *Fundamental Immunology*, Raven Press Ltd, New York, pp. 801–835.

Eisenberg, D.M. (1997) Advising patients who seek alternative medical therapies. *Ann. Int. Med.*, **127**, 61–69.

Elsässer-Beile, U., von Kleist, S., Sauther, W., Gallati, H., and Schulte Mönting, J. (1993a) Impaired cytokine production in whole blood cell cultures of patients with gynaecological carcinomas in different clinical stages. *British Jornal of Cancer*, **68**, 32–36.

Elsässer-Beile, U., von Kleist, S., Stähle, W., Schurhammer-Fuhrmann, C., Schulte Mönting, J., and Gallati, H. (1993b) Cytokine levels in whole blood cell cultures as parameters of the cellular immunologic activity in patients with malignant melanoma and basal cell carcinoma. *Cancer*, **71**, 231–236.

Finelli, A. and Limberg, R. (1998) Mistel-Lektin bei Patienten mit Tumorerkrankungen. *Diagnostik und Therapie im Bild* 1–8.

Friess, H., Beger, H.G., Kunz, J., Funk, N., Schilling, M., and Büchler, M.W. (1996) Treatment of advanced pancreatic cancer with mistletoe: results of a pilot trial. *Anticancer Research*, **16**, 915–920.

Grothey, A., Düppe, J., Hasenburg, A., and Voigtmann, R. (1998) Anwendung alternativmedizinischer Methoden durch onkologische Patienten. *Deutsche Medizinische Wochenschrift*, **123**, 923–929.

Gutsch, J., Berger, H., Scholz, G., and Denck, H. (1988) Prospektive Studie beim radikal operierten Mammakarzinom mit Polychemotherapie, Helixor und unbehandelter Kontrolle. *Deutsche Zeitschschrift für Onkologie*, **20**, 94–101.

Hagenah, W., Dörges, I., Gafumbegete, E., and Wagner, T. (1998) Subkutane Manifestationen eines zentrozystischen Non-Hodgkin-Lymphoms an Injektionsstellen eines Mistelpräparates. *Deutsche Medizinische Wochenschrift*, **123**, 1001–1004.

Hajto, T., Hostanska, K., Fornalski, M., and Kirsch, A. (1991) Antitumorale Aktivität des immunmodulatorisch wirkenden Beta-galaktosid-spezifischen Mistellektins bei der klinischen Anwendung von Mistelextrakten (*Iscador*). *Deutsche Zeitschrift für Onkologie*, **23**, 1–5

Hara, N., Ichinose, Y., Asoh, H., Yano, T., Kawasaki, M., and Ohta, M. (1992) Superoxide anion-generating activity of polymorphonuclear leukocytes and monocytes in patients with lung cancer. *Cancer*, **69**, 1682–1687.

Hassauer, W., Gutsch, J., and Burkhardt, R. (1979) Welche Erfolgsaussichten bietet die *Iscador*-Therapie beim fortgeschrittenen Ovarialkarzinom? *Onkologie* **2**, 3–11.

Howard, M.C., Miyajima, A., and Coffman, R. (1993) T-cell derived cytokines and their receptors. In W.E. Paul, (ed.), *Fundamental Immunology*, Raven Press Ltd, New York, pp. 763–800.

Kjaer, M. (1989) Mistletoe (*Iscador*) therapy in stage IV renal adenocarcinoma. *Acta Oncologica*, **28**, 489–494.

Lange Wantzin, G., Thomsen, K., and Nissen, N.I. (1983) Alvorlige bivirkninger efter mistelenekstraktbehandling.*Ugeskr Lærger*, **145**, 2223–2224.

Mahfouz, M.M., Ghaleb, H.A., Zawawy, A., and Scheffler, A. (1998) Significant tumour reduction, improvement of pain and quality of life and normalisation of sleeping patterns of cancer patients treated with a high dose of mistletoe. *Annals of Oncology*, **9** (Suppl 2), 129.

Mosman, T.R. and Coffman, R.L. (1989) Heterogeneity of cytokine secretion patterns and function of helper T-cells. *Advances in Immunology*, **46**, 111–147.

Ottenjann, R. (1992) Allergische Kolitis auf Mistelextrakt. *Selecta*, **9**, 29.

Pichler, W.J. and Angeli, R. (1991) Allergie auf Mistelextrakt. *Deutsche Medizinische Wochenschrift*, **116**, 1333–1334.

Schlodder, D. (1996) 75 Jahre additive Misteltherapie bei Krebspatienten. Eine kritische Zusammenfassung der ärztlichen Erfahrungen. In R. Scheer, H. Becker, P.A. Berg, (eds.), *Grundlagen der Misteltherapie. Aktueller Stand der Forschung und klinische Anwendung*. Hippokrates Verlag, Stuttgart, pp. 339–351.

Schultze, J.L., Stettin, A., and Berg, P.A. (1991) Demonstration of specifically sensitised lymphocytes in patients treated with an aqueous mistletoe extract (*Viscum album* L). *Klinische Wochenschrift*, **69**, 397–403.

Seidemann, W. (1984) Allergische Rhinitis durch Misteltee (*Viscum album*). *Allergologie*, **7**, 461–463.

Stein, G.M. and Berg, P.A. (1999) Characterisation of immunological reactivity of patients with adverse effects during treatment with an aqueous mistletoe extract. *European Journal of Medical Research*, **4**, 169–177

Stein, G.M. and Berg, P.A. (1998a) Modulation of cellular and humoral immune responses during exposure of healthy individuals to an aqueous mistletoe extract. *European Journal of Medical Research*, **3**, 307–314.

Stein, G. and Berg, P.A. (1994) Non-lectin component in a fermented extract from *Viscum album* L. grown on pines induces proliferation of lymphocytes from healthy and allergic individuals *in vitro*. *European Journal of Clinical Pharmacology*, **47**, 33–38.

Stein, G.M., Henn, W., von Laue, B., and Berg, P.A. (1998b) Modulation of the cellular and humoral immune responses of tumour patients by mistletoe therapy. *European Journal of Medical Research*, **3**, 194–202.

Stein, G.M., Stettin, A., Schultze, J., and Berg, P.A. (1997) Induction of anti-mistletoe lectin antibodies in relation to different mistletoe extracts. *Anticancer Drugs*, **8** suppl 1, S57–S59.

Stein, G.M., von Laue, H.B., Henn, W., and Berg, P.A. (1998c) Human anti-mistletoe lectin antibodies. In S. Bardocz, U. Pfüller, and A. Pusztai, (eds.), *COST 98. Effects of antinutritients on the nutritional value of legume diets*, European Community, Luxembourg, Vol. 5. pp. 168–175.

Stettin, A., Schultze, J.L., Stechemesser, E., and Berg, P.A. (1990) Anti-mistletoe lectin antibodies are produced in patients during therapy with an aqueous mistletoe extract derived from *Viscum album* L. and neutralise lectin-induced cytotoxicity *in vitro*. *Klinische Wochenschrift*, **68**, 896–900.

Stoss, M., van Wely, M., Musielsky, H., and Gorter, R.W. (1999) Study on local inflammatory reactions and other parameters during subcutaneous mistletoe application in HIV-positive patients and HIV-negative subjects over a period of 18 weeks. *Arzneimittel-Forschung/Drug Research*, **49**, 366–373.

Stoss, M., van Wely, M., Reif, M., and Gorter, R.W. Tolerability of a standardised, non-fermented aqueous *Viscum album* L. extract in immunocompromised and healthy individuals. *Am. J. Alt. Med.* [in press]

van Wely, M., Stoss, M., and Gorter, M. (1999) Toxicity of a standardised mistletoe extract in immunocompromised and in healthy individuals. *American Journal of Therapeutics*, **6**, 37–43

Weis, J., Bartsch, H.H., Hennies, F., Tietschel, M., Heim, M., Adam, G., *et al.* (1998) Complementary medicine in cancer patients: demand, patient's attitudes and physiological beliefs. *Onkologie* **21**, 144–149.

Zielinski, C.C., Mueller, C., Tyl, E., Tichatschek, E., Kubista, E., and Spona, J. (1990) Impaired production of tumour necrosis factor in breast cancer. *Cancer*, **66**, 1944–1948.

Zürner, P. (1992) Sarkoidose nach Misteltherapie (Helixor)? *Arznei-telegramm*, 5, 51.

12. OVERVIEW ON *VISCUM ALBUM* L. PRODUCTS

ARNDT BÜSSING

Krebsforschung Herdecke, Department of Applied Immunology, University Witten/Herdecke, Communal Hospital, 58313 Herdecke, Germany

INTRODUCTION

The major market for *Viscum album* L. products is at present Germany, Switzerland and Austria, while it is rarely used in other countries. The main indications of *Viscum album* uses are hypertension (Table 1), arthritis and arthrosis (Table 2), and cancer (Table 3).

Medicinal preparations containing *Viscum album* are listed in official pharmacopoeias and are produced in accordance with the permission by law either as a homeopathic, phytotherapeutic or anthroposophical remedy. According to the German Medicines Act (AMG), enforced in 1978, finished medicinal products need to be authorised by the Federal Institute for Drugs and Medicinal Devices (BfArM) before they are made available to the patient. This authorisation was formerly based on monographs prepared by the distinct commissions, i.e. commission C (anthroposophical medicine), commission D (phytotherapy), and commission E (homeopathy). The pharmaceutical companies must provide proof of efficacy, safety and adequate pharmaceutical quality of their products. However, also products that had already been on the market before the AMG was in force, are in the process called post-marketing approval. In several cases, pharmaceutical companies producing *Viscum album* containing drugs had withdrawn their application from an ongoing procedure or are still waiting for renewal of the authorisation by BfArM. Homeopathic products need simple registration only (without proofed efficacy and safety), provided that no indication claims are made for them and that adequate quality is demonstrated.

Due to differences in their authorisation or registration based on the distinct monographs, *Viscum album* containing products may strongly differ in regard of plant material and preparation. This chapter presents an overview of the German market and the products available. However, in several cases, the pharmaceutical companies were unable or unwilling to present details of their individual manufacturing process.

HARVEST AND MANUFACTURING

The plant, growing semi-parasitically on different host trees such as Abies, Acer, Amygdalum, Betula, Crataegus, Fraxinus, Malus, Pinus, Poplus, Salix, Tilia, Ulmus

Table 1 *Viscum album* containing remedies used to treat hypertension and/or arteriosclerosis

	Drug and application	*Extraction*	*Host tree*	*Harvest*	*Parts used*	*Other drugs**
Antihypertonicum-Tropfen N Schuck Schuck, Schwaig	♦ 100 ml: Viscum album D3 (according to HAB 2a) 10 ml	Ethanolic	Different host trees	autumn	Fresh leafy shoots, fruits	Barium carbonicum D6 25 ml, Crataegus D 3 10 ml
Antihypertonicum S Schuck Schuck, Schwaig	○ Herba Visci albi (6:1) 25 mg	Aqueous	Different host trees	autumn	Dried leafy shoots, fruits	Folia Betulae (6:1) 20 mg, Herba Crataegi sicc. 20 mg, Fructus Crataegi (2.5:1) 20 mg, Folia Oleae europae. (5.5:1) 30 mg, Folia Rhododendri (3:1) 12.5 mg
Antisclerosin[R] S Medopharm, Gräfeling	○ Herb. Visci albi (6:1) 40 mg	Aqueous	Deciduous trees	November until March	Fresh leafy shoots, fruits	Crataegus berries 100 mg, Crataegus leaves and blossoms 10 mg, nicotinic acid 15 mg, magnesium orotat 25 mg, rutoside $3H_2O$ 5 mg
Asgoviscum N Kapseln/Tropfen Rhein-Pharma/ Teneca, Schwetzingen	♦ 100 g: Herba Visci albi 75 mg ○ Herba Visci albi (5:1) 30 mg	Aqueous-ethanolic	Different host trees	October to December	Dried leafy shoots, fruits	Fruct. Crataegi oxyac. (75 g/2.5:1 6 mg), Bulb. Allii sat. (0.3 g/1 mg)
Doppelherz Knoblauch-Kapseln mit Mistel + Weißdorn Queisser-Pharma, Flensburg	○ Herb. Visci albi (1:1) 60 mg	Oil macerate	Different host trees	NN	NN	Allium sativum 150 mg, Crataegus 60 mg

Table 1 *continued*

	Drug and application	*Extraction*	*Host tree*	*Harvest*	*Parts used*	*Other drugs**
Hypercard Hotz, Insel Riems	♦ 10 ml: Herb. Visci albi (1:5) 20 mg	Ethanolic	NN	NN	NN	Fol. Oleae 100 mg, Rad. Ginseng 5 mg, Flor. Convalleriae 10 mg, Herba Convalleriae 10 mg, Rad. Rauwolia D4 1 ml, Silicea D8 750 mg, Sumbul. Moschat 25 mg, Conium D4 2 ml
Kneipp® Knoblauch-Kapseln mit Mistel + Weißdorn Kneipp, Würzburg	○ Herb. Visci albi (4:1) 25 mg	Cold macerate	NN	NN	NN	Crataegus and Allium sativum (separate ○)
Kneipp® Mistelpflanzensaft Kneipp, Würzburg	♦ 100 ml: Herb. Visci albi 100 ml	Pressed sap	NN	December	Leafy shoots, berries	No
Mistel Curarina® Harras-Pharma-Curarina, Munich	♦ 100 ml: Herb. Visci albi (1:5) 100 ml	Ethanolic	Different host trees	NN	Dried young leafy shoots, fruits	No
Mistel-Kräutertabletten Salushaus, Bruckmühl/Mangfall	○ Herb. Visci albi (6:1) 350 mg	Aqueous-ethanolic	Different host trees	NN	Fresh leafy shoots, fruits	No
Mistelöl-Kapseln Twardy, Flörsheim/Main	○ Herb. Visci albi (1:1) 270 mg	Oil macerate	NN	NN	NN	No
Plantacard N Madaus, Cologne	♦ 100 g: Viscum album (according to HAB 2a) 40 g	Ethanolic	Poplar	January	Fresh leafy shoots	Arnica D3 10 g, Crataegus D1 30 g

Table 1 *continued*

	Drug and application	*Extraction*	*Host tree*	*Harvest*	*Parts used*	*Other drugs**
Repowinon Truw Truw, Frechen	♦ 100 ml: Viscum album (according to HAB 2a) 20 ml	Ethanolic	NN	NN	NN	Crataegus 20 ml, Phosphorus D 5 20 ml, Rauwolfia serpentina D4 20 ml, Selenicereus grandifloris D2 20 ml
RauwolfiaViscomp Schuck/ RauwolfiaViscomp-Tab Schuck, Schwaig	♦ 100 ml: Viscum album ex herba recente 90 ml ○ Viscum album D1 (according to HAB 2a) 25 mg	Ethanolic	Different host trees	autumn	Fresh leafy shoots, fruits	Rauwolfia serpentina D2 (1 ml/2.5 mg), Allium sativum D3 (1 ml/2.5 mg), Barium chloratum D2 (1 ml/2.5 mg), Crataegus D3 (1 ml/2.5 mg), Equisetum arvense ex herba rec. (6 ml/15 mg, rutoside 10 mg, procyphyline 40 mg
Regivital Mistel-Tropfen Togal, München	♦ 100 g: Herb. Visci albi (1:5) 20 g	Ethanolic	NN	NN	NN	No
Salus Mistel-Tropfen Salushaus, Bruckmühl/Mangfall	♦ 100 g: Herb. Visci albi (1:5) 20 g	Ethanolic	Different host trees	NN	Fresh leafy shoots, fruits	No
Viscasan Bioforce, Konstanz	♦ 10 ml: Mother tincture (according to HAB 2a) 10 ml	Ethanolic	Different host trees	autumn	Fresh leafy shoots, fruits	No
Visconisan N Hanosan, Garbsen	♦ 100 g: Viscum album (according to HAB 2a) 16.5 g	Ethanolic	Pini	autumn	Fresh leafy shoots, fruits	Crataegus 25 g, Rauwolfia serpentina 0.1 g

Table 1 *continued*

	Drug and application	*Extraction*	*Host tree*	*Harvest*	*Parts used*	*Other drugs**
Viscophyll Krewel Meuselbach, Eitorf	♦ 1 ml: Herb. Visci albi (1:2) 0,4 g	Aqueous-ethanolic	Different host trees	NN	Dried young leafy shoots, fruits, flowers	Fucus 0.05 g
Viscum album H Pflüger, Rheda-Wiedenbrück	♦ 100 ml: Viscum album mother tincture (according to HAB 2a) 32 ml	Ethanolic	NN	autumn	Fresh leafy shoots, fruits	Aconitum napellus D4, Anamirta cocculus D3, Kalium phosphoricum D10, Rauwolfia serpentina D4, Veratrum album D4
Viscum/Crataegus Wala, Eckwälden	○ 100 g: Viscum album, Planta tota D3 (1:1.5) 1 g (according to HAB 33a) ↘ 1 ml: Viscum album, Planta tota D5 (1:1.5) 1 g (according to HAB 33a)	Aqueous fermentation	Tiliae	winter	Fresh leafy shoots, fruits, sinker	Crataegus e foliis et fructibus ferm D1 1 g and 0.1 g, respectively
Viscum comp. Wala, Eckwälden	○ 100 g: Viscum album, Planta tota D2 (1:1.5) 1 g (according to HAB 33a) ↘ 1 ml: Viscum album, Planta tota D5 (1:1.5) 0.1 g (according to HAB 33a)	Aqueous fermentation	Populi	winter	Fresh leafy shoots, fruits, sinker	Atropa belladonna e radice ferm. D11 1 g and 0.1 g, respectively
Viscysat[R] Bürger Ysatfabrik, Bad Harzburg	♦ 100 ml: Herb. Visci albi (2:1) 85 g	Ethanolic	Deciduous trees	December	Fresh leafy shoots, fruits	No

NN – no information available (host tree and harvest conditions in the responsibility of the provider); ♦ drops; ○ tablets; ↘ injection
* complex drugs containing > 8 different plant species were not taken into account for this table.

Table 2 *Viscum album* remedies used to treat arthrosis, chronic polyarthritis, rheumatoid arthritis etc.

	Drug and application	*Extraction*	*Host tree*	*Harvest*	*Parts used*	*Other drugs**
Hornerz/Cartilago comp. Wala, Eckwälden	○ 100 g: Viscum album, Planta tota D2 (1:2.25)0.1 g (according to HAB 33a) ↘ 1 ml: Viscum album, Planta tota D4 (1:2.25) 1 g (according to HAB 33a)	Aqueous fermentation	Mali	December	Fresh leafy shoots, fruits, sinker	Horn silver D5 0.1 and 1 g, respectively
Plenosol N Madaus, Cologne	↘ 1 ml: Herb. Visci albi (1:1.3) 1 mg	Aqueous	Populi	January	Fresh leafy shoots	No
Syviman[R] N pasture Müller Göppingen, Göppingen	✋ Tinctura Symphyti 6.5 g, Tinctura Visci albi 0.2 g (according to HAB 2a)	Ethanolic	Different host trees	Autumn	Fresh leafy shoots, fruits	No
Viscum Mali e planta tota 3%, Unguentum Wala, Eckwälden	✋ 30 g: Viscum album, Planta tota (1:2.25) 3 g (according to HAB 33a)	Aqueous fermentation	Mali	December	Fresh leafy shoots, fruits, sinker	No
Viscum Mali ex herba W 5%, Oleum Wala, Eckwälden	✋ 50, 100 ml: Viscum album, Planta tota (1:20) 20 g (according to HAB 12 g)	Oil extraction	Mali	December	Fresh leafy shoots, fruits, sinker	No

↘ ampoule (injection); ○ tablets; ✋ Topic application of pasture/oleo.

Table 3 *Viscum album* remedies used to treat cancer

Extracts	*Drug*	*Extraction*	*Host tree*	*Harvest*	*Parts used*
Abnoba Abnoba, Pforzheim	↘ 1 ml: Viscum album, Planta tota/Herba rec. 15, 1.5, 0.15, 0.015 mg; D6, D10, D20, D30	Pressing	A, Ac, Am, B, C, F, M, P, Qu	summer and winter	Fresh leafy shoots, fruits
Cefalektin Cefak, Kempten	↘ 1 ml: Herb. Visci albi (1:10) 10 mg	Aqueous	Po and others	autumn/winter	Dried leafy shoots
Eurixor Biosyn, Fellbach	↘ 1 ml: Herb. Visci albi (1:1.3) 1 mg	Aqueous	Po	January	Fresh leafy shoots
Helixor Helixor, Rosenfeld	↘: Viscum album, Herba rec. (1:19) 1, 5, 10, 20, 30, 50, 100 mg	Aqueous	A, M, P	summer and winter	Fresh leafy shoots, fruits
Iscador Weleda AG, Schwäbisch Gmünd	↘: Viscum album, Herba rec. 50, 30, 20, 10, 1, 0.1, 0.01, 0.001, 0.0001 mg	Aqueous fermentation	M, P, Qu, U	summer and winter	Fresh leafy shoots, fruits
Iscador spezial Weleda AG, Schwäbisch Gmünd	↘: Viscum album, Herba rec. 5 mg (mistletoe lectins in M: 250 ng/ml, in Q: 375 ng/ml)	Aqueous fermentation	Qu, M	summer and winter	Fresh leafy shoots, fruits
Iscucin Wala, Eckwälden	↘ 1 ml: Viscum album, Planta tota (according to HAB 38) (1:7)	Aqueous[1]	A, C, M, P, Po, Qu, S, T	summer and winter[2]	Dried leafy shoots, fruits, sinker[3]

Table 3 *continued*

Extracts	*Drug*	*Extraction*	*Host tree*	*Harvest*	*Parts used*
Isorel/Vysorel Novipharm, Pörtschach	↘ 1 ml: Viscum album, Planta tota (1:16.5) 991.54 mg	Aqueous	A, M, P	summer and winter	Fresh leafy shoots, fruits, sinker
Lektinol Madaus, Cologne	↘ 0.5 ml: Herb. Visci albi (1:1.1–1.5) 0.02–0.07 mg (Mistletoe lectins: 30 ng/ml)	Aqueous	Po	January	Fresh leafy shoots
Plenosol N Madaus, Cologne	↘ 1 ml: Herb. Visci albi (1:1.3) 1 mg	Aqueous	Po	January	Fresh leafy shoots

Host trees: A – Abietis; Ac – Aceris; Am – Amygdali; B – Betulae; C – Crataegi; F – Fraxini; M – Mali; P – Pini; Po – Poplar; S – Salicis; T – Tiliae; U – Ulmi; Qu – Quercus.

[1] there are also fermented extracts from mistletoes grown on defined host trees (such as Abietis, Mali and Pini) available: Fresh leafy shoots, fruits and sinker harvested in winter (according to HAB 33a).

[2] The mistletoe used to produce Iscucin Salicis is harvested only in winter.

[3] not in Iscucin Quercus.

↘ ampoule (injection).

and Quercus all over Europe, is harvested in defined harvest resources in France, Germany, Austria, Switzerland and Eastern-Europe countries. To a minor extend, the plant is propagated artificially by spreading its berries during winter to defined host trees, especially oak trees (see Ramm *et al.*, this book). Vegetative propagation of the plant, however, was never been successful. Eight month after the spreading of the berries, either by birds or artificially, and successful connection of the haustorium to the host branch, the first leaves may appear. The first small blossoms, and later on the berries, will appear after 3 to 5 years. After this phase of plant growth, the plant material may be harvested by the manufacturers.

In contrast to the drugs used in phytotherapy (Table 1), the processed plant material, host tree and time of harvest is highly defined in the extracts used in anthroposophical medicine (Table 3). As the relevant compounds within the plant, such as proteins and carbohydrates, strongly differ within a year, the anthroposophical manufacturers combine the saps from both, the summer and winter harvest, while in phytotherapy, only plant material harvested in winter from poplars is used. The drugs used to treat arthrosis are produced from fresh plant material of the winter (and autumn) harvest only (Table 2).

Mistletoe in Homeopathy

Homeopathic remedies are products received from various sources suggested to exert therapeutic properties at very small dosages. The mother tincture might be diluted several hundred times by shaking, even though all remaining molecules of the original substance may be absent. This agitation is thought to impart the homeopathic action to the remedy. Remedies are chosen which at pharmacologic or toxic doses cause symptoms that mimic those which are the subject of treatment. Among a bunch of symptoms, application of *Viscum album* as a homeopathic drug in healthy individuals may produce symptoms of increased blood pressure, pain similar as in cases of neuralgia, arthrosis and arthritis, but is also used to treat epilepsy and asthma (Boericke, 1992).

In several preparations, *Viscum album* is combined with different other plants which have a reputation for decreasing blood pressure or strengthen the heart activity. The route of application is in general orally. The plant material is extracted either by ethanol, aqueous fermentation or oil, as handed down in the German Homeopathic Pharmacopoeia (HAB). However, in several cases the host tree of the processed mistletoe is not defined and pooled saps of mistletoes from different host trees are provided by commercial companies to the manufacturers.

The mother tinctures manufactured by Method 2a HAB 1 are produced by macerating the fresh leafy shoots and fruits from *Viscum album* collected in autumn using ethanol (approximately 43%). The plants or parts of the plants are cut up finely. A sample is used to determine loss on drying. Subsequently, at least half of the amount by weight of ethanol 86% is added to the cut-up plant material. The mixture is left to stand for at least 10 d at a temperature not exceeding 20°C, shaking repeatedly, and expressed and filtered later on. Mother tinctures made in accordance with Method 2b are manufactured as per Method 2a, using ethanol

62% (ethanol content approximately 30%). For potentization, the first decimal dilution (D1) is made with 2 parts of the mother tincture and 8 parts of ethanol 30%, while for the second decimal dilution (D2), 1 part of the D1 dilution and 9 parts of ethanol 15% are used. Subsequent dilutions are produced in the same way as for the D2 dilutions.

According to Method 33a, crushed plant material (100 parts) is fermented with honey (0.75 parts), lactose (0.75 parts) and water (50 parts). The mixture is kept in a water bath at 37°C (warming phase), and only in the morning and in the evening for 2 h in ice water (cooling phase). After the decrease of the pH, with the exception of the cooling phases in the morning and in the evening, the mixture is kept at room temperature. Three days later the mixture is pressed within a cooling phase. The pressed sap again is stored at room temperature but kept for 2 h in ice water in the morning and in the evening. Three days later, the extract is filtered. To form the mother tincture, 50 mg ash of air-dried residues from the fermented plant material is added to 100 ml of the filtered extract.

An external oleo is produced according to Method 12 g HAB 1. One part of dried plant material is mixed with 20 parts of plant oil (peanut, olive or sesame oil) and kept at 37°C for 7 d. Later on the extract is pressed and filtered.

Mistletoe in Phytotherapy

According to the definition of the European Scientific Cooperative on Phytotherapy (ESCOP), phytomedicines, or herbal medicinal products, are medicinal products containing as active ingredients only plants, parts of plants or plant materials, or combinations thereof, whether in the crude or processed state. Plant materials include juices, gums, fixed oils, essential oils, and any other directly derived crude plant product. They do not include chemically defined isolated constituents, either alone or in combination with plant materials.

In phytotherapy, *Viscum album* is used mainly to treat hypertension (Table 1), but also arthritis and arthrosis (*Plenosol* N; Table 2), and to stimulate the immune system of cancer patients (*Plenosol* N, *Lektinol, Eurixor, Cefalektin*; Table 3) (DAB-Kommentar, 1997). Table 1 gives an indication of the large number of commercial products which are available in Germany to treat hypertension, even in the absence of a positive monograph; many medicines are on the market due to their long traditionally use rather than scientifically evaluated. Due to the process of post-approval, it is unclear whether all these products will still be available on the German market. In most cases, the plant material is extracted with ethanol from dried or fresh plant material. In this case, the most relevant compounds are the flavonoids, lignans and amines. Also teas containing *Viscum album* are used to treat hypertension and to enhance coronary blood flow. Two representatives are given below (according to Weiß, 1991):

Herb. Visci albi	33.3 g		
Fol. et Flor. Crataegi	33.3 g	Viscum album conc.	100.0 g
Fol. Melissae	33.3 g		

2 small spoons of plant material with 1 cup hot water (5–10 min), to be drunken in the morning and in the evening

2–4 small spoons of plant material with 250 ml cold water (overnight), to be drunken in the morning and in the evening

The phytotherapeutic *Viscum album* extracts (VA-E) used to treat cancer patients are aqueous extracts from fresh plant material harvested from poplars during the winter season. The preparations *Lektinol* and *Eurixor* state that the content of mistletoe lectins (ML) is of major importance for an effective treatment and thus, defined amounts of these proteins are declared. Both preparations are produced under controlled temperature conditions from aqueous VA-E of fresh plant material. To avoid loss of ML, *Lektinol* is stabilised with polyvidon by a special pharmaceutical formulation which is under patent.

Mistletoe in Anthroposophical Medicine

Anthroposophically extended medicine does not regard illness as a chance occurrence or mechanical breakdown, but rather as something intimately connected to the biography of the human being, and thus, treatment integrates conventional dietary and nutritional therapy, rhythmical massage, hydrotherapy, art therapy, and counselling. To treat cancer patients, several manufacturers produce VA-E in accordance with recommendations of Rudolf Steiner (1924), the founder of anthroposophy. However, the interpretation of his instructions given in 1923/24 is rather difficult and thus, different manufacturing methods are used.

The manufacturing process of these drugs is in accordance with the methods handed down in the HAB (32, 33, 38). In contrast to the dried plant material mainly used in phytotherapy (in accordance with DAB 10), anthroposophical VA-E contain fresh material (with the exception of *Iscucin* which follows HAB 1, Method 38).

Table 4 Extraction of *Viscum album* at different pH values

	pH 4	*pH 8*
proteins (nMol/ml)	90	482
amino acids (nMol/ml)	1721	2178
Mistletoe lectin I (ng/ml)	0.431	9.983
carbohydrates (mg/ml)	189	198

The frozen plant material (mistletoe from apple tree harvested in winter) was crushed and extracted for 1 h in water supplemented with NaCl. The results were kindly provided by Helixor, Rosenfeld.

One and two-year old leaves and stems of male and female plants, and the short-tribe from the summer harvest, and the ripened berries from the winter harvest, and in some cases also the sinker, are harvested from defined harvest resources in France, Germany, Austria and Switzerland. According to recommendations of Rudolf Steiner, plants from different host trees will be processed separately to form distinct drugs. In fact, the content of toxic proteins from *Viscum album* strongly differs in VA-E from different host trees (Scheer *et al.*, 1995; Schaller *et al.*, 1996; Büssing *et al.*, 1999). The crushed plants may be stored either as fresh material in liquid nitrogen or stored as a sterile extract from fresh plants at 4 to 8°C until its use. The processed plant material is strictly protected against oxidation.

As shown in Table 4, variations of the pH during extraction will have a major impact on the relevant compounds. The most relevant changes are an increase of the protein content, and thus, ML content (and cytotoxicity) at higher pH level. In contrast, a decrease of the extraction temperature (10°C) will slightly increase the content of proteins (5%), while the content of amino acids decreases (3%). Thus, variations of the temperature will not significantly influence the content of relevant compounds.

- *Abnoba* extracts are produced by aqueous maceration (with ascorbate-phosphate buffer) of fresh plant material pressed in a special squeezer at room temperature. The sap is rich in ML, viscotoxins, polysaccharides, and contains membrane lipids which form liposome-like vesicles, which are micelles from the cell membranes of the plant.
- *Helixor* extracts are produced by cold water extraction (1 h at 14–20°C) of fresh plant material and are rich in ML, poly- and oligosaccharides, while the viscotoxins are below the detection level of 5 μg/ml. However, due to a special filtration step, ML III is the dominating mistletoe lectin in these extracts, while ML I is nearly missing.
- *Iscador* extracts are produced by fermentation of fresh plant material at 20–23°C using *Lactobacillus plantarum*. Within two or three days, lactic acid production reduces the pH and ML content (Ribéreau-Gayon *et al.*, 1986), while the content of viscotoxins raises. Later on, the fermented mistletoe is pressed to remove solid and insoluble residues from the extract.
- *Iscucin* is produced in accordance with the HAB 1, Method 38. Lyophilised plant material harvested in winter is mixed with a 6 fold amount of water (8.8 parts NaCl, 0.2 parts NaOH and 991 parts H_2O) and kept at 4°C for 14 d.
- *Isorel* extracts are produced by cold water extraction (2 h at 6–12°C) of fresh plant material and contain ML, poly- and oligosaccharides and few amounts of viscotoxins. However, also the ML III is the dominating lectin.

In contrast to industrial companies which favour the concept of a single biologically active substance within the plant (such as mistletoe lectin I), anthroposophical manufacturers state that several components are active (and may interact), and thus their extracts are recommended to be applied at increasing concentrations. Further, anthroposophical manufacturers mix the saps of the summer and winter harvest by complicated procedures (Koehler, 1992) which strongly differ between the manufacturers. Generally, the plant extracts are mixed using high speed homogenisation on a rotating disc or by whirling within an egg-shaped vessel. As a result of this process, both saps may form a new entity with additional qualities. To identify these additional qualities, some researcher observed differences in the germination rate and growth of cress shoots (*Lepidium sativum*) germinated in high dilutions of VA-E, which are produced on a rotating titanium disc (10,000 rounds per minute; centrifugal force approximately 55,000 g), as compared to hand-mixed saps (Gorter, 1998). However, the clinical impact of these differences is unclear and needs evaluation with adequate methods.

The mixed saps may be diluted with isotonic saline and sterile-filtered. To ensure consistent quality of the drugs, the time of harvest, processed plant material and the manufacturing process is highly standardised. Later on, the extracts go through extensive tests (specific constituents, biological activity etc.; see Lorch and Tröger, this book). To exclude contamination with pesticides, heavy-metals or bacteria, the plant material is under strict control.

REFERENCES

Boericke, W. (1992) *Manual der Homöopathischen Materia Medica*. Karl F. Haug Verlag, Heidelberg.

Büssing, A. and Schietzel, M. (1999) Apoptosis-inducing properties of *Viscum album* L. extracts from different host trees correlate with their content of toxic mistletoe lectins. *Anticancer Research*, **19**, 23–28.

German Pharmacopoeia, "*Deutsches Arzneibuch*" (DAB). 10. Auflage (1993) Deutscher Apotheker Verlag, Stuttgart, and Govi-Verlag, Frankfurt.

German Pharmacopoeia, "*Deutsches Arzneibuch*" (DAB 10; M82) (1998) Deutscher Apotheker Verlag, Stuttgart.

German Homeopathic Pharmacopoeia, "*Homöopathisches Arzneibuch*" (HAB), 1. Ausgabe (1978), and 4. Nachtrag (1985), Deutscher Apotheker Verlag, Stuttgart.

Gorter, R. (1998) *Isacdor*. Mistletoe preparations used in anthroposophically extended cancer treatment. Verlag für GanzheitsMedizin, Basel.

Koehler, R. (1992) Mistelbildung und Strömungsverfahren. *Elemente der Naturwissenschaft*, **57**, 3–19.

Ribéreau-Gayon, G., Jung, M.L., Di Scala, D., and Beck, J.P. (1986) Comparison of the effects of fermented and unfermented mistletoe preparations on cultured tumor cells. *Oncology*, **43** (Suppl. 1), 35–41.

Schaller, G., Urech, K., and Giannattasio, M. (1996) Cytotoxicity of different viscotoxins and extracts from the European subspecies of *Viscum album* L. *Phytotherapy Research*, **10**, 473–477.

Scheer, R., Errenst, M., and Scheffler, A. (1995) Wirtsbaumbedingte Unterschiede von Mistelpräparaten. *Deutsche Zeitschrift für Onkologie*, **27**, 143–149.

Steiner, R. (1985) *Physiologisch-Therapeutisches auf Grundlage der Geisteswissenschaft* (GA 314). 22. April 1924. Rudolf Steiner Verlag, Dornach, pp. 294–295.

Weiß, R.F. (1991) *Lehrbuch der Phytotherapie*. Hippocrates Verlag, Stuttgart, pp. 217.

13. PHARMACEUTICAL QUALITY CONTROL OF MISTLETOE PREPARATIONS

ELMAR LORCH AND WILFRIED TRÖGER

Verein Gemeinschaft Fischermühle, Department of Holistic Cancer and Immune Therapy, Rosenfeld, Germany

INTRODUCTION

When applying medicinal products, the quality, the efficacy and the safety have to be considered with respect of drug laws. According to the German Drug Law (AMG, 1998), the quality of remedies is defined as "the property of a medicinal product being determined according to identity, content, purity, other chemical, physical, biological properties or by the manufacturing procedure". The manufacturer is obliged by law to take the appropriate measures for the control of quality, constant efficacy and safety of the drug (EU-GMP-Guideline, 1989).

The present contribution deals with quality control in the production of mistletoe extracts. Mistletoe extracts are plant extracts containing several substances and substance groups. They have to fulfil quality requirements in order to attain a batch-to-batch consistency. Therefore, methods of standardization and norming are established. Norming means that a multi-component mixture is adjusted by dilution to a defined content of substance or substance group. If this substance is characterised and can clearly be detected in the whole plant extract, it may be used as tracer substance for the extract. It would be optimal, if the tracer substance is also the main effective component. Some mistletoe producers suggest the mistletoe lectins (ML) to define efficacy of the drug, although no clear proofs are available yet. In the past, the viscotoxins were regarded as the effective principle (Winterfeld, 1942). At present, the lectins are discussed to be the main active substance (Hajto, 1989). In the future, oligosaccharides or an up to now unknown 5 kD protein (Kuttan, 1988) may be in the centre of interest. As recent researches prove interactions between mistletoe substances, it may be possible that not a single substance defines the efficacy of mistletoe preparations, and thus, standardization of whole plant extracts from mistletoe seems to be appropriate. Standardization considers all substances to be responsible for the efficacy of the whole plant extract. Methods to proof its quality gain importance, e.g. by measuring the *in vitro* effects of the product. The constant quality of the herbal material and also in-process-controls during the production procedure in each and every step leads to the desired batch-to-batch consistency.

Mainly two aspects have to be considered: (1) the quality requirements of the drug law, and (2), the standardization with respect to a multicomponent extract.

This chapter focussed on quality control of the plants and the produced herbal remedies. The manifold other aspects of quality control of a GMP-conform drug production according to the guidelines of "Good Manufacturing Practices" (GMP) as e.g. controls during manufacturing process, documentation, standard operation procedures, qualification and validation as well as the special requirements in the manufacturing of sterile injection preparations under aseptic conditions shall not be mentioned in detail; please see the relevant literature (EU-GMP-Guidelines 1989; FDA-Guideline, 1987).

IDENTITY

Plant Material

Botanical aspects

The basic material for the production of mistletoe preparations is the white-berry mistletoe, *Viscum album* L. (Viscaceae) domestic in Europe. Its three most important botanical subspecies differ in their host tree specification: ssp. *abietis* (fir mistletoe), ssp. *album* (leafwood mistletoe), and ssp. *austriacum* (pine mistletoe). If dried plant material is used for the manufacturing of extracts, the quality test can be performed according to the German pharmacopoiea, which describes in detail macroscopic and microscopic identity test (DAB, 1999). In most cases, however, fresh plants are taken for the production of mistletoe extracts used to treat tumour patients. Furthermore, the anthroposophic manufacturers use plants grown on different host tree, e.g. mistletoe grown on fir or apple-tree.

For the identity of the plants as well as the identity of the host trees, it is important that the plants are harvested by trained company personnel, as suggested in a draft of a Guideline for the commercial collection of wild plants for medical purposes (GHP = Good Harvesting Practices) (Harnischfeger, 1999]. Harvest date, harvest place, weather conditions and condition of the material (development, pest infestation, etc.) are observed and documented. After picking and washing, the plants are cut and deep-frozen.

Chromatographic analysis

The determination of identity of the deep-frozen mistletoe material is preferably carried out by high-pressure-liquid-chromatography (HPLC) or thin-layer-chromatography (TLC). The phenylpropan-glycosides syringin and syringenin-4′-O-apiosylglucoside are suited for characterisation by these methods (Wagner *et al.*, 1984). Also, flavonoidglycosides can be identified. The different plant subspecies can be differentiated according to their distinct flavonoid pattern (Hamacher, 1996; Lorch, 1993, and own results). A host tree specific distinction is possible by the determination of the distribution of the viscotoxins. For this, a suitable sample work-up with subsequent HPLC-determination was described (Jordan *et al.*, 1986a; Schaller *et al.*, 1996a). Meanwhile, six different viscotoxins were detected and characterized in the European mistletoe (Samuelsson, 1958, 1961; Schaller *et al.*, 1996a,b, 1998).

Herbal Products

The identity test in the herbal remedies can be performed in different ways depending on the distinct extract. As already described above, the mistletoe typical phenylpropanglycosides syringin and syringenin-4′-O-apiosylglucoside as well as flavonoidglycosides can be proven by DC or HPLC in aqueous extracts of fresh plants. In fermented mistletoe extracts and pressed juices, a host tree specific distinction is possible by means of the viscotoxin pattern by HPLC (Schaller *et al.*, 1996a,b), whereas they are hardly detectable in aqueous extracts. The proof of mistletoe lectins (ML) by means of specific methods (ELLA, ELISA) that will be discussed in detail in paragraph CONTENT, may serve as a further identity control.

PURITY

According to nature, the purity of plant extracts depends on the purity of the basic material, as in the further course of the manufacturing process a contamination during the production process can be strictly excluded by monitoring measurements during the production and a validated filling process under aseptic conditions (EU-Guideline, 1998).

Plant Material

Directly after harvesting, foreign substances and discoloured and vermin-ridden or moldy plant parts will be separated by hand picking. After crushing of the picked plant material, it will be tested as to drying loss, sulphate ash, heavy metals, pesticides and microbiological pollution. Although increased cadmium values in *Viscum album* tea drugs were reported in literature (Nagell *et al.*, 1997), however, no increased cadmium values are observed in the ready to use medicinal products (Lorch *et al.*, unpublished results).

In order to limit the content of bacterial organisms it is important to carefully handle the material after harvesting. For this, the careful picking of the plants, washing with sterile *aqua purificata*, and quick freezing is necessary. Only plant material with test results conforming with the specifications may be used for further processing.

Medicinal Products

Mistletoe preparations for the adjuvant and palliative tumour therapy are used as sterile solution for parenteral application. The purity test of the ready to use product has to include all tests for this kind of application according to the drug law, i.e. the test of contamination, sterility as well as to bacterial endotoxins by limulus-amoebocyte-lysate (LAL) test. However, it has to be considered that mistletoe components may influence the LAL test, and thus may give false positive results (Scheer, 1993). The release specification for bacterial endotoxins in the ready to use drugs has to be fixed in that way that the required limit of 5 E.U./BW/h

(E.U. = international endotoxin unit) for parenteral preparations will not be exceeded (European Pharmacopoea, 1998).

CONTENT

The determination of components is useful for characterising the plant extracts and for the examination of the batch consistency. Depending on the pharmacologically effective components being known or not, two procedures are possible for the quality control (Hamacher, 1996):

1. The pharmacologically effective component or components of the extract are known (e.g. digitalis, fructus cardui mariae): The determination of contents of components can be performed by analytic determination of the effective substances. In most cases, a norming is done as to the component determining the efficacy.
2. Several components or component groups are responsible for the efficacy or the principle determining the efficacy is not known: In these cases, a chromatographic fingerprint-analysis can be combined with a determination of the content of so-called tracer substances, i.e. chemically defined components of the relevant plant extract without claim to therapeutic relevance.

Although ML I is considered by some groups as one of the efficacy determining principle of the mistletoe treatment in adjuvant cancer therapy, anthroposophic manufacturers of mistletoe extracts suppose mistletoe extracts as medicinal products that may be related to the second group. This is supported by the detection of several pharmacological active compounds, i.e. the phenylpropanglycosides, oligo- and polysaccharides, viscotoxins, and lectins. Dependent on the manufacturing procedure, these components can be used for the determination of content in the sense of a constant quality from batch to batch. The chances and limits are discussed in the following.

Phenylpropanglycosides

Up to now it is not known that phenylpropanglycosides attribute to the efficacy of mistletoe preparations in subcutaneous application. It is, however, possible to use this substance group, especially syringenin-4′-O-apiosylglucoside, as tracer substances for the standardization of extracts according to a suggestion of Wagner (1984). Their suitability for the identity test of plant material and extracts can be used in non-fermented extracts for testing the constant extract quality. Dependent on the relevant host tree, the values can be kept within a constant level from batch to batch.

Oligo- and Polysaccharides

Although there are some reports on oligo- and polysaccharides from *Viscum album* (Jordan *et al.*, 1986b; Heine, 1987; Müller *et al.*, 1990; Stein *et al.*, 1999), no suitable analytic procedure is available in order to use the determination of content of this substance group in mistletoe extracts for the quality control.

Viscotoxins

In mistletoe extracts produced by fermentation or squeezing, and thus containing a provable content of viscotoxins, these compounds can be used for determination of content (see also paragraph IDENTITY).

Lectins

The detection of lectins in mistletoes was reported already at the end of the seventies (Franz, 1977; Luther, 1980), but only in recent times these substances became the centre of interest in the research on the immunological effects of mistletoe extracts. In the European mistletoe, three isolectin groups (ML I, ML II and ML III) are detected (Franz, 1981) which differ in regard of sugar specificity (see Pfüller, this book). Their analytic determination by immunological test procedures (ELLA, ELISA) is suitable for the quality of mistletoe preparations; the described methods are specific and very sensitive. At least two procedures for the analytical detection of ML in the extracts are used (detection limit is 10 ng/ml):

- An enzyme-linked lectin assay (ELLA) as described (Vang *et al.*, 1986; Schöllhorn, 1993; Jäggy *et al.*, 1995). Briefly, asialofetuin (ASF) or galactose (GAL) is fixed to a microtiter plate. The ML present in the test sample bind to this layer according to their sugar-specificity and binding capacity; accompanying substances are washed out. The next step is to add anti-ML antibodies to the plate which bind to the fixed ML. For quantitative evaluation, Streptavidin-peroxidase-marked IgG-antibodies bind to the anti-ML-antibodies and are determined photometrically. As this test is based on the sugar binding capacity of the ML, this assay measures "active ML". As all three ML bind to ASF, the total content of ML is determined in the given test sample. If galactose is fixed to the plate, a specific determination of ML I can be carried out in this way (Schöllhorn, 1994).
- An enzyme-linked immunosorbent assay (ELISA) as described (Musielski *et al.*, 1996; Temyakov *et al.*, 1997). Here, monoclonal anti-ML antibodies fixed to the microtiter plate will bind ML present in the test sample. Subsequently, biotinylated anti-ML antibodies detecting another epitope of the ML bind to the fixed ML. After binding to Streptavidin-POD, reaction follows with a suitable substrate. In case of antibodies specific to the isolectins are available, a differentiation of the ML pattern can be made. The antibodies bind to amino-acid sequences of the lectin B chain or the toxic A chain. As these sequences are not necessarily identical with the sugar-binding domain of the lectin, also not "sugar-binding" – or inactive ML and fragments which might also induce immunological raections – are detected. The test procedures, therefore, deliver different results of the concentration of lectins in a preparation.

The ELLA as well as the ELISA test require a defined ML standard. When comparing different mistletoe preparations and test systems, the use of the same standards is essential. However, up to now there is no generally accepted ML standard available. At the moment, a ML standard is established by the group of Prof. Uwe Pfüller, Institute of Phytochemistry at the University of Witten/Herdecke (Eifler *et al.*, 1994). Also, the anti-ML antibody and the ASF quality require attention. According to our studies, the commercial available ASF preparations show different cross reactivities with the three ML subunits (Lorch, unpublished results).

Moreover, results of the investigations of different mistletoe preparations with one test system and the same standard are not comparable. The different composition of the preparations influence the test result, especially the rate of recovery of ML, and thus, the ML concentrations in different preparations measured with different methods cannot easily be compared. Consequently, there is a need of a specially validated ML test for each and every mistletoe preparation.

Using a specially validated ELLA test, in 10 subsequent batches (produced within a period of 9 month) of a commercially available process-standardized mistletoe extract (*Helixor* A 50 mg), the mean ML content was 165 ng/ml, with a maximal value of 183 ng/ml (+11%) and a minimal value of 135 ng/ml (–18%). Compared to the guaranteed ML range of 50–70 ng/ml (± 16%) in a ML-normed mistletoe extract (*Eurixor*), the results of the standardized product are comparable to those of the normed drug.

In this context, the declaration of ML contents in mistletoe preparations should, however, not give the impression that the ML are the sole effective component. Even the dosage recommendation of 1 ng ML I per kg bodyweight is based on only a few experimental settings (Hajto *et al.*, 1989, 1990) and lacks reliable dosage finding studies with purified substance. The immune response of the cancer patients to the subcutaneously applied mistletoe extract is not depending on the body weight, but turns out to be very individual (Büssing *et al.*, 1999). Last but not least other effective components in mistletoe extracts have to be considered to be effective (Berg and Stein, 1995; Stein and Berg, 1994, 1996; Stein *et al.*, 1996, 1999), as the patient develops anti-ML-antibodies during the treatment phase which block the ML activity *in vitro* and probably also *in vivo* (Stettin *et al.*, 1990).

Keeping these facts in mind, test methods suitable for quality control of the whole plant extract gain importance, especially the determination of the biological activity *in vitro* by means of cell culture methods.

CELL CULTURE METHODS FOR QUALITY CONTROL

The biological activity of mistletoe extracts can be determined by testing the cytotoxicity to cell cultures, i.e. leukaemic Molt-4 cells (Minowada *et al.*, 1972; Ribéreau-Gayon *et al.*, 1986). The cells are incubated with different concentrations of the mistletoe preparation, and the growth inhibition is measured, e.g. the ID_{50} value (see Figure 1). The cytotoxicity of mistletoe extracts of different tumour cell lines correlates with certain mistletoe components. It was shown that the ML exert

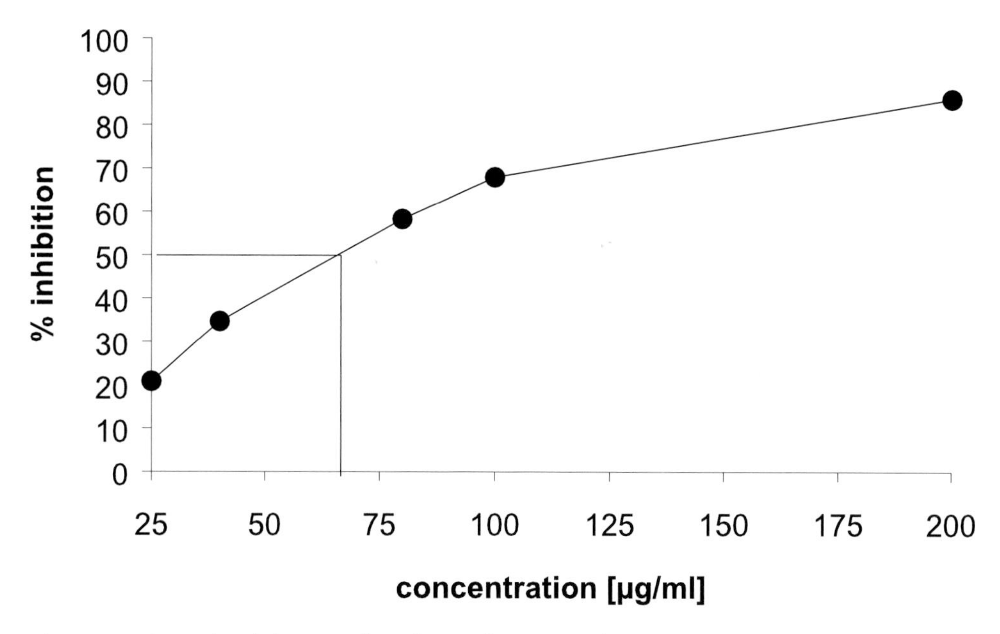

Figure 1 Growth inhibition of Molt-4 cells treated for 72 h with an aqueous mistletoe extract (Helixor M 50 mg/ml, according to fresh plant material; lot number 970 433). The ID_{50} value represents the concentration of the drug leaving about 50% viable cells. Results are mean values of triplicated experiments (SD < 3%).

the main inhibitory effect in Molt-4 cell cultures (Doser *et al.*, 1989; Jung *et al.*, 1990; Urech *et al.*, 1995), while Yoshida sarcoma cells are more sensitive to the viscotoxins and, therefore, may be used for the quality control of viscotoxin-rich mistletoe extracts (Urech *et al.*, 1995).

Another method for the quality control of mistletoe extracts is the measurement of distinct immune responses. In contrast to tumour cell lines, it is difficult to keep normal immune cells in culture for a long period. In an *in vitro* human skin bioassay (Joller *et al.*, 1996) used to analyse the induction of cytokines such as interleukin (IL) 1α and IL-6 in response to applied drugs, a crude mistletoe extract stimulated the cytokines twice as high as purified ML (Table 1). Surprisingly, the reproducibility of the stimulation by the mistletoe extract was better than by the pure ML. It

Table 1 Induction of cytokines in a human skin bioassay.

	not standardised/normed mistletoe extract [referred to 6 ng ML/ml]	*pure ML I [6 ng ML/ml]*
IL-1α (pg/ml)	530.6 ± 10.6 (corr. ± 2%)	265.9 ± 9.0 (corr. ± 3%)
IL-6 (ng/ml)	178.2 ± 27.4 (corr. ± 15%)	120.7 ± 26.7 (corr. ± 22%)

Results are from Joller *et al.* (1996).

remains to be shown whether test systems based on such *in vitro* assays have an impact on quality control procedures for immunostimulating herbal remedies.

STABILITY

An essential aspect referring to the quality of plant extracts is their stability within the expiry time. Especially the content of components with proven therapeutic effect should be stable during the period of applicability. Repeated measurements of the inhibitory effects of commercially available mistletoe extracts in the Molt-4 cell assay approve the stability of mistletoe preparations (Figure 2).

Problems referring to storage stability are reported only for mistletoe preparations standardised to their ML content. They require stabilisation by the addition of different auxiliary substances and cool storage (Wächter *et al.*, 1997). In contrast, also the stabilizing effect of concomitant compounds to the main effective components in plant extracts was reported (Eder *et al.*, 1998). Regarding their activity in the Molt-4 bioassay, there is evidence that the ML are more stable in whole plant extracts as compared to pure lectins in aqueous solution (Lorch, unpublished results).

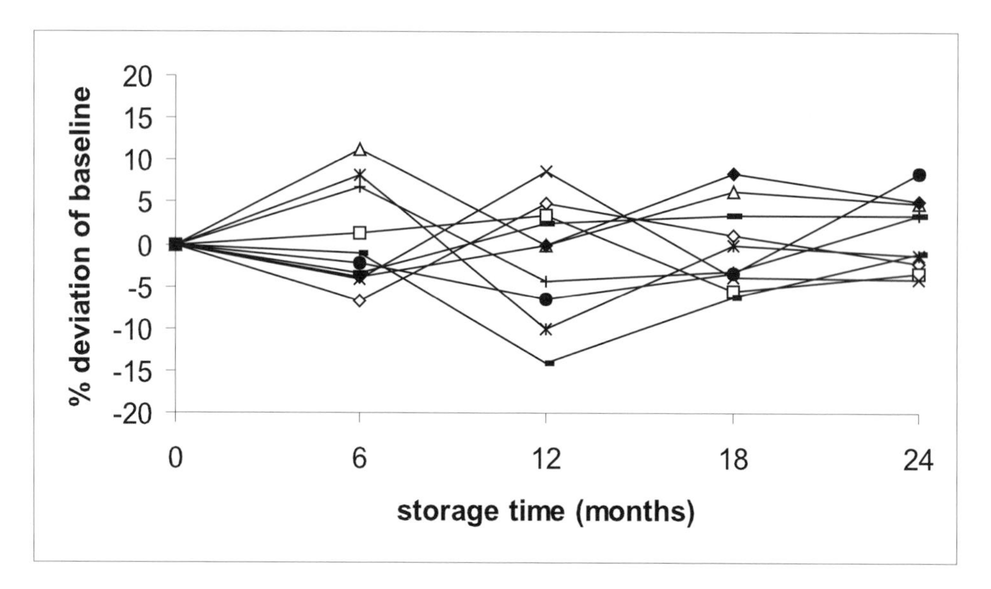

Figure 2 Growth inhibition of Molt-4 cells treated for 72 h with 10 different batches of an aqueous mistletoe extract (Helixor M 50 mg/ml), stored at a temperature of 20 ± 5° C over a period of 2 years. Results are expressed as relative changes of growth inhibition at 0.2 mg/ml of the extract. Obviously, the cytotoxic effect is maintained over the whole storage period of 2 years.

FURTHER MEASURES OF QUALITY CONTROL

Selection of Plants

In contrast to phytotherapeutic manufacturers, anthroposophic manufacturers process mistletoes harvested in summer and winter separately, keeping in mind that the component spectrum relies to seasonal rhythms (Scheer *et al.*, 1992). To take into account also the different distribution of components in the single plant parts (Scheffler, 1996), only certain parts (mainly leafy shoots and fruits) are used. The final product is produced by mixing the distinct extracts by defined methods which differ from one manufacturer to the other.

The Manufacturing Process

In spite of detailed knowledge of component groups in mistletoe preparations, it is not yet possible to define their contribution to the effect of the whole plant extract. The exact fixing and description of every single step in the manufacturing process is therefore essential for obtaining a constant composition of the extract. Each production step has to be monitored by in-process-controls, and thoroughly documented. Even unknown components or component relations might be guaranteed from batch to batch by this process standardization. This reliability cannot be achieved by norming the extract to a single substance.

Storage

As described above, the cool storage of ML-normed preparations is recommended. Standardized whole plant extracts, however, show sufficient storage stability at room temperature (Figure 2). According to recent unpublished results, the ampoules with mistletoe extracts should be stored under strict light protection in a closed package.

SUMMARY

The pharmaceutical quality control of mistletoe preparations has to be performed in accordance with the legal prescriptions. The identity of the plant material, the purity, content and activity of components from batch to batch as well as the stability have to be examined. To obtain a constant high quality of the medicinal product, first of all, the main effective components have to be defined. Supposed that a certain substance or a certain substance group is responsible for the efficacy of the drugs, the quality of the drug may improve by norming the extract to the suggested substance or substance group. In the case that there are legitimate suppositions that the composition of the substance groups or still unknown components are decisive to efficacy, one may consider a standardization of the production process. In mistletoe preparations, especially the ML are discussed to define the efficacy. However, several findings indicate the diversity of effective components and even synergistic effects. Therefore, the standardization of mistletoe preparations is preferred by anthroposophical manufacturers. Subsequent analysis of mistletoe extracts approves that standardization is equivalent to norming regarding the batch to batch consistency.

REFERENCES

AMG (German Drug Law): *Gesetz über den Verkehr mit Arzneimitteln* (AMG), 8. Gesetz zur Änderung des Arzneimittelgesetzes vom 07. Sept. 1998 (BGBl. I, S. 2649)

Berg, P.A. and Stein, G. (1995) Ein Inhaltsstoff allein genügt nicht. *Zeitschrift für Phytotherapie,* **16**, 282.

BMG (Federal Ministry of Health): *Bekanntmachung von Empfehlungen für Höchstmengen an Schwermetallen bei Arzneimitteln pflanzlicher und tierischer Herkunft*, Entwurf des Bundesministeriums für Gesundheit vom 17.10.1991.

Büssing, A., Rosenberger, A., Stumpf, C., and Schietzel M. (1999) Verlauf lymphozytärer Subpopulationen bei Tumorpatienten nach subkutaner Applikation von Mistelextrakten. *Forschende Komplementärmedizin,* **6**, 196–204.

DAB (German Pharmacopoeia): *Deutsches Arzneibuch* (DAB 1999) Deutscher Apotheker Verlag, Stuttgart.

Doser, C., Doser, M., Hülsen, H., and Mechelke, F. (1989) Influence of carbohydrates on the cytotoxicity of an aqueous mistletoe drug and of purified mistletoe lectins tested on human T-leukemia cells. *Arzneimittel-Forschung/Drug Research*, **39**, 647–651.

Eder, M. and Mehnert, W. (1998) Bedeutung pflanzlicher Begleitstoffe in Extrakten. *Pharmazie*, **53**, 285–293.

Eifler, R., Pfüller, U., Göckeritz, W., Pfüller, K., Gelbin, M., and Tonevitsky, A.G. (1994) Verfahren zur Gewinnung von Lektinen aus Mistelpflanzen, *Offenlegungsschrift DE 42 29 876 A1* (10.03.1994)

EU-GMP-Guideline (1989). *Guidelines of a Good Production Practice for drugs*. III/2244/87, Rev. 3-01/89

EU-Guideline (1998) Ergänzende und überarbeitete Leitlinien für die Herstellung steriler Arzneimittel (September 1996). In G. Auterhoff, (ed.), *EG-Leitfaden einer Guten Herstellungspraxis für Arzneimittel.* Editio Cantor Verlag, Aulendorf, pp. 73–88.

European Pharmacopoeia (1998) *Europäisches Arzneibuch* 1997, Nachtrag 1998, Deutscher Apotheker-Verlag Stuttgart.

FDA-Guideline (1987) *Guideline on sterile drug products produced by aseptic processing.* US Food and Drug Administration.

Franz, H., Haustein, B., Luther, P., Kuropka, U. and Kindt, A. (1977) Isolierung und Charakterisierung von Inhaltsstoffen der Mistel (*Viscum album* L.). *Acta Biol. Med. Germ.*, **36**, 113–117.

Franz, H., Ziska, P. and Kindt, A. (1981) Isolation and properties of three lectins from mistletoe (*Viscum album* L.). *Biochem. J.*, **195**, 481–484.

Hajto, T., Hostanska, K., and Gabius, H.J. (1989) Modulatory potency of the β-galactoside-specific lectin from mistletoe extract (Iscador®) on the host defence system *in vivo* in rabbits and patients. *Cancer Research*, **49**, 4803–4808.

Hajto, T., Hostanska, K., Gabius, H.-J. (1990) Zytokine als Lektin-induzierte Mediatoren in der Misteltherapie. *Therapeutikon*, **4**, 136–145.

Hamacher, H. (1996) Standardisierung komplexer Naturstoffgemische. In R. Scheer, H. Becker, and P.A. Berg, (eds.), *Grundlagen der Misteltherapie. Aktueller Stand der Forschung und klinische Anwendung*. Hippokrates Verlag, Stuttgart, pp. 119–138.

Harnischfeger, G. (1999) Vorschlag zu Richtlinien für die kommerzielle Wildsammlung von Pflanzenmaterial für medizinische Zwecke (GHP) *Journal of Herbs, Spices and Medicinal Plants* [in press]

Heine, H. (1987) Antitumorpolysaccharide der Mistel. *Zeitschrift für Phytotherapie*, **8**, 122–124.

Jäggy, C., Musielski, H., Urech, K., and Schaller, G. (1995) Quantitative Determination of Lectins in Mistletoe Preparations. *Arzneimittel-Forschung/Drug Research*, **45**, 905–909.

Joller, P.W., Menrad, J.M., Schwarz, T., Pfüller, U., Parnham, M.J., and Weyhenmeyer, R. (1996) Stimulation of cytokine production via a special standardized mistletoe preparation in an *in vitro* human skin bioassay. *Arzneimittel-Forschung/Drug Research*, **46**, 649–653.

Jordan, E. and Wagner, H. (1986a) Nachweis und quantitative Bestimmung von Lektinen und Viscotoxinen in Mistelpräparaten. *Arzneimittel-Forschung/Drug Research*, **36**, 428–433.

Jordan, E. and Wagner, H. (1986b) Structure and properties of polysaccharides from *Viscum album* L. *Oncology*, **43** (Suppl 1), 8–15.

Jung, M.L., Baudino, S., Ribéreau-Gayon, G., and Beck, J.P. (1990) Characterization of cytotoxic proteins from mistletoe (*Viscum album* L.). *Cancer Letters*, **51**, 103–108.

Kuttan, G., Vasudevan, D., and Kuttan, R. (1988) Isolation and identification of a tumour reducing component from mistletoe extract (Iscador®). *Cancer Letters*, **41**, 307–314.

Lorch, E. (1993) Neue Untersuchungen über Flavonoide in *Viscum album* L. ssp. *abietis, album* und *austriacum. Zeitschrift für Naturforschung*, **48c**, 105–107.

Luther, P., Theise, H., Chatterjee, B., Karduck, D., and Uhlenbruck, G. (1980) The lectin from *Viscum album* L. – Isolation, characterization, properties and structure. *Journal of Biochemistry*, **11**, 428–435.

Minowada, J., Ohnuma, T., and Moore, G.E. (1972) Rosette forming human lymphoid cell-lines. 1. Establishment and evidence for origin of thymus derived lymphocytes. *Journal of the National Cancer Institute*, **49**, 892–895.

Müller, E.A. and Anderer, F.A. (1990) A *Viscum album* oligosaccharide activating human natural cytotoxicity is an interferon γ inducer. *Cancer Immunology and Immunotherapy*, **32**, 221–227.

Musielski, H. and Rüger, K. (1996) Verfahren zur quantitativen Bestimmung von Mistellektin I und Mistellektin II und/oder Mistellektin III in Mistelextrakten unter Verwendung monoklonaler Antikörper, die spezifisch mit Mistellektin reagieren. In R. Scheer, H. Becker, and P.A. Berg, (eds.), *Grundlagen der Misteltherapie. Aktueller Stand der Forschung und klinische Anwendung*. Hippokrates Verlag, Stuttgart, pp. 95–104.

Nagell, A. and Grün, T.A. (1997) Reinheitssprüfung an pflanzlichen Rohstoffen und daraus hergestellten Zubereitungen. *Pharmazeutische Industrie*, **59**, 706–711.

Ribéreau-Gayon, G., Jung, M.L., Baudino, S., Sallé, G., and Beck, J.P. (1986) Effects of mistletoe (*Viscum album* L.) extracts on cultured tumor cells. *Experientia*, **42**, 594–599.

Samuelsson, G. (1958) Phytochemical and pharmacological studies on *Viscum album* L., I. Viscotoxins, its isolation and properties. *Svensk Farmaceutisk Tidskrift*, **8**, 169–189.

Samuelsson, G. (1961) Phytochemical and pharmacological studies on *Viscum album* L., V. Further improvements in the isolation methods for viscotoxin. *Svensk Farmaceutisk Tidskrift*, **19**, 481–494.

Schaller, G., Urech, K., Giannattasio, M., and Jäggy, C. (1996a) Viscotoxinspektren von *Viscum album* L. auf verschiedenen Wirtsbäumen. In R. Scheer, H. Becker, and P.A. Berg, (eds.), *Grundlagen der Misteltherapie. Aktueller Stand der Forschung und klinische Anwendung*. Hippokrates Verlag, Stuttgart, pp. 105–110.

Schaller, G., Urech, K., and Giannattasio, M. (1996b) Cytotoxicity of different viscotoxins and extracts from the European subspecies of *Viscum album* L. *Phytotherapy Research*, **10**, 473–477.

Schaller, G., Urech, K., Grazi, G., and Giannattasio, M. (1998) Viscotoxin composition of the three European subspecies of *Viscum album*. *Planta Medica*, **64**, 677–678.

Scheer, R. (1993) Beeinflussung des Limulus-Amöbozyten-Lysat-Tests durch Mistellektine. *Arzneimittel-Forschung/Drug Research*, **43**, 795–800

Scheer, R., Scheffler, A., and Errenst, M. (1992) Two harvesting times, summer and winter: Are they essential for preparing pharmaceuticals from mistletoe (*Viscum album*)? *Planta Medica*, **58** (Suppl. 1), 594.

Scheffler, A., Richter, C., Beffert, M., Errenst, M., and Scheer, R. (1996) Differenzierung der Mistelinhaltstoffe nach Zeit und Ort. In R. Scheer, H. Becker, and P.A. Berg, (eds.), *Grundlagen der Misteltherapie. Aktueller Stand der Forschung und klinische Anwendung*. Hippokrates Verlag, Stuttgart, pp. 49–76.

Schöllhorn, V. (1993) An ELLA-system to quantify mistletoe I and II isolectins. In E. van Driessche, H. Franz, S. Beeckmans, U. Pfüller, A. Kallikorm, and T.C. Bog-Hansen, (eds.), *Lectins: Biology, Biochemistry, Clinical Biochemistry*. Vol. 8. Textop, Hellerup, Denmark, pp. 14–20.

Schöllhorn, V. (1994) *Verfahren zur selektiven quantitativen Bestimmung der Konzentration von Lektinen, insbesondere Mistellektinen*. Patentschrift DE 4123263 C2 (03.11.1994).

Stein, G.M. and Berg, P.A (1994) Non-lectin component in a fermented extract from Viscum album L. grown on pines induces proliferation of lymphocytes from healthy and allergic individuals *in vitro*. *European Journal of Clinical Pharmacology*, **47**, 33–38

Stein, G.M. and Berg, P.A (1996) Evaluation of the stimulatory activity of a fermented mistletoe lectin-1 Free Mistletoe Extract on T-Helper Cells and Monocytes in Healthy Individuals in vitro. *Arzneimittel-Forschung/Drug Research*, **46**, 635–369.

Stein, G.M., Meink, H., Durst, J. and Berg, P.A. (1996) Release of cytokines by a fermented lectin-1 (ML-1) free mistletoe extrakt reflects differences in the reactivity of PBMC in healthy and allergic individuals and tumour patients. *European Journal of Clinical Pharmacology*, **151**, 247–252.

Stein, G.M., Edlund, U., Pfüller, U., Büssing, A., and Schietzel, M. (1999) Influence of polysaccharides from *Viscum album* L. on human lymphocytes, monocytes and granulocytes *in vitro*. *Anticancer Research*, **19** [in press]

Stettin, A., Schultze, J.L., Stechemesser, E., and Berg, P.A. (1990) Anti-Mistletoe Lectin Antibodies Are Produced in Patients During Therapy with Aqueous Mistletoe Extract Derived from Viscum album L. and Neutralize Lectin-Induced Cytotoxicity *in vitro*. *Klinische Wochenschrift*, **68**, 896–900.

Temyakov, D.E., Agapov, I.I., Moisenovich, M.M., Prokofev, S.A., Malyuchenko, N.V., Egorova, S.E., *et al.* (1997) Heterogeneity of mistletoe lectin catalytic subunits assessed with monoclonal antibodies. *Molecular Biology*, **31**, 448–453.

Urech, K., Schaller, G., Ziska, P., and Giannattasio, M. (1995) Comparative study on the cytotoxic effect of viscotoxin and mistletoe lectin on tumour cells in culture. *Phytotherapy Research*, **9**, 49–55.

Vang, O., Pii Larsen, K., and Bog-Hansen, T.C. (1986) A new quantitative and highly specific assay for lectinbinding activity. In T.C. Bog-Hansen and E. van Driessche, (eds.), *Lectins:*

Biology, Biochemistry, Clinical Biochemistry, Vol. 5, Walter de Gruyter Verlag, Berlin, New York, pp. 637–644.

Wächter, W. and Witthohn, K. (1997) Immunologisch wirksame Mistelextrakt-Zubereitungen. *Offenlegungsschrift DE 19548 367 A1* (03.07.1997)

Wagner, H., Feil, B., and Bladt, S. (1984) *Viscum album* – Die Mistel. *Deutsche Apotheker-Zeitung*, **124**, 1429–1432.

Winterfeld, K. and Bijl, L.H. (1942). Viscotoxin, ein neuartiger Inhaltsstoff der Mistel (*Viscum album* L.) *Archiv der Pharmazie*, **280**, 23–26.

14. NATURAL *VERSUS* RECOMBINANT MISTLETOE LECTIN-1

Market Trends

JOSEF BEUTH

Institute for Scientific Evaluation of Naturopathy, University Cologne, Robert-Koch Strasse 10, 50931 Cologne, Germany

INTRODUCTION

Mistletoe preparations are widely used in traditional medicine as immunomodulating agents and biological response modifiers, respectively. They were introduced into oncological treatment by Rudolf Steiner (about 1920) and still belong to the so called "non-conventional/non-proven medications". For evidence-based medicine main problems of mistletoe extract treatment comprise 1. the variable composition of clinically available extract preparations leading to an anticipated lack of reproducibility of experimental and clinical effects, and 2. the missing (or inadequate) prove of efficacy in experimental or clinical settings. Although research efforts were intensified recently, and due to their special authorisation by the Federal Institute for Drugs and Medicinal Devices (BfArM), mistletoe extracts warrant precise declaration of compounds and scientific investigations on the influence of well-defined components on basic mechanisms, experimental (*in vitro/in vivo*) activities and clinical studies (in accordance with Good Clinical Practice). Recently, clinically approved mistletoe extracts were biochemically separated in monocomponents (e.g. mistletoe lectins 1–3, viscotoxins, vesicles, carbohydrates) and experimentally/clinically investigated. Especially the plant-derived, natural mistletoe lectin-1 (ML-1) yielded promising results and may be considered as one of the main relevant immunoactive extract component. After genome analysis recombinant mistletoe lectin (rML) was expressed in *Escherichia coli* and is currently under investigation in pre-clinical settings. However, other mistletoe lectins and components are also being investigated and might gain therapeutical relevance in future.

MISTLETOE TREATMENT IN COMPLEMENTARY ONCOLOGY

Complementary Medicine: a Scientific Approach to Comprehensive Therapy in Oncology

The concept of scientific complementary medicine arose from the growing awareness that cytotoxic tumour destructive therapies (e.g. radiotherapy, chemotherapy) obviously fail to provide a reasonable benefit for patients suffering from advanced carcinomas. Although the toxicity of chemo/radiotherapeutic regimens ultimately increased, demanding stem cell transplantation and other cost-intensive supports, no statistically evaluable benefit on overall survival could be observed for most patients with advanced carcinomas. These disappointing data and missing therapeutical options finally resulted in the definition of criteria other than survival to suggest a therapeutical benefit. Accordingly, remission rate (further specified in complete/partial remission) was postulated to correlate with therapeutical success, however, biometric meta-analysis totally neglected this correlation (Abel, 1995; Moss (1995). Some studies even demonstrated inverse correlations between remission and patient survival, however, it is still established as a marker of therapeutical success.

Whereas conventional cytotoxic tumour destructive strategies (chemo-/radiotherapy) are well appreciated for most paediatric tumour entities as well as for primarily systemic neoplastic diseases (e.g. leukaemia, lymphoma) and defined non-carcinomatous neoplasms (e.g. of testicular origin), most epithelial derived cancers (= carcinoma) do not reasonably respond. Especially asymptomatic patients with advanced carcinoma after surgical treatment do not profit from adjuvant cytotoxic treatment, except for defined tumour entities and stages. This conclusion can be drawn from biometric analyses by Abel (1995) and Moss (1995) who demonstrated currently that most standard chemotherapeutic schedules lack adequate scientific evaluation. Obviously, most cytotoxic treatment modalities were not evaluated in accordance to Good Clinical Practice, however, influential industry driven interests delay an important change of paradigma.

Immunomodulatory Activity of Natural Mistletoe Lectin-1 (ML-1)

The scientific evaluation of the immunomodulatory efficacy of the galactoside-specific ML-1 was recently initiated. Promising preliminary results have been discovered *in vitro*. These include the upregulation of immune-cell activation markers and cytokine release (Hajto *et al.*, 1997; Heiny and Beuth, 1994), downregulation of tumour cell proliferation and tumour spheroid growth (Lenartz *et al.*, 1997) as well as pronounced dose-dependent cytotoxicity towards various cell lines (Staak *et al.*, 1998). *In vivo*, the regular subcutaneous administration of a small but defined ML-1 dosage (1 ng/kg body weight, twice a week) yielded enhanced thymocyte proliferation, maturation and emigration (Beuth *et al.*, 1993), significantly increased peripheral blood immune cell counts and activities (Beuth *et al.*, 1993), immunorestoration after steroid application (Beuth *et al.*, 1994), as well as significant antimetastatic (Beuth *et al.*, 1991) and antibacterial (Stoffel *et al.*, 1996) effects in different murine models. These promising experimental data encouraged the administration of ML-1 standardised mistletoe extract to cancer patients outside the setting of a formal study. All patients were subcutaneously injected with a mistletoe extract standardised on ML-1 at a dose of 1 ng per kg body weight, twice a week over a 4 to 5 weeks' period. This ML-1 concentration had been shown to be the optimal dose in the preceding experiments (Beuth *et al.*, 1994). The time schedule was fixed after recommendations from a scientific board. However, the dosage and application schedule of ML-1 for the immunopotentiation of cancer patients is somewhat provisional. Experimental *in vivo* models showed that a prolongation of treatment intervals as well as the administration of higher concentrations of ML-1 yielded improved effects (Lenartz *et al.*, 1997). In cancer patients the clinical experience yielded the following observations. Patients in the aggregate experienced:

1) significantly increased counts and activities of peripheral blood lymphocytes and natural killer (NK)-cells (Beuth *et al.*, 1992);
2) increased serum levels of acute phase reactants (Beuth *et al.*, 1993);
3) enhanced delayed type of hypersensitivity after intracutaneous challenge (Beuth *et al.*, 1995):

Clinical Trials

These data, derived from the non-study observations, formed the basis for more advanced controlled clinical studies following prospectively randomised designs. The types of cancer that are currently treated with ML-1 standardised mistletoe extract (as a complement to more established therapies), include glioblastoma multiforme, breast and colorectal carcinoma. The primary aims of these trials, as delineated in the study protocols, are improvement in quality of life (as determined by questionnaires and confirmed by measurements of β-endorphin plasma levels) and immunoprotection. Secondary aims, evaluated during and after the administration of this form of immunotherapy include the influence on predictable side effects of tumour-destructive treatments (particularly chemotherapy), on the metastasis, relapse rate and on overall survival.

GLIOBLASTOMA MULTIFORME

Astroglial brain tumours (e.g. glioblastoma multiforme) are the most common primary brain tumours. Neurosurgery and radiotherapy represent the only tumour destructive therapeutical approaches and are still lacking convincing evidence of their benefit. The administration of (neo)adjuvant chemotherapy is still controversial since it has not resulted in biometrically evaluable prolongation of survival time (Levin *et al.*, 1990). In spite of progress in these techniques only a moderate improvement of overall survival has been achieved over the past decade.

Generally, tumour destructive therapies induce immunosuppression. A prospectively randomised clinical trial with patients suffering from malignant glioma stage III/IV (n = 35) was performed to evaluate a) the immunosuppressive effects of standard tumour destructive therapy b) the immunoprotective efficacy of complementary immunotherapy with ML-1 standardised mistletoe extract and c) the benefit of immunotherapy to quality of life.

As recently shown for other tumour entities (e.g. colorectal and breast carcinoma; Heiny *et al.*, 1998a,b), tumour destructive therapy of stage III/IV malignant glioma (neurosurgery, perioperative cortisone treatment, local radiation postoperatively) proved to be immunosuppressive, especially down regulating peripheral blood lymphocyte counts but not granulocytes and monocytes. Flow cytometry revealed that counts of T-cell subsets (e.g. $CD3^+$, $CD4^+$, $CD8^+$, $CD3^+$ $CD16^+/CD56^+$), B-cells, and NK-cells as well as T-cell activation markers (e.g. CD25, HLA-DR) were significantly down regulated after primary treatment.

During the study, patients of the control and study groups were regularly monitored concerning the cellular immune system. The significant postoperative down-regulation of counts and activation markers of lymphatic cells was followed by recovery to almost preoperative values in the control group after 3–6 months. However, regular subcutaneous administration of the extract with the defined immunomodulating dosage of ML-1 (1 ng/kg body weight, twice a week, for 3 months) induced a considerable upregulation of lymphocyte counts and activities which was statistically significant (as compared to preoperative control values) for $CD3^+$, $CD4^+$, $CD8^+$, $CD25^+$ and $HLA\text{-}DR^+$ T-cells after 3 months of treatment (Lenartz *et al.*, 1996).

In an attempt to demonstrate the clinical benefit of immunotherapy with ML-1 standardised mistletoe extract, quality of life was assessed by standard questionnaire (Spitzer). Although no obvious difference between control and study group could be shown initially (3 months postoperatively) patients of the ML-1 treated *verum* group presented a considerably higher questionnaire score (correlating with an improved quality of life) after a 6 months follow up period, as compared to patients of the control group (Lenartz *et al.*, 1996). Accordingly, a co-stimulation of the neuro-immuno-endocrine system was anticipated and further confirmed for breast carcinoma patients, as shown below.

Breast carcinoma

Breast carcinoma is the most frequent female malignancy. Treatment strategies include surgery, chemo/radio/hormonal therapy, however, convincing evidence for biometrically verifiable benefits (e.g. increased overall survival) is still lacking for most patients. This disappointing therapeutic success resulted in conception of high dose chemotherapy regimes demanding stem cell transplantation and other cost-intensive supports; however, no proof of efficacy is available for this option so far (Minckwitz *et al.*, 1997). Criteria other than survival were recently suggested to demonstrate therapeutical benefit, e.g. remission rate which was postulated to correlate with prolongation of overall survival. However, biometric meta-analyses totally neglected this correlation, some studies even demonstrated inverse correlations between remission and patient survival (Abel, 1995).

The aim of an initial study with breast carcinoma patients was to assess whether complementary treatment with ML-1 standardised mistletoe extract can favourably affect immunological/neuroendocrinological parameters. Accordingly, patients (control group $n = 32$, therapy group $n = 36$) with histologically verified breast carcinoma (TNM stages III, IV) were enrolled in this study. All patients were surgically treated and hospitalised for chemotherapy (both according to standard protocols).

To correlate critically empirical clinical observations (stabilisation of mood, perception of pain) with the administration of mistletoe extract standardised for

ML-1, β-endorphin plasma levels of the patients were determined and compared to non-immunomodulated patients. Prior to ML-1 treatment, the mean β-endorphin plasma level of non ML-1 treated patients (6.32 pg/mL), and ML-1 treated patients (7.46 pg/mL) were comparable, both within the normal range (3–10 pg/mL).

To further analyse the neuro-immunological activity of ML-1 standardised mistletoe extract, breast carcinoma patients were divided after treatment into therapeutical responders (n = 25) and non-responders (n = 11). This procedure proved to be favourable since ML-1 responders presented an evidently improved quality of life (as determined by standard questionnaire). Furthermore, therapeutical responders presented an enhanced activity of defined immune parameters (cytokine release, peripheral blood lymphocyte counts) and a positive skin reaction (rubor, infiltration) at the injection site whereas non-responders did not present any of these reactions. Separation of responders/non-responders to ML-1 standardised mistletoe extract treatment appeared to be of relevance since after 6/12 weeks of application the mean β-endorphin plasma levels of responders (13.6/14.6 pg/mL) were statistically significantly ($p < 0.005$) different from 1) basic β-endorphin plasma levels of this group of patients (7.46 pg/mL) 2) β-endorphin plasma levels of control patients without ML-1 standardised mistletoe extract treatment (6.03/7.32 pg/mL) 3) β-endorphin plasma levels of non-responders (6.22/6.46 pg/mL).

The increased β-endorphin plasma levels in the responder group after ML-1 standardised extract administration correlated positively with enhanced *in vitro* cytokine release (interleukin-2, tumour necrosis factor-α, interferon-γ) by mononuclear cells of these patients. Further, peripheral blood lymphocyte subset counts also correlated with β-endorphin plasma levels after ML-1 standardised mistletoe extract treatment. As compared to the control group of patients, complementary ML-1 standardised mistletoe extract application presented with a less pronounced chemotherapy induced downregulation of lymphocyte subpopulations (Heiny and Beuth, 1994).

In the course of another prospectively randomised clinical trial with breast carcinoma patients (n = 47, histologically verified, TNM stages III/IV), β-endorphin plasma levels were correlated to NK-cell and T-lymphocyte activities by analysis of SPEARMAN correlation coefficient (Heiny *et al.*, 1998a). This biometrical procedure is adequate even with limited numbers of patients and events, respectively, since more than 95% of all data are under evaluation. This investigation definitely suggested a close correlation of defined cellular immune parameters (NK-cell, T-lymphocyte activities) and plasma β-endorphin levels and further demonstrates the close correlation of the immuno-neuro-endocrine axis (Heiny *et al.*, 1998a,b). Obviously, complementary ML-1 standardised mistletoe extract treatment modulates defined immune functions (involved in antitumour/antimicrobial resistance) and neuroendocrine functions (determining the quality of life) in cancer patients and may thus be beneficial for those patients.

Colorectal carcinoma

Colorectal carcinoma is one of the commonest cancers in the developed countries, affecting more males than females (Sherman, 1990). In the treatment of this malignancy and its metastatic spread, including surgery, chemotherapy and radiotherapy, little progress has been achieved over the last decade. Approximately 50% of patients who develop colorectal carcinoma do not survive 5 years, although surgery with curative intent is possible in about 80% of all cases. In order to reduce the high mortality rate, complementary treatment modalities are warranted to improve the prognosis for these patients (Schumacher *et al.*, 1998).

A prospectively randomised clinical study was initiated to investigate the efficacy of ML-1 standardised mistletoe extract application on defined effects on patients with advanced colorectal carcinoma. A total of 79 patients were enrolled into this study and treated on standard protocol with 5-FU (Fluorouracil) and FA (Folinic acid). Patients were randomised into control group (n = 41, no complementary treatment) and *verum* group (n = 38; complementary treated with ML-1 standardised mistletoe extract, 1 ng ML-1/kg body weight, twice a week for 8 weeks followed by a 4 weeks break) following the "matched pairs" design. Analysis of peripheral blood cells (including lymphocyte subsets and activities by flow cytometry), therapy/disease-induced side effects, length of remission, overall survival and quality of life (FACT: Functional Assessment of Cancer Therapy Scale V 3.0) was regularly accomplished.

Concerning the primary aim of this study the quality of life (assessed in 6 weeks turns), a significant improvement was established for ML-1 standardised mistletoe extract treated patients as compared to control group patients. Since this beneficial effect of complementary immunotherapy reached statistical significance not earlier than 12 weeks, a placebo effect can be ruled out. A non therapy-induced improvement of the quality of life apparently would have been detectable in the early phase of treatment (Heiny *et al.*, 1998b). Although the late onset of improvement of quality of life is a strong indication for the beneficial efficacy of ML-1 standardised mistletoe extract treatment, a placebo-controlled confirmative study is necessary for definite proof.

Evaluation of therapy-induced side effects demonstrated an evident benefit for complementary ML-1 treated patients. As compared to rate and severity of side effects in the control group of patients, those of the *verum* group suffered significantly less from leukopenia and mucositis WHO grade III (Heiny *et al.*, 1998b). Furtheron, duration of severe mucositis was significantly reduced; however, no significant effect of ML-1 standardised mistletoe extract administration could be verified on frequency and length of remission, relapse-free interval and overall survival. Since improvement of quality of life and reduction of side effects were the primary variables of the design of this prospectively randomised clinical study, the benefit for complementary ML-1 treated colorectal carcinoma patients is obvious.

Expression, Characterisation and Activity of Recombinant Mistletoe Lectin (rML)

The recombinant mistletoe lectin (rML) is a new biological response modifier developed for cancer treatment. To clone different fragments of the ML gene from mistletoe genomic DNA, a polymerase chain reaction strategy was initiated and performed (Langer *et al.*, 1997). All fragments were shown to belong to a particular gene in the mistletoe genome. The full length sequences of both A- and B-chains were established by alignment of the fragments. Expression vectors (A- and B-chain coding region) were constructed and the single chains were expressed in *E. coli* separately. After renaturation of the inclusion bodies, functional A- and B-chains were obtained and associated *in vitro* to obtain ML-holoprotein (Langer *et al.*, 1997).

Experimental investigations on the activity of rML were promising (Möckel *et al.*, 1997) and confirmed 1. the cytotoxic activity against cell lines (e.g. the human lymphoblastic leukemia cell line MOLT-4), 2. the induction of apoptosis (e.g. peripheral blood cells), 3. the increased NK-cell mediated cytotoxicity towards target tumour cells, 4. the immune cell activating potency (e.g. cytokine release, expression of activation markers). Current analysis of the *in vivo* activity of rML (e.g. immunomodulating potency and antitumoural/antimetastatic capacity) in murine models may further clarify the value of rML for clinical settings in oncology.

MARKET TRENDS

Critical analysis of multicomponent mistletoe extract *vs.* scientifically-based monocomponent (ML-1) respectively ML-1 standardised mistletoe extract treatment is currently being performed. Since both therapeutic directions (multicomponent *vs.* monocomponent) seem to have advantages/disadvantages, a profund analysis is warranted. Whereas conventional (scientific-based) medicine obviously tends to prefer defined molecules of multicomponent extracts/agents, anthroposophical medicine is more holistic relying on interactive potencies of diverse components. The recombinant technique now offers pure molecules (e.g. rML); however, the prove of efficacy of the non-glycosylated proteins (expressed in *E. coli*) is still lacking. Since oncology nowadays follows a more mechanistic approach to cancer, monocomponent treatment (ML-1) or even more administration of recombinant molecules (rML) may be speculated to be more successful in the market race, at least in conventional circles. The clear cut demand of evidence-based therapy, however, may be the great chance for holistic anthroposophical treatment modalities, since evaluation of therapeutic effects may stabilise the acceptance.

REFERENCES

Abel, U. (1995) *Chemotherapie fortgeschrittener Karzinome*, Hippokrates, Stuttgart.

Beuth, J., Ko, H.L., Tunggal, L., Steuer, M.K., Geisel, J., Jelaszewicz, J., *et al.* (1993) Thymocyte proliferation and maturation in respone to galactoside-specific mistletoe lectin. *In Vivo*, 7, 407–410.

Beuth, J., Ko, H.L., Tunggal, L., and Pulverer, G. (1993) Das Lektin der Mistel als Immunmodulator in der adjuvanten Tumourtherapie. *Deutsche Zeitschrift für Onkologie*, **25**, 73–76.

Beuth, J., Ko, H.L., Tunggal, L., Buss, G. Jeljaszewicz, J., Steuer, M.K. *et al.* (1994) Immunoprotective activity of the galactoside-specific mistletoe lectin in cortisone treated BALB/c-mice. *In Vivo*, **8**, 989–992.

Beuth, J., Ko, H.L., Gabius, H.-J., and Pulverer, G. (1991) Influence of treatment with the immunomodulatory effective dose of the ß-galactoside-specific lectin from mistletoe on tumour colonization in BALB/c-mice for two experimental models. *In Vivo*, **5**, 29–32.

Beuth, J., Ko, H.L., Tunggal, L, Buss, G., Jeljaszewicz, J., Steuer, M.K., *et al.* (1994) Immunaktive Wirkung von Mistellektin-1 in Abhängigkeit von der Dosierung. *Arzneimittel Forschung/Drug Research*, **11**, 1255–1258.

Beuth, J., Ko, H.L., Gabius, H.-J., Burrichter, H., Oette, K., and Pulverer, G. (1992) Behavior of lymphocyte subsets and expression of activation markers in response to immunotherapy with galactoside-specific lectin from mistletoe in breast cancer patients. *Clinical Investigator*, **70**, 658–661.

Beuth, J., Gabius, H.J., Steuer, M.K., Geisel, J., Steuer, M., Ko, H.L., *et al.* (1993) Einfluß der Mistellektintherapie auf den Serumspiegel definierter Serumproteine (Akutphaseproteine) bei Tumourpatienten. *Medizinisch. Klinik*, **88**, 287–290.

Beuth, J., Ko, H.L., and Pulverer, G. (1995) Immunreaktion vom verzögerten Typ unter lektinnormierter Misteltherapie bei Tumourpatienten. *Deutsche Zeitschrift für Onkologie*, **27**, 130–133.

Hajto, T., Hostanska, K, Fischer, J., and Saller, R. (1997) Immunomodulatory effects of Viscum album agglutinin-1 on natural immunity. *Anti-Cancer Drugs*, **8**, 43–46.

Heiny, B.M. and Beuth, J. (1994) Misteltoe extract standardized for the galactoside-specific lectin (ML-1) induces ß-endorphin release and immunopotentation in breast cancer patients. *Anticancer Research*, **14**, 1339–1342.

Heiny, B.M., Albrecht, V., and Beuth, J. (1998a) Correlation of immune cell activities and β-endorphin release in breast cancer patients treated with galactoside-specific lectin standardized mistletoe extract. *Anticancer Research*, **18**, 583–586.

Heiny, B.M., Albrecht, V., and Beuth, J. (1998b) Lebensqualitätsstabilisierung durch Mistellektin-1 normierten Extrakt beim fortgeschrittenen kolorektalen Karzinom. *Onkologe*, **4**, 35–39.

Langer, M., Zinke, H., Eck, J., Möckel, B., and Lentzen, H. (1997) Cloning of the active principle of mistletoe: the contributions of mistletoe lectin single chains to biological functions. *European Journal of Cancer*, **33**, 24.

Lenartz, D., Herrman, S., Pietch, T., Rommel, T., Menzel, J., and Beuth, J. (1997) Cytotoxic activity of the galactoside-specific lectin from mistletoe on anaplastic glioma cell spheroids and cell growth. *Zeitschrift für Onkol./Journal of Oncology*, **29**, 11–15.

Lenartz, D., Andermahr, J., Menzel, J., and Beuth, J. (1998) Efficiency of treatment with galactoside-specific lectin from mistletoe against rat glioma. *Anticancer Research*, **18**, 1011–1014.

Lenartz, D., Stoffel, B., Menzel, J., and Beuth, J. (1996) Immunoprotective activity of galactoside-specific lectin from mistletoe after tumour destructive therapy in glioma patients. *Anticancer Research*, **16**, 3799–3802.

Levin, V.A., Silver, P., Hannigan, J., Wara, W.M., Gutin, P.H., Davis, R.L., *et al.* (1990) Superiority of postradiotherapy adjuvant chemotherapy with CNNU, procarbazine, and vincristin (PVC) over BCNU for anaplastic glioma. *Int. J. Radiat. Oncol. Phys. Biol.*, **18**, 321–326.

Minckwitz von, G., Da Costa, S., and Kaufmann, M. (1997) Hochdosis-Chemotherapie beim Mammakarzinom. *Deutsches Ärzteblatt*, **94**, 2835–2837.

Möckel, B., Schwarz, T., Eck, J., Langer, M., Zinke, H., and Lentzen, H. (1997) Apoptosis and cytokine release are biological responses mediated by recombinant mistletoe lectin in vitro. *European Journal of Cancer*, **33**, 35.

Moss, R. (1995) *Questioning chemotherapy*. Equinox, Brooklyn.

Schumacher, K., Beuth, J., and Uhlenbruck, G. (1998) Prophylaxe von Lebermetastasen durch Blockade von Adhäsionsmolekülen bei kolorektalen Karzinomen. *Onkologe*, **4**, 28–34.

Sherman, C.D. (1990) Cancer of the gastrointestinal tract. In D.K. Hossfeld, C.D. Sherman, and R.R. Love, (eds.), *Manual of Clinical Oncology*, Springer Verlag, Heidelberg, pp. 228–252.

Staak, O., Stoffel, B., Wagner, H., Pulverer, G., and Beuth, J. (1998) In vitro-Zytotoxizität der Viscum album-Agglutinine I und II. *Zeitschrift für Onkol./Journal of Oncology*, **30**, 29–33.

Stoffel, B., Beuth, J., and Pulverer, G. (1996) Effect of immunomodulation with galactoside-specific mistletoe lectin on experimental listeriosis. *Zentralblatt für Bakteriologie*, **284**, 439–442.

15. THE MAGIC POTION BECOMES SERIOUS: WHOLE PLANT EXTRACTS *VS.* DEFINED COMPONENTS

GERBURG M. STEIN AND MICHAEL SCHIETZEL

Krebsforschung Herdecke, Communal Hospital, University Witten/Herdecke, Herdecke, Germany

INTRODUCTION

In the preceding review articles of this book, botanical, biochemical, and biological/pharmacological characteristics of the different plant species of the genus *Viscum* have been described with respect to their use in public medicinal science and clinic. In order to refer especially the immunological properties of commercially available whole plant extracts to major constituents, single components have been isolated, highly purified and at least partially characterised.

Already in 1952, Winterfeld speculated that a freshly prepared extract from *Viscum album* L. might be more effective after oral application with respect to the influence on heart functions as compared to the purified viscotoxins. Probably, other extract components may exert an influence on the resorption of the viscotoxins which are suggested to be degraded in stomach and intestine.

Later, Selawry and colleagues (1961) demonstrated differences in the tumour inhibiting effect of extracts from distinct parts of *Viscum album*. Thus, in an animal system, using sarcoma 180 cells in swiss mice, pressed sap from the whole plant exerted the strongest tumour inhibition after short term application as compared to the extracts from leaves, stems, berries or buds. Due to different purification steps, they found at least 3 tumour inhibiting components. Here, the combination of the different components within the whole plant extracts might have been responsible for the strongest effect, although characterisation of these substances is still unclear.

From these data and also from recent findings, there is growing evidence that the whole plant extracts may exert different effects than purified substances, especially the mistletoe lectins (ML). Because the ML are the best characterised components and major constituents of *Viscum album* extracts (VA-E) exerting immunomodulatory properties, these molecules are propagated to be responsible for most immunological effects mediated by VA-E. However, there is a great need to characterise in further detail the effects of other components (Stein *et al.*, 1998b). In addition, only few attempts were made yet to disclose interactions between the different extract components. In the following chapter an overview on the current knowledge of direct interactions between *Viscum album* components is presented as far as it is not performed in the preceding articles. In this respect, it is noteworthy that in these

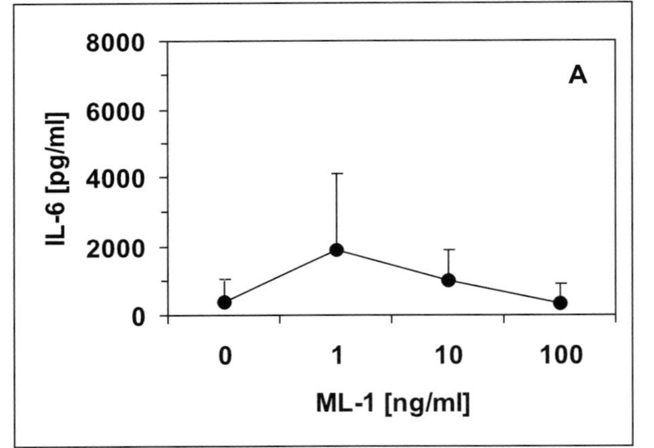

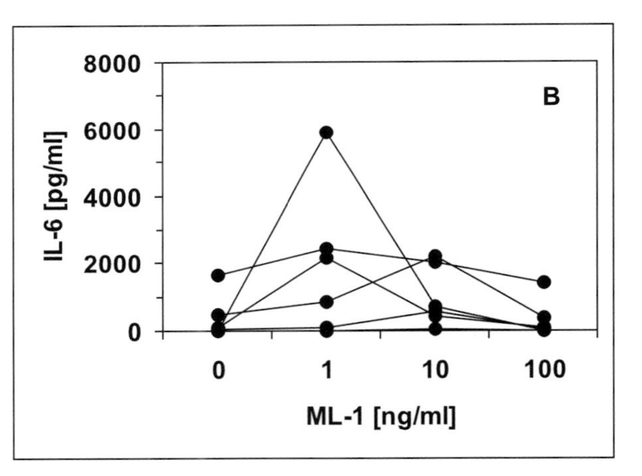

Figure 1 Comparison of the mean value and standard deviation of the ML-1 induced release of Interleukin (IL)-6 (A) with the individual concentrations of IL-6 (B). Cytokines were detected in the supernatants of cultures of PBMC from 6 different controls after *in vitro* stimulation for 6 days by ELISA.

in vitro investigations not a single reactivity pattern of the immune responses towards the different components and their combinations became obvious, but mainly individual immune responses occurred. Highly individual immune responses were also found to be induced by *Viscum album* and their purified components (Stein *et al.*, 1994, 1996b). Although it has been shown that ML-1 and ML-3 induce the release of different cytokines like IL-6 (Hajto *et al.* 1990), not all individuals produce this cytokine after stimulation *in vitro*, and, as shown in Figure 1, they do respond at different concentrations of the ML used. Finally, the conclusions, which have to be drawn from these data and the future aspects for the use of VA-E are discussed.

INTERACTIONS OF DIFFERENT EXTRACT COMPONENTS

Interactions of Different Mistletoe Lectins

Since the ML were suggested to be the main relevant components in VA-E and standardisation of the extracts on the content of the ML was demanded, it is of importance to determine the exact content of at least the different ML within the whole plant extracts. This, however, is very difficult and comparison of the content of the ML obtained by the different methods (haemagglutination, sugar binding activity, cytotoxicity etc.) is impossible (Tröger, 1992; Lorch and Tröger, this book). In addition, there is evidence for an influence of one ML on the detection of another ML. Schöllhorn (1993) studied different ELLA [enzyme-linked lectin binding assay] methods in order to reproducibly detect concentrations of ML-1 and ML-2/3. He found that there is no problem to determine the ML-1 content at < 500 ng/ml in the presence of ML2/3 (galactose-BSA ELLA), however, at a concentration of more than 500 ng/ml ML-1 within a solution of ML-2/3, no real specification was possible (N-acetyl-galactosamine-BSA ELLA). Probably, this effect may be related to the low content of ML-2 within the isolated ML-1, which may be large enough to be detected by the ELLA method at these high concentrations. Furthermore, using an asialofetuin ELLA, ML-1 values slightly decreased when ML-2 was added to the ML-1 solution as compared to the pure ML-1 solution. Since both ML bind to asialofetuin, rather an increase was expected. The decrease, however, suggested an interaction of the ML at the sugar binding sites.

In addition to this methodical problems, also a mutual influence of the different ML was documented for their cytotoxic effects. Kopp *et al.* (1993) reported an experiment testing cytotoxicity of a mixture of the three different ML against activated human T-cells. The measured IC_{50} (inhibitory concentration) value was about threefold higher (and, thus, cytotoxicity decreased) than expected from the experiments performed with the single lectins. Probably, the less toxic ML-2 might have blocked the receptors for ML-1 and ML-3, which in turn could not exhibit full cytotoxicity.

Already from these few experiments, there is obviously a great requirement to study in further detail the influence of the different ML on each other with respect to biochemical and immunological properties.

Interactions of Mistletoe Lectins and Polysaccharides

Mistletoe lectins have been shown to bind to a variety of different carbohydrates (Franz *et al.*, 1981; Gabius *et al.*, 1990; Ziska *et al.*, 1993). Although oligo- and polysaccharides (PS) are present in VA-E (Franz, 1989; Jordan and Wagner, 1986; Klett *et al.*, 1989; Luther and Becker, 1986; Müller and Anderer, 1990a, b), and ML-mediated effects might be altered in the presence of these *Viscum album* carbohydrates, only few investigations were engaged in this topic.

Büssing and co-workers did not observe changes of ML-mediated cytotoxicity on human peripheral blood mononuclear cells (PBMC) in the presence of a *Viscum album*-PS (unpublished own data). In contrast, there is some evidence for an interaction of ML-1 and ML-3 with PS-induced effects (Stein *et al.*, 1999a). Table 1 summarises the influence of ML-1 and ML-3 on the PS-stimulated proliferation of PBMC from healthy controls using an arabinogalactan from *Viscum album* as stimulatory agent *in vitro*. Proliferation was measured by incorporation of the thymidine analogue 5-Bromo-2′-deoxyuridine (BrdU) into the cells following the method described by Carayon and Bord (1995). Although only few controls were tested, these studies revealed an individual immune response demonstrating a synergistic effect in one and a slight inhibitory effect in another individual. These data may indicate, that not only the individual reactivity towards a single component has to be considered, but also the reactivity towards further components, which may modify the response.

Interactions of Mistletoe Lectins and Vesicles

A synergistic effect of ML-1 with vesicles, which are genuine membrane systems from the cell membranes of *Viscum album* obtained by pressing the fresh plant

Table 1 Interaction of ML-1 or ML-3 with the polysaccharide-induced proliferation of human PBMC*.

Individual	*Control*	*Polysaccharides*	*ML-1*	*ML-3*	*ML-1*	*ML-3*	*PWM*
					+ Polysaccharides		
	% CD4⁺ BrdU⁺ CD25⁺ cells						
1	0.1	*4.6*	1.2	0.1	*4*	*4.3*	*6.1*
2	0.1	0.5	0.1	0.1	0.7	1.6	*15.7*
3	0.1	*3.4*	0.7	0.1	*9.9*	*6.8*	*28.4*
4	0.8	*8.1*	0.1	0.9	*4.8*	*5.3*	*7*
5	0.1	1.2	0.1	n.d.	2.1	n.d.	n.d.
6	0.2	1.3	0.7	0.1	0.2	1	*8.7*

* proliferation was measured by incorporation of BrdU and subsequent staining with FITC-conjugated anti-BrdU antibody and PE-conjugated anti-CD25 (interleukin-2 receptor) antibodies. PBMC from healthy controls were cultured for 7 days in autologous plasma (10%) with the polysaccharides (100 μg/ml) in the absence or presence of ML 1 or ML 3 at 1 ng/ml (individuals 1 + 2), 5 ng/ml (individuals 3 + 4) or 10 ng/ml (individuals 5 + 6). Pokeweed mitogen (PWM, 2.5 μg/ml) served as a positive control. n.d. = not determined.

material (Scheffler, 1990), was shown by Scheffler *et al.* (1995). Binding of ML-1 to the vesicles depended upon the pH and ionic strength of the solution with a stronger binding at low pH and low ionic strength. The binding of ML-1 to the vesicles resulted in an enhanced haemagglutination activity of ML-1. In contrast, re-elution of the ML-1 by affinity chromatography revealed only 1/5 of the ML-1 content, demonstrating stability of this ML-vesicle binding. In addition, ML-concentration of the solution containing ML-1 and vesicles appeared to be diminished when tested by ELISA (enzyme-linked immunosorbent assay) using an A-chain and a B-chain specific anti-ML-1 antibody as compared to the same ML-1 concentration in the absence of the vesicles. These data point to the influence of attendant molecules for the correct determination of the lectin content within VA-E. This may be also of importance for the *in vivo* activity of the ML.

In vitro, cytotoxicity of the ML to human PBMC (Fischer *et al.*, 1996b) and also towards different human tumour cells implanted into nude mice (Scheer *et al.*, 1995) was abolished in the presence of the vesicles. Purified ML-1 was more cytotoxic than a combination of ML-1 with the vesicles. Furthermore, using different VA-E, the toxicity could not be related to the lectin content, suggesting a modulation of the ML-associated effects by further extract components.

Another aspect of the investigations concerning the influence between ML and vesicles was the estimation of the proliferation of PBMC/CD4$^+$ T-cells from VA-E (*Abnobaviscum*) treated patients. Proliferation was synergistically enhanced in the presence of both, the ML and the vesicles as compared to the effect mediated by either component (Fischer *et al.*, 1996b).

Interactions of Mistletoe Lectins and Viscotoxins

Cytotoxicity of different *Viscum album* fractions, which were obtained by FPLC (fast protein liquid chromatography) of an unfermented experimental extract and of a fermented commercially available VA-E grown on oak trees were compared by Jung *et al.* (1990) in a short term assay (24 h) by incorporation of [^{3}H]-thymidine into the ML-sensitive human T-cell leukaemia cell line MOLT-4. Inhibition of [^{3}H]-thymidine uptake into MOLT-4 cells treated with the unfermented extract was mainly mediated by the ML-associated fractions and was abolished by the presence of anti-ML-antibodies. In contrast, the viscotoxin (VT)-containing fractions did not exert strong cytotoxic effects in this test system in the concentrations used. In the fermented preparation, however, cytotoxicity of the fractions changed, in that the cytotoxic fractions now corresponded to the viscotoxin fractions, while no such toxicity was observed in the fractions eluted at the position of the ML. Since this toxic effect of the altered VT fraction was also blocked by anti-ML antibodies, the authors concluded that fermentation led to an alteration of the ML, which might form complexes with the VT but maintain cytotoxicity of the ML. It is noteworthy, that in contrast to the above mentioned experiments recent investigations on MOLT-4 cells pointed to an expression of the apoptosis related molecules caspase 3 and Apo2.7 due to incubation with VT for 24 h (Büssing *et al.*, 1999). Probably, the differences might be explained by the high concentrations of the VT used by Büssing and colleagues.

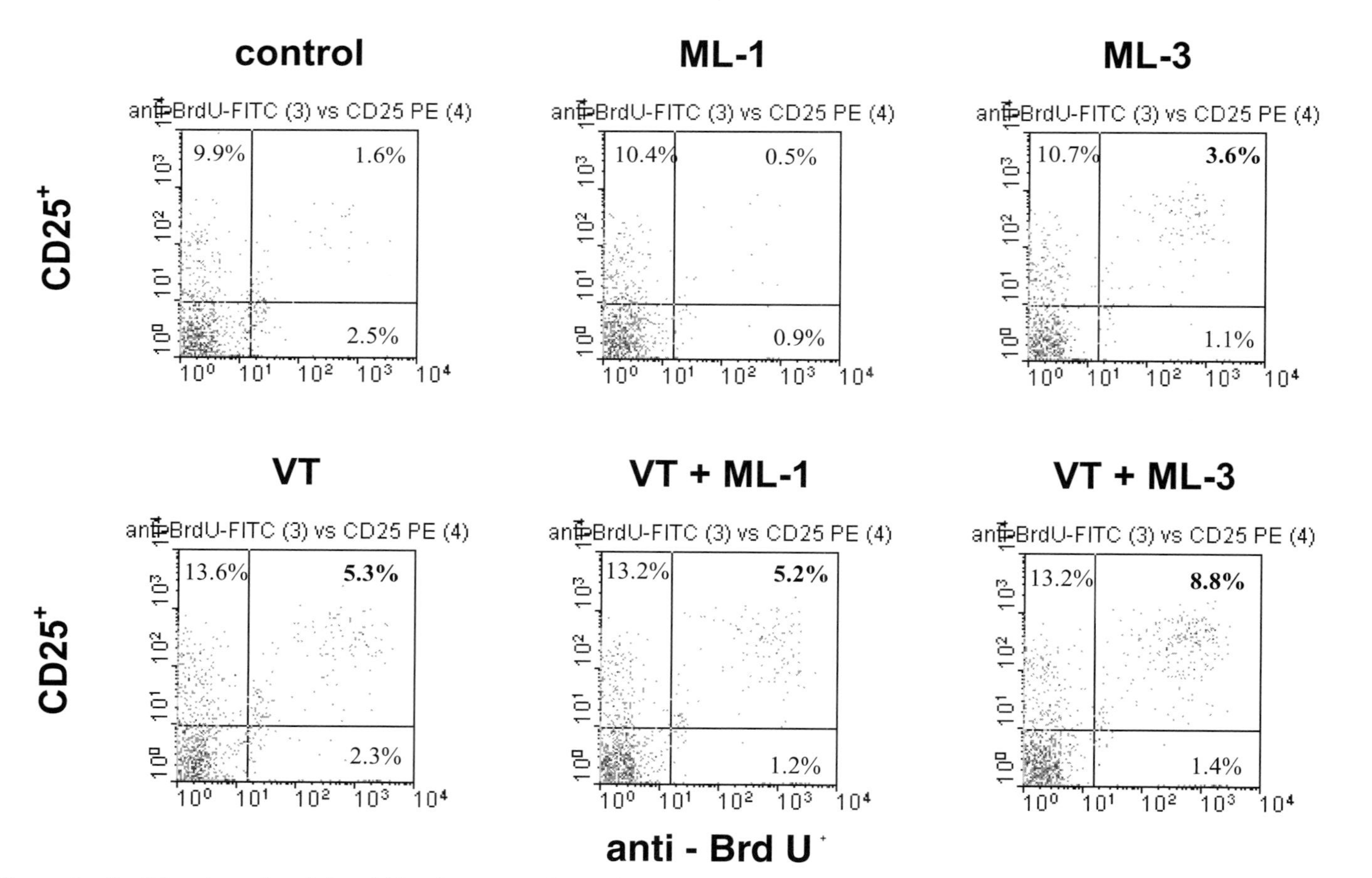

Figure 2 Proliferation of peripheral blood mononuclear cells (PBMC) from a healthy control stimulated with 1 μg/ml viscotoxins (VT) and/or 5 ng/ml mistletoe lectin (ML)-1 or ML-3 for 7 days. Proliferation was measured as BrdU-incorporation. BrdU was stained with a fluorescent dye-conjugated anti-BrdU antibody and subsequently measured by flow cytometry. Cells were $CD4^+$ T-cells.

Table 2 Interaction of ML-1 or ML-3 with the viscotoxins on the proliferation of human PBMC*.

Individual	*Control*	*Viscotoxins (μg/ml)*		*ML-1*	*ML-3*	*ML-1*	*ML-3*	*ML-1*	*ML-3*	*PWM*
						+ Viscotoxins (μg/ml)				
		1	*10*			*1*	*1*	*10*	*10*	
	% CD 4$^+$BrdU$^+$ 25$^+$ cells									
1	0.1	0.1	0.1	0.2	0.1	0.1	0.1	0.3	0.1	17.2
2	1.0	0.9	0.2	0.2	0.6	0.9	0.8	0.2	0.3	28.4
3	0.4	0.8	1.2	0.8	3.1	1.3	0.3	0.4	0.4	9.2
4	1.6	5.3	0	0.5	3.6	5.2	8.8	0.1	0.1	25.2

* proliferation was measured by incorporation of BrdU and subsequent staining with FITC-conjugated anti-BrdU antibody and PE-conjugated anti-CD25 antibodies by flow cytometry. PBMC from healthy controls were cultured for 7 days in autologous plasma (10%) with the viscotoxins (1 or 10 μg/ml) in the absence or presence of ML 1 or ML 3 at 5 ng/ml. Pokeweed mitogen (PWM; 2.5 μg/ml) served as a positive control.

[1] Not used in analysis.

Recently, it was shown by Stein *et al.* (1999b) that VT-mediated enhanced phagocytosis of obsonised *Escherichia coli* and *E. coli*-stimulated respiratory (oxidative) burst of human granulocytes was not influenced by ML-1 or ML-3 at a non-toxic concentration of 1.25 ng/ml. In contrast, an influence of VT and ML on the proliferation of human PBMC was found in single individuals, as measured by flow cytometry *via* incorporation of BrdU (Figure 2). Although only four individuals were studied, in two cases an influence could be visualised with an additive effect in one and an inhibitory effect in another individual (Table 2).

Interactions of Viscotoxins and Polysaccharides

No influence of an arabinogalactan isolated from *Viscum album* on the VT-enhanced phagocytotic activity of human granulocytes was observed (Stein and Edlund, unpublished observations). With respect to the influence of VT and the PS on the proliferation of PBMC from healthy controls, in one out of four cases studied a slight synergistic effect was induced *in vitro* only in the presence of both molecules, while the PS and VT at these concentrations remained without any effect (Figure 3).

NON-ML-RELATED IMMUNOLOGICAL EFFECTS OF WHOLE PLANT EXTRACTS

As demonstrated previously, some immunological properties of VA-E are obviously not related to the ML. This was especially true for effects observed with the fer-

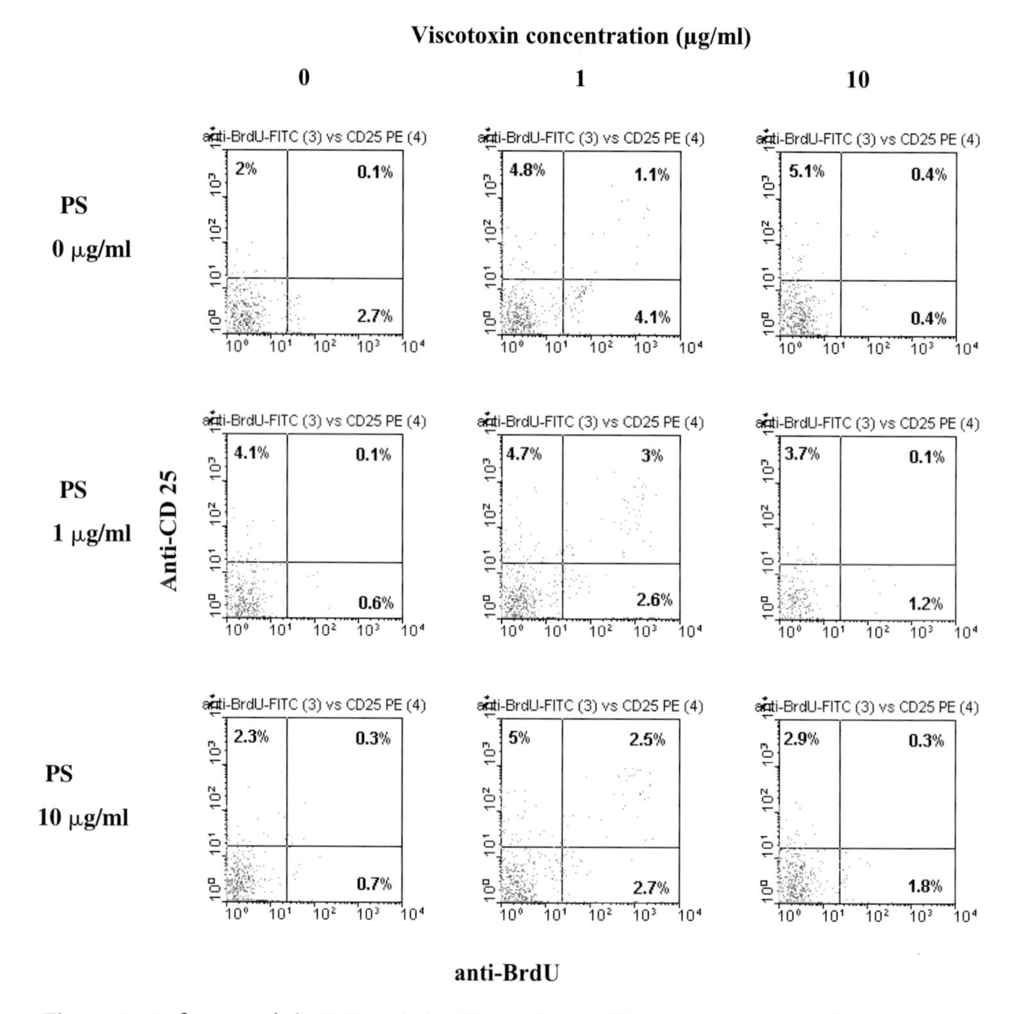

Figure 3 Influence of the VT and the PS on the proliferation of peripheral blood mononuclear cells (PBMC) from a healthy control stimulated with viscotoxins (VT) and polysaccharides (PS). Proliferation was measured as BrdU-incorporation. BrdU was stained with a fluorescent dye-conjugated anti-BrdU antibody and subsequently measured by flow cytometry. Cells were CD4+ T-cells.

mented VA-E from pines (*Iscador* Pini). This extract does not contain ML-1 and only minor amounts of ML2/ML-3 were detectable (Stein and Berg, 1994). Stein *et al.* published a set of experiments dealing with that extract. VA-E-stimulated proliferation of PBMC especially from untreated allergic/atopic individuals could not be related to the ML (Stein and Berg, 1994). Figure 4 compares the effects of this extract and of different components on the proliferation of human PBMC from

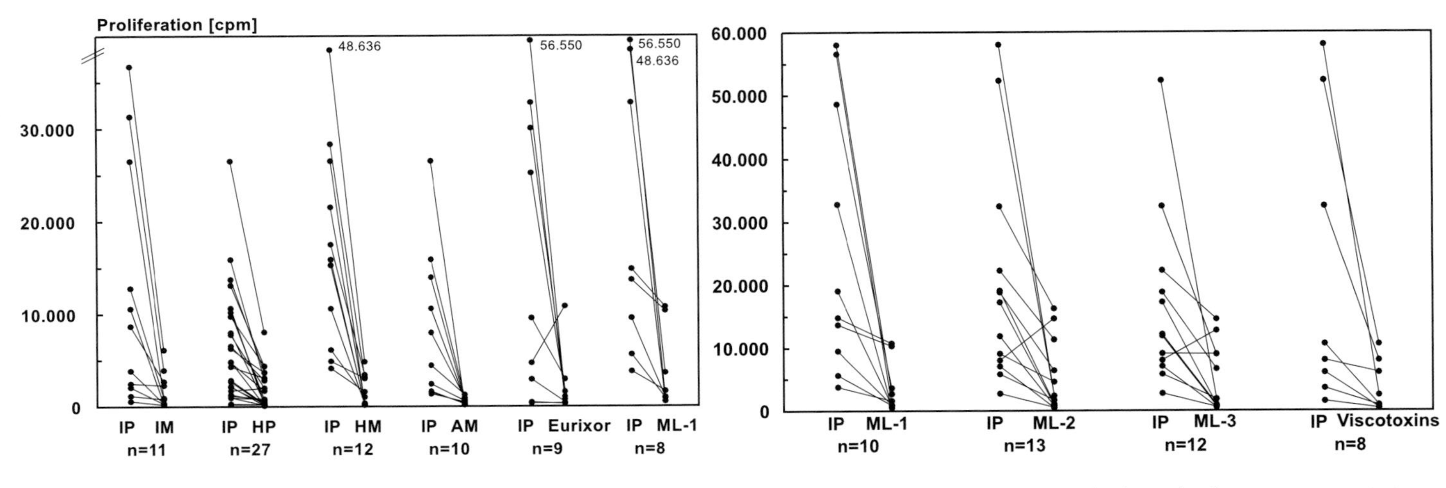

Figure 4 Comparison of the stimulatory activity of a fermented ML-1 free VA-E (*Iscador* Pini, IP) with that of other extracts, mistletoe lectins (ML) and the viscotoxins (VT). Proliferation was measured in cultures of human PBMC after 7 days *via* incorporation of [^{3}H]-thymidine. AM: *Abnobaviscum* Mali, cpm: counts per minute, HP: *Helixor* Pini, HM: *Helixor* Mali, IM: *Iscador* Mali.

untreated controls as measured by [^{3}H]-thymidine incorporation (Stein and Berg, 1997). None of these single purified components exerted comparable effects than this extract and also other VA-E did not induce similar proliferation, irrespective of the host tree they were derived from.

Furthermore, this extract induced the release of different cytokines, especially of TNF-α and IL-6, and in some individuals also IFN-γ or IL-4/IL-5 (Stein *et al.*, 1996b). In addition, expression of different activation markers (Stein and Berg, 1996a) and co-stimulatory signals (Stein and Berg, 1998a) were also increased, indicating a stimulation of cells of the monocyte/macrophage lineage, which might serve as antigen presenting cells for the activated CD4$^+$ T-helper cells. Although the relevant antigen has not been characterised yet, there is clear evidence that these effects were not lectin-associated.

Differences in the effects of a VA-E (*Isorel*) and its fractions of low (< 30 kD) and high (> 30 kD) molecular weight were demonstrated by Zarkovic *et al.* (1998). Proliferation of murine B16F10 melanoma cells but not Con-A-stimulated murine lymph node lymphocytes was inhibited by the whole plant extract as effectively as by ML-1 and by the low and the high molecular weight fractions. From these data, it is suggested that not only the ML but also low molecular weight substances (< 30 kD) are active. Probably, interactions of different components may be of further importance as indicated by the observation that addition of both fractions to the melanoma cells did not exert the same effect as the whole plant extract especially at higher concentrations of 0.1–1 μg/ml.

Reduction of sister chromatid exchange (SCE) frequency in PHA-stimulated PBMC from healthy individuals by whole plant extracts from VA-E (*Helixor*) has not been related to a single extract component yet (Büssing *et al.*, 1994). ML-3 seems not to be responsible for this phenomenon, since no significant influence was observed (Büssing *et al.*, 1998) and similar data were obtained for ML-1 (Büssing, personal communication).

Comparison of the *in vitro* cytotoxicity of mistletoe grown on oak trees using Korean mistletoe (*Viscum album* var. *coloratum*) and the European mistletoe (*Viscum album* var. *album* or *platyspermum*), against different tumour cell lines revealed similar results (Choi *et al.*, 1996). However, the lectin content of *Viscum album* var. *coloratum* was reduced and also the lectin pattern was changed with a dominating Korean-ML- (K-ML) 2/3 pattern in the extract from Korean mistletoe and predominantly ML-1 in the European mistletoe, both grown on oak trees (Choi *et al.*, 1996). Yet, it is unclear whether K-ML-1, K-ML-2 or K-ML-3 are identical with ML-1, ML-2 and ML-3 from European mistletoe.

Further non-ML-related immunological effects like those induced by defined *Viscum album* oligo- and polysaccharides (Klett *et al.*, 1989; Müller and Anderer, 1990a,b) and a still undefined component, a peptide of about 5 kDa used by the group of Kuttan (1992,1997) are described by Büssing in a preceding chapter.

CONCLUDING REMARKS/FUTURE ASPECTS

Conventional medicine favours application of single components. Thus, major constituents of plant extracts, e.g. *Convallaria, Taxus* etc., were isolated and provided the basis for new drug design. Similar efforts are on the way for the characterisation of ML-1. In the meantime, recombinant ML-1 (rML-1) has been expressed in *E. coli*, as described by Beuth in the preceding chapter. However, as for all remedies, efficacy and safety of the molecule has to be proven. There are some interesting data providing evidence for immunological activity of the rML-1, although today too little experimental data (and no clinical studies at all) do not allow a final statement.

Reflecting all available data on the current knowledge of *Viscum album*, as reviewed in this book, it is, naturally, a great need to standardise remedies containing VA-E applicated to patients. However, there are some important questions, which might be answered at least in part, and will justify further research in this topic. These most important questions are: Is the standardisation of VA-E on ML-1, the postulated main relevant component, really the conclusion we have to draw? Is the application of a single component to tumour patients superior than the application of a whole plant extract?

With respect to the standardisation procedure, a similar situation as it appears now for VA-E, occurred with extracts derived from St. John's Wort (*Hypericum perforatum*) which is used for treatment of mild to moderate depression. Until recently, standardisation on the content of hypericin was state of the art, however, there is now evidence that hyperforin is the major component exerting antidepressive effects and other components contribute to the pharmacological efficacy of whole plant extracts (Bhattacharya *et al.*, 1998; Butterweck *et al.*, 1998; Chatterjee *et al.*, 1998).

As reviewed in this book, the current knowledge of the biochemical, biological, pharmacological and/or immunological properties of the different *Viscum album* components does not allow a restriction of the standardisation on one component like ML-1. Obviously, there is a great need of further investigations to better characterise the effects of all *Viscum album* components including ML, VT and PS but also molecules like the *Viscum album* chitin-binding lectin (VisalbCBL), Viscumamid, the so-called "45 kDa epitope" (still uncharacterised protein, of 45 kDa demonstrated by Ribéreau-Gayon *et al.*, 1993; Stettin *et al.*, 1990; Stein *et al.*, 1994), the 5 kDa peptide used by Kuttan *et al.* (1992, 1997), the vesicles (Fischer *et al.* 1996a, 1997) flavonoids etc. Although only limited data on these and further undefined components are available, one may not exclude them as ineffective. Until recently, VT were regarded to be without relevant immunological effects but to exert mainly strong cytotoxicity. However, detailed experiments now demonstrate most interesting properties with respect to induction of necrosis, apoptosis and also stimulatory properties especially on granulocytes (for review see Büssing, this book).

In addition, one has to keep in mind that a single component may not be made responsible for all effects observed with a variety of preparations. Therefore, especially the differences between the various commercially available extracts have to be studied in more detail. Probably, it may be possible to better define the most suitable extracts for different tumour entities and/or tumour stages implying a far deeper knowledge of basic tumour immunology. Probably, there will be a rationale for the use of VA-E derived from different host trees or different manufacturing processes.

A major problem for therapy with VA-E as for all immunomodulating therapies is the highly individual response. In contrast to the postulation, that a single dose of one component (i.e., 1 ng/kg b.w. of ML-1), found in experiments performed with only few animals, is optimal, clinical experience points to the fact that tumour patients respond to different concentrations and may require higher concentrations during treatment. Interestingly, no clinical data are available dealing with long term treatment with the "ML-optimised" therapy with VA-E for several years. In addition, animal experiments confirm that suggestion, since much higher concentrations also exerted immunostimulatory and antimetastatic effects (Mengs *et al.*, 1998; Weber *et al.*, 1998).

For classical medicine, therapy with recombinant or natural ML-1 may provide one possibility, however, therapy with different extracts provide further alternatives for immunotherapy of the malignancies.

REFERENCES

Bhattacharya, S.K., Chakrabarti, A., and Chatterjee, S.S. (1998) Activity profiles of two hyperforin-containing *hypericum* extracts in behavioral models. *Pharmacopsychiatry*, **31** (Suppl 1), 22–29.

Büssing, A., Azhari, T., Ostendorp, H., Lehnert, A., and Schweizer, K. (1994) *Viscum album* L. extracts reduce sister chromatid exchanges in cultured peripheral blood mononuclear cells. *Eur. J. Cancer*, **30A**, 1836–1841.

Büssing, A., Multani, A.S., Pathak, S., Pfüller, U., and Schietzel, M. (1998) Induction of apoptosis by the N-acetyl-galactosaimne-specific toxic lectin from *Viscum album* L. is associated with a decrease of nuclear p53 and Bcl-2 proteins and induction of telomeric associations. *Cancer Let.t*, **130**, 57–68.

Büssing, A., Verwecken, W., Wagner, M., Wagner, B., Pfüller, U., and Schietzel, M. (1999) Expression of mitochondrial Apo2.7 molecules and caspase-3 activation in human lymphocytes treated with the ribosome-inhibiting mistletoe lectins and the cell membrane permeabilizing viscotoxins. *Cytometry*, **37**, 133–139.

Butterweck, V., Petereit, F., Winterhoff, H., and Nahrstedt, A. (1998) Solubilized hypericin and pseudohypericin from *Hypericum perforatum* exert antidepressant activity in the forced swimming test. *Planta Med.*, **64**, 291–294.

Carayon, P. and Bord, A. (1992) Identification of DNA-replicating lymphocyte subsets using a new method to label the bromo-deoxyuridine incorporated into the DNA. *J. Immunol. Methods*, **147**, 225–230.

Chatterjee, S.S., Nöldner, M., Koch, E., and Erdelmeier, C. (1998) Antidepressant activity of *hypericum perforatum* and hyperforin: the neglected possibility. *Pharmacopsychiatry*, **31** (Suppl 1), 7–15.

Choi, O.B., Yoon, T.J., Drees, M., Scheer, R., and Kim, J.B. (1996) Inhaltsstoffe und *in vitro* Zytotoxizität eines Extraktes aus *Viscum album* L. ssp. *coloratum* (Koreanische Mistel) – Konsequenzen für die Standardisierung von Mistelpräparaten. *Z. Onkol.*, **28**, 77–81.

Fischer, S., Scheffler, A., and Kabelitz, D. (1996a) Activation of human $\gamma\delta$ T-cells by heat treated mistletoe plant extracts. *Immunol. Letters*, **52**, 69–72.

Fischer, S., Scheffler, A., and Kabelitz, D. (1997) Oligoclonal *in vitro* response of CD4 T cells to vesicles of mistletoe extracts in mistletoe-treated cancer patients. *Cancer Immunol. Immunother.*, **44**, 150–156.

Fischer, S., Scheffler, A., and Kabelitz, D. (1996b) Reaktivität von T-Lymphozyten gegenüber Mistel-Inhaltsstoffen. In R. Scheer, H. Becker, P.A. Berg, (eds.), *Grundlagen der Misteltherapie. Aktueller Stand der Forschung und klinische Anwendung*. Hippokrates Verlag, Stuttgart, pp. 213–223.

Franz, H. (1989) Viscaceae lectins. *Advances in Lectin Research*, **2**, 28–59.

Franz, H., Ziska, P., and Kindt, A. (1981) Isolation and properties of three lectins from mistletoe (*Viscum album* L.). *Biochem. J.*, **195**, 481–484

Gabius, H.J. (1990) Influence of type of linkeage and spacer on the interaction of β-galactoside-binding proteins with immobilized affinity ligands. *Analyt. Biochem.*, **189**, 91–94.

Hajto, T., Hostanska, K., Frei, K., Rordorf, C., and Gabius, H.J. (1990) Increased secretion of Tumour Necrosis Factor α, Interleukin 1, and Interleukin 6 by human mononuclear cells exposed to ß-galactoside-specific lectin from clinically applied mistletoe extract. *Cancer Res.*, **50**, 3322–3326.

Jordan, E. and Wagner, H. (1986) Structure and properties of polysaccharides from *Viscum album* (L.). *Oncology*, **43** (Suppl 1), 8–15.

Jung, M.L., Ribéreau-Gayon, G., Beck, J.P., and Baudino, D. (1990) Characterisation of cytotoxic proteins from mistletoe (*Viscum album* L). *Cancer Lett.*, **51**, 103–108.

Klett, C.Y. and Anderer, F.A. (1989) Activation of Natural-Killer cytotoxicity of human blood monocytes by a low molecular weight component from *Viscum album* extract. *Arzneimittel-Forsch./Drug Res.*, **39**, 1580–1585.

Kopp, I., Koerner, I.J., Pfüller, U., Göckeritz, W., Eifler, R., Pfüller, K., *et al.* (1993) Effects of mistletoe lectins I, II and III on normal and malignant cells. In E. van Driessche, H. Franz, S. Beeckmans, U. Pfüller, A. Kallikom and T.C. Bøg-Hansen, (eds.), *Lectins: Biology, Biochemistry, Clinical Biochemistry*, Vol. 8, Textop, Hellerup, Denmark, pp. 41–47.

Kuttan, G. and Kuttan, R. (1992) Immunomodulatory activity of a peptide isolated from *Viscum album* extract (NSC 635 089). *Immunol. Invest.*, **21**, 285–296.

Kuttan, G., Menon, L.G., Antony, S., and Kuttan, R. (1997) Anticarcinogenic and antimetastatic activity of Iscador, *Anticancer Drugs*, **8** (Suppl. 1), S15–S16.

Luther, P. and Becker, H. (1986) *Die Mistel-Botanik, Lektine, medizinische Anwendung*. VEB Verlag/Springer Verlag, Heidelberg.

Mengs, U., Weber, K., Schwarz, T., Hajto, T., Hostanska, K., and Lentzen, H. (1998) Effects of standardised mistletoe preparation Lektinol on granulopoiesis and pulmonary metastases in mice. In S. Bardocz, U. Pfüller and A. Pusztai, (eds.), *COST 98. Effects of antinutritients on the nutritional value of legume diets*. European Commission, Luxembourg, Vol. 5, pp. 194–201.

Müller, E.A. and Anderer, F.A. (1990a) Chemical specificity of effector cell/tumour cell bridging by a *Viscum album* rhamnogalacturonan enhancing cytotoxicity of human NK cells. *Immunopharmacology*, **19**, 69–77.

Müller, E.A. and Anderer, F.A. (1990b) Synergistic action of a plant rhamnogalacturonan enhancing antitumour cytotoxicity of human Natural Killer and lymphokine-activated killer cells: chemical specificity of target cell recognition. *Cancer Res.*, **50**, 3646–3651.

Ribéreau-Gayon, G., Jung, M.L., Dietrich, J.B., and Beck, J.P. (1993) Lectins and viscotoxins from mistletoe (*Viscum album* L.) extracts: development of a bioassay of lectins. In E. van Driessche, H. Franz, S. Beeckmans, U. Pfüller, A. Kallikorm and T.C. Bøg-Hansen, (eds.), *Lectins: Biology, Biochemistry, Clinical Biochemistry*, Vol. 8, Textop, Hellerup, Denmark, pp. 21–28.

Scheer, R., Fiebig, H.H., and Scheffler, A. (1995) Synergisms between lectins and vesicles of *Viscum album* L. (2) direct cytotoxic effects. Proc 1st World meeting APGI/APV Budapest 09.–11.05.1995, pp. 873–874

Scheffler, A. (1990) Neue Aspekte zur Herstellung von Mistelpräparaten. *Therapeutikon,* **4**, 16–22.

Scheffler, A., Musielski, H., and Scheer, R. (1995) Synergismus zwischen Lektinen und Vesikeln von *Viscum album* L. *Dtsch Zschr. Onkol.*, **27**, 72–75.

Schöllhorn, V. (1993) An ELLA system to quantify mistletoe I and II isolectins. In E. van Driessche, H. Franz, S. Beeckmans, U. Pfüller, A. Kallikorm and T.C. Bøg-Hansen, (eds.), *Lectins: Biology, Biochemistry, Clinical Biochemistry*, Vol. 8, Textop, Hellerup, Denmark, pp. 14–20.

Selawry, O.S., Vester, F., Mai, W., and Schwartz, M.R. Zur Kenntnis der Inhaltsstoffe von *Viscum album*, II Tumorhemmende Inhaltsstoffe. *Hoppe Seyler's Z Physiol. Chem.*, **324**, 262–281.

Stein, G. and Berg, P.A. (1994) Non-lectin component in a fermented extract from *Viscum album* L. grown on pines induces proliferation of lymphocytes from healthy and allergic individuals *in vitro*. *Eur. J. Clin. Pharmacol.*, **47**, 33–38.

Stein, G.M. and Berg, P.A. (1996a) Evaluation of the stimulatory activity of a fermented mistletoe lectin-1 free mistletoe extract on T-helper cells and monocytes in healthy individuals *in vitro*. *Arzneim-Forsch./Drug Res.*, **46**, 635–639.

Stein, G.M., Meink, H., Durst, J., and Berg, P.A. (1996b) Release of cytokines by a fermented lectin-1 (ML-1) free mistletoe extract reflects differences in the reactivity of PBMC in healthy and allergic individuals and tumor patients. *Eur. J. Clin. Pharmacol.*, **51**, 247–252.

Stein, G.M. and Berg, P.A. (1997) Mistletoe extract-induced effects on immunocompetent cells: *in vitro* studies. *Anticancer Drugs*, **8** (Suppl 1), S 39-S 42.

Stein, G.M. and Berg, P.A. (1998a) Flow cytometric analyses of the specific activation of peripheral blood mononuclear cells from healthy donors after *in vitro* stimulation with a fermented mistletoe extract and mistletoe lectins. *Eur. J. Cancer*, **34**, 1105–1110.

Stein, G.M., Edlund, U., Pfüller, U., Büssing, A., and Schietzel, M. (1999a) Influence of polysaccharides from *Viscum album* L. on human lymphocytes, monocytes and granulocytes *in vitro*. *Anticancer Res.*, **19**, 3907–3914.

Stein, G.M., Schietzel, M., and Büssing, A. (1998b) Mistletoe in immunology and the clinic (short review). *Anticancer Res.*, **18**, 3247–3249.

Stein, G.M., Schaller, G., Pfüller, U., Schietzel, M., and Büssing, A. (1999b) Thionins from *Viscum album* L.: influence of viscotoxins on the activation of granulocytes. *Anticancer Res.*, **19**, 1037–1042.

Stettin, A., Schultze, J.L., Stechemesser, E. and Berg, D.A. (1990) Anti-mistletoe lectin antibodies are produced in patients during therapy with an aqueous mistletoe extract derived from *Viscum album* L. and neutralise lectin-induced cytotoxicity *in vitro*. *Klin. Wochensehr.*, **68**, 896–900.

Tröger, W. (1992) Nachweismethoden von Mistellektinen. *Der Merkurstab*, **6**, 456–460.

Weber, K., Mengs, U., Schwarz, T., Hajto, T., Hostanska, K., Allen, T.R., *et al.* (1998) Effects of standardised mistletoe preparation on metastatic B16 melanoma colonisation in murine lungs. *Arzneimittel-Forsch./Drug Res.*, **48**, 497–502.

Winterfeld, K. (1952) Die Wirkstoffe der Mistel (*Viscum album* L.). *Pharm. Ztg.*, **88**, 573–574.

Zarkovic, N., Kališnik, T., Loncaric, I., Borovi'c, S., Mang, S., Kissel, D., *et al.* (1998) Comparison of the effects of *Viscum album* lectin ML-1 and fresh plant extract (Isorel) on the cell growth *in vitro* anf tumourigenicity of melanoma B16F10. *Cancer Biother Radiopharmaceuticals*, **13**, 121–131.

Ziska, P., Gelbin, M., and Franz, H. (1993) Interaction of mistletoe lectins ML-I, ML-II, and ML-III with carbohydrates. In E. van Driessche, H. Franz, S. Beeckmans, U. Pfüller, A. Kallikom and T.C. Bøg-Hansen, (eds.), *Lectins: Biology, Biochemistry, Clinical Biochemistry*, Vol. 8, Textop, Hellerup, Denmark, pp. 10–13.

INDEX

Other volumes in preparation in Medicinal and Aromatic Plants – Industrial Profiles

Allium, edited by K. Chan
Aloes, edited by A. Sweck and R. George
Artemisia, edited by C. Wright
Cardamom, edited by P.N. Ravindran and K.J. Madusoodanan
Chamomile, edited by R. Franke and H. Schilcher
Cinnamon nad Cassia, edited by P.N. Ravindran and S. Ravindran
Colchicum, edited by V. Simánek
Curcuma, edited by B.A. Nagasampagi and A.P. Purohit
Eucalyptus, edited by J. Coppen
Hypericum, edited by K. Berger Büter and B. Büter
Illicium and Pimpinella, edited by M. Miró Jodral
Kava, edited by Y.N. Singh
Licorice, edited by L.E.Craker, L. Kapoor and N, Mamedov
Narcissus, edited by G. Hanks
Plantago, edited by C. Andary and S. Nishibe
Stevia, edited by A.D. Kinghorn
Thymus, edited by W. Letchamo, E. Stahl-Biskup and F. Saez
Trigonella, edited by G.A. Petropoulos
Urtica, by G. Kavalali

This book is part of a series. The publisher will accept continuation orders which may be cancelled at any time and which provide for automatic billing and shipping of each title in the series upon publication. Please write for details.